High-Impact Weather Events over the SAARC Region

Kamaljit Ray • M. Mohapatra
B.K. Bandyopadhyay • L.S. Rathore
Editors

High-Impact Weather Events over the SAARC Region

Editors
Kamaljit Ray
Nowcasting Unit
India Meteorological Department
Mausam Bhawan, Lodhi Road
New Delhi, India

M. Mohapatra
Cyclone Warning Division
India Meteorological Department
Mausam Bhawan, Lodhi Road
New Delhi, India

B.K. Bandyopadhyay
India Meteorological Department
Mausam Bhawan, Lodhi Road
New Delhi, India

L.S. Rathore
India Meteorological Department
Mausam Bhawan, Lodhi Road
New Delhi, India

Co-published by Springer International Publishing, Cham, Switzerland, with Capital Publishing Company, New Delhi, India.

Sold and distributed in North, Central and South America by Springer, 233 Spring Street, New York 10013, USA.

In all other countries, except SAARC countries—Afghanistan, Bangladesh, Bhutan, India, Maldives, Nepal, Pakistan and Sri Lanka— sold and distributed by Springer, Haberstrasse 7, D-69126 Heidelberg, Germany.

In SAARC countries—Afghanistan, Bangladesh, Bhutan, India, Maldives, Nepal, Pakistan and Sri Lanka—printed book sold and distributed by Capital Publishing Company, 7/28, Mahaveer Street, Ansari Road, Daryaganj, New Delhi 110 002, India.

ISBN 978-3-319-10216-0   ISBN 978-3-319-10217-7 (eBook)
DOI 10.1007/978-3-319-10217-7
Springer Cham Heidelberg New York Dordrecht London

Library of Congress Control Number: 2014955791

© Capital Publishing Company 2015
This work is subject to copyright. All rights are reserved by Capital Publishing Company, whether the whole or part of the material is concerned, specifically the rights of translation, reprinting, reuse of illustrations, recitation, broadcasting, reproduction on microfilms or in any other physical way, and transmission or information storage and retrieval, electronic adaptation, computer software, or by similar or dissimilar methodology now known or hereafter developed. Exempted from this legal reservation are brief excerpts in connection with reviews or scholarly analysis or material supplied specifically for the purpose of being entered and executed on a computer system, for exclusive use by the purchaser of the work. Duplication of this publication or parts thereof is permitted only under the provisions of the Copyright Law of the Publisher's location, in its current version, and permission for use must always be obtained from Capital Publishing Company. Permissions for use may be obtained through Capital Publishing Company. Violations are liable to prosecution under the respective Copyright Law.

The use of general descriptive names, registered names, trademarks, service marks, etc. in this publication does not imply, even in the absence of a specific statement, that such names are exempt from the relevant protective laws and regulations and therefore free for general use.

While the advice and information in this book are believed to be true and accurate at the date of publication, neither the authors nor the editors nor the publisher can accept any legal responsibility for any errors or omissions that may be made. The publisher makes no warranty, express or implied, with respect to the material contained herein.

Printed on acid-free paper

Springer is part of Springer Science+Business Media (www.springer.com)

Preface

High impact weather events are an inherent aspect of the climate system and are of different spatial and temporal scales. They have the potential to cause significant loss of life and property and a major disruption in communication and transport. Understanding the patterns of extreme weather events has assumed even more importance in recent years in the scenario of global climate change. Because of the significance of the extreme weather events in India, the India Meteorological Department, Ministry of Earth Sciences, Government of India, in collaboration with SMRC, Bangladesh, took the initiative to organize the SAARC Seminar on 'High Impact Weather Events over SAARC Region' during 2–4 December, 2013, in New Delhi, India, in order to shed light on the scientific basis and the complexities inherent in combating these events.

The objective of this seminar was to create a forum for discussion on the causes and consequences of high impact weather events in the SAARC member countries, to promote research activities with a view to make better understanding of the high impact weather phenomena and improve their forecasting to minimize loss of lives and properties of this region.

The broad thematic areas of the seminar were:

1. Climatology of high impact weather events
2. The dynamics of extreme events—improving forecasts in the current climate
3. High impact weather events/extreme events under changing climate
4. Consequence of high-impact weather events on the economy, infrastructure and society in various SAARC countries

Papers were received from scientists and National Hydrometeorological Service representatives from SAARC countries and a number of institutions like IIT, IITM, NCMRWF, IISc etc. from India. About 70 delegates from different SAARC countries participated in the seminar. During the seminar, there were nine technical sessions, a panel discussion and the concluding session. There were nine lead talks by eminent scientists in the field of heavy rains, thunderstorms, cyclones and temperature extremes. A number of recommendations emerged after the seminar in each area of specialization.

The panelists agreed that, research being an important component of IMD, it should give special emphasis to high impact weather events, particularly impacts and prediction of heat waves and thunderstorms. More research work was required to make use of DWR data through calibration, validation and networking. Based on remote sensing data, flood and drought hazard proneness needs to be evaluated for SAARC region. There was a need for hazard and vulnerability analysis and climatology of HIWE for SAARC region. In case of heavy rains, there is the need to use rainfall forecasts in a hydrological model to generate surface run-off and thus chance of flooding. The need of high resolution mesoscale models with an interactive land surface model and data assimilation to generate heavy rainfall forecasts was discussed. The panel felt that more sensitivity studies on regional meso scale models was needed to understand the basic mechanism of rainfall over different regions as a result of interaction of monsoon circulation, transient systems, orography and mid-latitude interactions. A number of papers were presented on tropical cyclones and the committee stressed on application of DT to microwave imageries, microwave sounders to estimate the intensity of TC, augmentation of observational network in SAARC region, including surface and upper air observations. The committee stressed on a standard operational mechanism for exchange of data and information among SAARC member countries. Regarding the lack of ground-based observations, space-based observation through satellite was to be utilized maximum for monitoring of high impact weather events including rainfall, temperature extremes, winds etc. Need of a structured system of forecasting and warnings over SAARC region using High Resolution Ensemble for Short Range NWP models for nowcasts, regional cooperation through Severe Weather FDP, and standard operating procedure for all elements of monitoring, prediction and warning was stressed upon.

Considering the significant findings presented in the seminar by various delegates and the recommendations made in the seminar, it was decided to publish the selected papers presented during the seminar as a book after the peer review of the manuscripts.

This book deals with recent advances in our understanding and prediction of cyclone, severe thunderstorms, squalls, heat and cold waves and heavy rainfall, based on the latest observational and NWP modeling platform. The chapters are based on four broad high impact weather events i.e. thunderstorms, cyclones, heavy rains, and drought and temperature. They are authored by leading experts both in research and operational fields.

The book reviews research work, future needs, forecasting skills and societal impacts of above extreme weather events and is relevant to weather forecasters, managers, graduate students and provides high-quality reference material for the users.

As editors of this volume, we are highly thankful to all the authors for their efforts and cooperation in bringing out this publication. We are thankful to SMRC,

Dhaka, for approving publishing of this book. We are also thankful to the Advisory Council, National Organizing Committee, SMRC Organizing Committee and Local Organizing Committee for successfully organizing the SAARC Seminar during 2–4 December in New Delhi, India. Authors also thank the Ministry of Earth Sciences for facilitating the organization of the seminar.

New Delhi, India
Kamaljit Ray
M. Mohapatra
B.K. Bandyopadhyay
L.S. Rathore

Contents

Part I Thunderstorms

**Study of Severe Thunderstorms over Bangladesh
and Its Surrounding Areas During Pre-monsoon Season
of 2013 Using WRF-ARW Model** 3
Md. Abdul Mannan, Md. Nazmul Ahasan, and Md. Shah Alam

**Assimilation of Doppler Weather Radar Data Through Rapid
Intermittent Cyclic (RIC) for Simulation of Squall Line Event
over India and Adjoining Bangladesh** 23
Kuldeep Srivastava, Vivek Sinha, and Rashmi Bharadwaj

**Impact of Data Assimilation in Simulation of Thunderstorm
Event over Bangladesh Using WRF Model** 35
Nazlee Ferdousi, Sujit K. Debsarma, Md. Abdul Mannan,
and Md. Majajul Alam Sarker

**Numerical Simulation of a Hailstorm Event over Delhi, India
on 28 Mar 2013** 49
A. Chevuturi and A.P. Dimri

**Simulation of Mesoscale Convective Systems Associated
with Squalls Using 3DVAR Data Assimilation over Bangladesh** 63
Mohan K. Das, Someshwar Das, and Md. Mizanur Rahman

**Simulation of Severe Convective Weather Events
over Southern India Using WRF Model** 73
S. Stella and Geeta Agnihotri

Part II Tropical Cyclones

Early Warning Services for Management of Cyclones over North Indian Ocean: Current Status and Future Scope 87
M. Mohapatra, B.K. Bandyopadhyay, Kamaljit Ray, and L.S. Rathore

Development of NWP-Based Cyclone Prediction System for Improving Cyclone Forecast Service in the Country 111
S.D. Kotal, Sumit Kumar Bhattacharya, S.K. Roy Bhowmik, and P.K. Kundu

Interannual and Interdecadal Variations in Tropical Cyclone Activity over the Arabian Sea and the Impacts over Pakistan 129
Wash Dev Khatri, Zhi Xiefei, and Zhang Ling

Impact of Cloud Microphysics and Cumulus Parameterisation on Meso-scale Simulation of TC Sidr over the Bay of Bengal Using WRF Model 147
Md. Mahbub Alam

Eddy Angular Momentum Fluxes in Relation with Intensity Changes of Tropical Cyclones Jal (2010) and Thane (2011) in North Indian Ocean 165
S. Balachandran and B. Geetha

Impact of Initial and Boundary Conditions on Mesoscale Simulation of Bay of Bengal Cyclones Using WRF-ARW Model 179
K.S. Singh and M. Mandal

Performance of Global Forecast System for the Prediction of Intensity and Track of Very Severe Cyclonic Storm 'Phailin' over North Indian Ocean 191
V.R. Durai, S.D. Kotal, S.K. Roy Bhowmik, and Rashmi Bhradwaj

Part III Heavy Rains

Observational Analysis of Heavy Rainfall During Southwest Monsoon over India 207
Pulak Guhathakurta

Long Term Trends in the Extreme Rainfall Events over India 229
D.S. Pai and Latha Sridhar

Diagnostic Study of Heavy Rainfall Events in Monsoon Season over Northern Part of Bangladesh Using NWP Technique 241
Md. Abdul Mannan and Md. Mahbub Alam

Analysis of Increasing Heavy Rainfall Activity over Western India, Particularly Gujarat State, in the Past Decade 259
Manorama Mohanty, Kamaljit Ray, and Kalyan Chakravarthy

Simulation of Heavy Rainfall Event over Gujarat During September 2013 277
S.I. Laskar, S.D. Kotal, S.K. Bhattacharya, and S.K. Roy Bhowmik

Simulation of Rainfall over Uttarakhand, India, in June Using WRF-ARW Model and Impact of AIRS Profiles 287
Sqn Ldr Prabodh Shukla, Wg Cdr Anil Kumar Devrani, and Sqn Ldr Himanshu Singh

Incessant Rainfall Event of June 2013 in Uttarakhand, India: Observational Perspectives 303
M.R. Ranalkar, H.S. Chaudhari, G.K. Sawaisarje, A. Hazra, and S. Pokhrel

Convergence of Synoptic and Dynamical Conditions Responsible for Exceptionally Heavy Rainfall over Uttarakhand, India 313
Charan Singh

Changes in Rainfall Concentration over India During 1871–2011 325
Naresh Kumar and A.K. Jaswal

Identifying the Changes in Rainfall Pattern and Heavy Rainfall Events During 1871–2010 over Cherrapunji 335
Pulak Guhathakurta, Preetha Menon, and N.B. Nipane

Part IV Drought and Temperature

Agricultural Drought Assessment: Operational Approaches in India with Special Emphasis on 2012 349
S.S. Ray, M.V.R. Sesha Sai, and N. Chattopadhyay

Trends in Extreme Temperature Events over India During 1969–2012 365
A.K. Jaswal, Ajit Tyagi, and S.C. Bhan

Analysis of Extreme High Temperature Conditions over Uttar Pradesh, India 383
Ramesh Chand and Kamaljit Ray

Use of Remote Sensing Data for Drought Assessment: A Case Study for Bihar State of India During Kharif, 2013 399
K. Choudhary, Inka Goel, P.K. Bisen, S. Mamatha, S.S. Ray, K. Chandrasekar, C.S. Murthy, and M.V.R. Sesha Sai

Index 409

About the Editors

Dr. Kamaljit Ray is presently the Incharge of Nowcasting Unit of India Meteorological Department. She was instrumental in starting the All India Nowcast of thunderstorms in 2013 for 120 cities covered by Doppler weather radar network of IMD. She was also a faculty member in the Gujarat University, Department of Physics Electronics and Space Science. She has more than 21 years of experience in meteorological services, research and forecasting and has about 25 research publications in peer-reviewed journals and proceedings.

Dr. M. Mohapatra is Head of Cyclone Warning Division of India Meteorological Department and also looks after the activities of WMO recognized Regional Specialised Meteorological Centre for Tropical Cyclones at IMD, New Delhi. His main research interests include high impact weather events including tropical cyclones. He has 21 years of experience in meteorological services and research and is the author of 50 research papers published in peer-reviewed journals. He has received a number of recognitions including 25th Biennial Mausam Award and Young Scientist Award of Ministry of Earth Sciences (MoES), Government of India, for his research contributions in the field of atmospheric sciences.

B.K. Bandyopadhyay is Deputy Director General of Meteorology (Services), India Meteorological Department. He joined as a Research Scholar in Indian Institute of Tropical Meteorology, Pune, and during next 3 years, he was associated with research on microphysical characteristics of clouds. He joined India Meteorological Department in 1981 and was engaged in operational weather forecasting for past 30 years which mainly included cyclone and heavy rainfall warning services and allied meteorological research. He has made significant research contributions mainly on tropical cyclones. He has about 40 research publications in national and international journals.

Dr. L.S. Rathore is Director General of Meteorology, India, Meteorological Department and Permanent Representative of India with World Meteorological Organization. He is Co-Vice Chairman of Intergovernmental Board of Climate Services (IBCS) and former Vice President of Commission for Agriculture Meteorology, WMO, and presently on its management board. He is former chairman of SAARC Meteorological Research Centre, Dhaka, and also former President of Indian Meteorological Society and President of Association of Agro-meteorologists. He made significant contribution in setting up Integrated Agro-meteorological Service in India. He has 33 years of experience in meteorological services and research and has published about 100 research papers and seven books. He is recipient of Dr Lakhi Ram Memorial Award 2011 constituted by Society for Recent Development in Agriculture. He has been conferred Fellowship by Indian Meteorological Society.

Part I
Thunderstorms

Study of Severe Thunderstorms over Bangladesh and Its Surrounding Areas During Pre-monsoon Season of 2013 Using WRF-ARW Model

Md. Abdul Mannan, Md. Nazmul Ahasan, and Md. Shah Alam

1 Introduction

Bangladesh is located in the northeastern part of the Indian subcontinent and faces the Bay of Bengal in the south and the Meghalaya plateau in the northeast. Almost the entire country is less than 10 m above sea level and on a flat plane. Severe Thunderstorms (henceforth referred to simply as STS) frequently occur in Bangladesh during the pre-monsoon season from March to May, causing deaths and damage to property every year. In Bangladesh, STSs are classified depending on the magnitude of wind speed. The ones producing wind gusts above 42 m s^{-1} are defined as tornadoes, while those producing wind gusts ranging from 11 to 42 m s^{-1} are defined as 'nor'westers'. The term 'nor'wester' means that STS come mostly from the northwestern direction. Despite being highly arbitrary, such criteria for classifying STSs have been used in a number of climatological studies addressing STSs in Bangladesh (Yamane et al. 2008).

Chowdhury and Karmakar (1986) investigated the climatology of nor'westers and reported that nor'westers occurred most frequently in the north central region of Bangladesh during the pre-monsoon season, peaking in April. Yamane and Hayashi (2006) showed the seasonal variation of Convective Available Potential Energy (CAPE) and the vertical wind shear between the surface and the midlevel of the troposphere in Bangladesh using ERA-40. They showed that both CAPE and vertical wind shear are high during the pre-monsoon season with a peak in April. Brooks

---

Md.A. Mannan (✉) • Md.S. Alam
SAARC Meteorological Research Centre (SMRC), Dhaka, Bangladesh

Bangladesh Meteorological Department (BMD), Dhaka, Bangladesh
e-mail: mannan_u2003@yahoo.co.in

Md.N. Ahasan
Bangladesh Meteorological Department (BMD), Dhaka, Bangladesh

et al. (2003) depicted that the atmospheric conditions displaying high CAPE and strong vertical wind shear are favourable for convective storms.

In Bangladesh, there has been little research on the environmental conditions of STS. Although some case studies of STS have been performed, but their environmental conditions have not been comprehensively studied. Convective parameters (e.g., CAPE) are useful tools for forecasting of STSs. Many statistical studies of convective parameters in the outbreak of STS have been conducted. Rasmussen and Blanchard (1998) showed statistical climatology of convective parameters in the outbreak of tornadic supercells in the United States using rawinsonde data. But STSs are one of the least predictable weather phenomena, especially if they are severe. They may cause damage to property and electric utilities, and endanger humans and livestock (Schemetis et al. 2008). Karmakar and Alam (2006) showed the statistics of convective parameters associated with nor'westers during the pre-monsoon season in Bangladesh using rawinsonde data at 0000 UTC in Dhaka. They provided critical values indicating the likelihood of occurrence of nor'westers for each parameter. However, the critical values provided in their study are subjectively determined.

According to Bangladesh Meteorological Department (BMD), a number of severe thunderstorms have occurred over Bangladesh and its surrounding areas in the pre-monsoon season (March–May) of 2013. They are associated with high winds and moderate to heavy rainfalls. But the events of 22 March, 16, 19 and 27 April, 4, 8, 13 and 16 May are significant and have considerable impact on the life and livelihood of the affected areas. The prediction process of the severe thunderstorms over SAARC region including Bangladesh is still inadequate and demands further study for its improvement. The present study comprehensively examines the environmental conditions of STS occuring during 19 April in Bangladesh. Simulated parameters are investigated. It is believed that the present study greatly contributes to the understanding and forecasting of the environmental conditions of STS in Bangladesh.

1.1 Observed Weather

On 19 April 2013, STSs were recorded at different places of western, southwestern regions and some parts of central region of Bangladesh with the maximum winds of 70 km/h at Ishurdi. Some of the information related to this is given in Table 1. Light to moderate rainfall was recorded at different places of Bangladesh in addition to the places where severe thunderstorms were recorded. Significant amounts of rainfall were recorded at Ishurdi (57 mm), Jessore (47 mm), Rangpur (39 mm), Srimongal (21 mm) and Chuadanga (20 mm). Spatial distribution depicts that the light to moderate rainfall was recorded over west-central and extreme northern parts and light rainfall over central and northern parts of Bangladesh (Fig. 1a). Similar signatures were found from TRMM (version 7) product (Fig. 1b).

Table 1 Recorded significant winds and rainfall on 19 April 2013 in Bangladesh

Station	Gusty/Squally wind			Significant weather	
	Direction (°)	Speed (km/h)	Time (UTC)	Rainfall (mm)	Hail
Chuadanga	NW'ly	67	1400–1430	20	–
Faridpur	W'ly	50	1630	7	–
Khulna	SW'ly	46	1330–1400	17	–
Ishurdi	NW'ly	70	1430–1440	57	–
Dhaka	W'ly	41	1500–1530	9	–

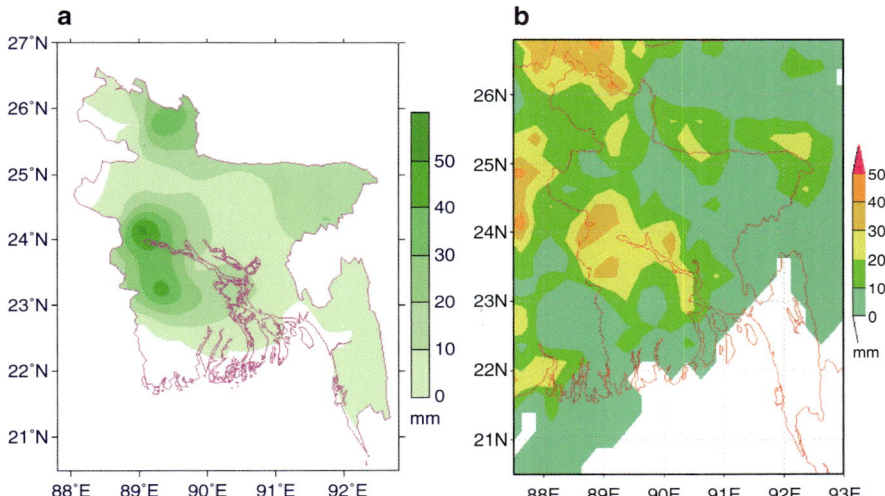

Fig. 1 Spatial distribution of rainfall for (**a**) observed and (**b**) TRMM over Bangladesh on 19 April 2013

1.2 Synoptic Condition on 19 April 2013 over Bangladesh and Adjoining Areas

NCEP reanalysis data set reveals a low pressure area lay over Bihar and adjoining area with its trough extending to West Bengal at 0000 UTC. Another part of the trough extended to Jharkhand, Chattisgarh and Orissa in India. At 0600 UTC, the distribution of MSLP was similar to the trough but it extended to central part of Bangladesh and covered some more parts over eastern ghat of India. At 1200 UTC, the low pressure system and its associated trough strengthened and covered Bangladesh and its adjoining areas including some more parts over India. At 1800 UTC, the low pressure system weakened (Fig. 2). The surface winds were southerly flowing from the Bay of Bengal and bringing moisture over central and eastern parts of India and Bangladesh. The positive vorticity field over this region at surface level was also established (Fig. 3). Accordingly, a strong CAPE field persisted over

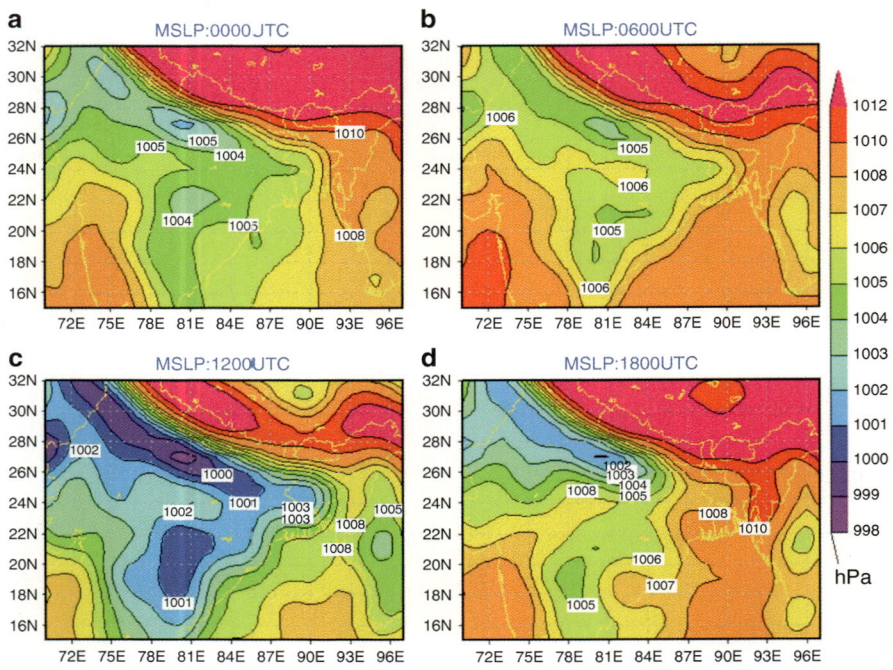

Fig. 2 MSLP distribution over Bangladesh and adjoining areas at (**a**) 0000, (**b**) 0600, (**c**) 1200 and (**d**) 1800 UTC on 19 April 2013 derived from NCEP data

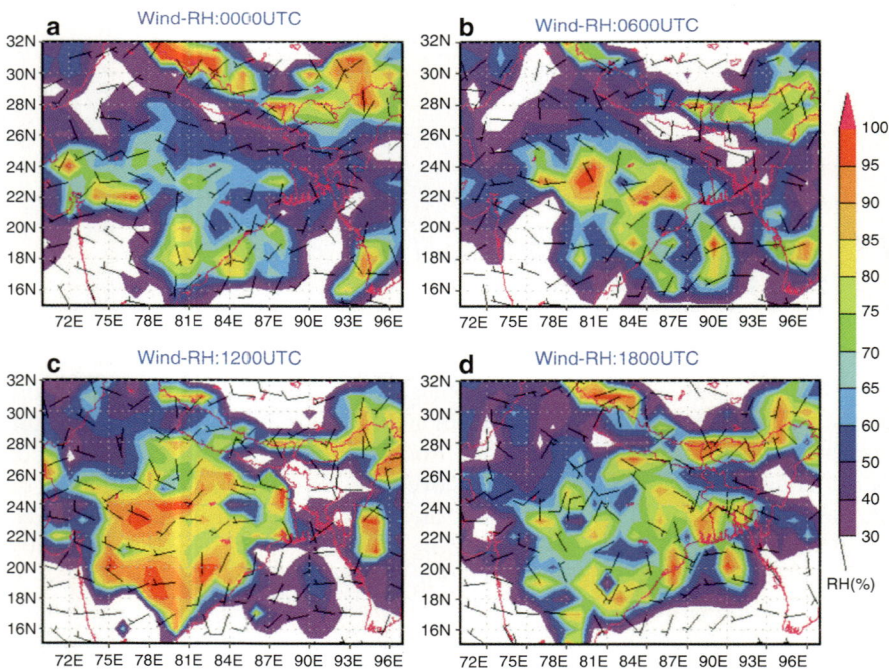

Fig. 3 RH distribution over Bangladesh and adjoining areas at (**a**) 0000, (**b**) 0600, (**c**) 1200 and (**d**) 1800 UTC on 19 April 2013 derived from NCEP data

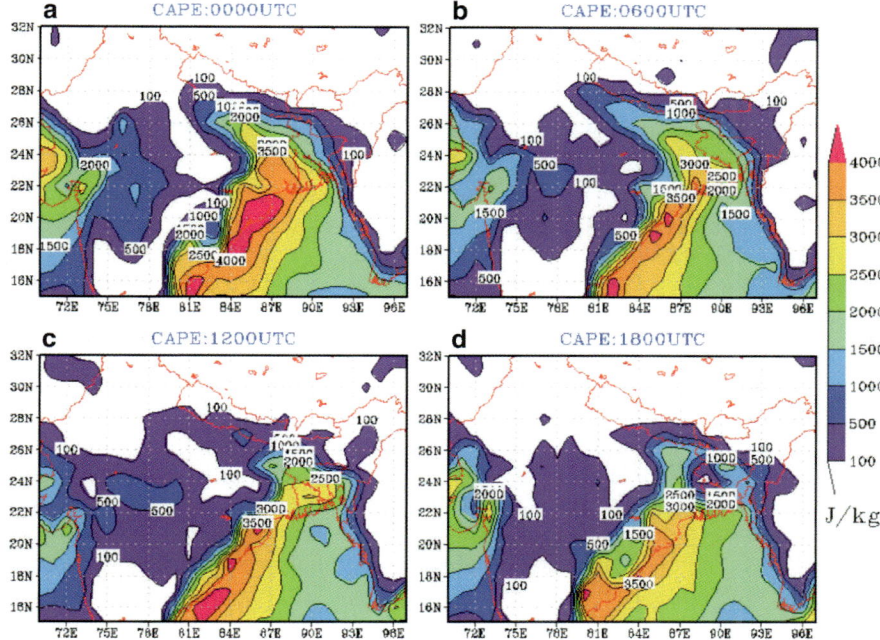

Fig. 4 CAPE distribution over Bangladesh and adjoining areas at (**a**) 0000, (**b**) 0600, (**c**) 1200 and (**d**) 1800 UTC on 19 April 2013 derived from NCEP data

eastern Ghat of India and adjoining Bay of Bengal which extended upto West Bengal of India, Bangladesh and adjoining areas and became strong with the progress of the day and weakened after the occurrence of severe thunderstorm over Bangladesh and adjoining areas (Fig. 4).

2 Methodology

WRF-ARW model (version 3.2.1) with the grid resolution of 9 km is used to diagnose the event using Ferrier (FR), Kessler (KS), Lin et al. (LN), WRF Single-Moment 5 Class (WSM5), WRF Single-Moment 6 Class (WSM6) and Thompson Graupel (TH) microphysics (MPs) schemes with the combination of Betts-Miller-Janjic (BMJ), Grell-Devenyi ensemble (GD), Kain-Fritsch (KF) and New Grell (NG) cumulus scheme (CPs) for extracting the physical processes. Therefore, the combination of MPs and CPs are: BMJFR, BMJKS, BMJLN, BMJTH, BMJWSM5, BMJWSM6, GDFR, GDKS, GDLN, GDTH, GDWSM5, GDWSM6, KFFR, KFKS, KFLN, KFTH, KFWSM5, KFWSM6, NGFR, NGKS, NGLN, NGTH, NGWSM5 and NGWSM6. The coverage area of the model domain is 12–30°N and 80–100°E.

The topography in the model is obtained from USGS land cover data set. NCEP data have been provided at every 6 h as initial and boundary conditions. The model has been run with 19 sigma levels in the vertical direction from the ground to the 100 hPa level to simulate and analyse the parameters of sea level pressure (SLP), relative humidity (RH), wind at 10 m, u and v-components of upper wind, vorticity, convergence and divergence, convective rain and non-convective rain etc. extracted by GrADS. Kringing method facilitated by Win Surfer (version 7.0) software has been used for preparation of all kinds of spatial distributions.

Convective parameters are essential to calculate/quantify the environmental conditions in the outbreak of STS. Commonly used convective parameters are K Index (KI), Lifted Index (LI), Total Totals Index (TTI), Showalter Index (SI), Precipitable Water (PW), Convective Available Potential Energy (CAPE), Convective Inhibition (CIN), Mean Shear (MS), Storm Relative Environmental Helicity (SREH), Vorticity Generation Parameter (VGP), Energy Helicity Index (EHI), and Bulk Richardson Number (BRN). MS, SHEAR and SREH are measures of vertical wind shear of the atmosphere. The vertical wind shear is important for organisation of convection like supercell and multicell (Weisman and Klemp 1982). Supercells and multicells have greater potential to produce severe weather such as tornadoes than ordinary cells (Houze 1993). Therefore, the vertical wind shear is an important ingredient for the outbreak of STS (Brooks et al. 2003). VGP, EHI and BRN are the combinations of the thermal instability and vertical wind shear parameters and common indicators of the formation of supercells and tornadoes. In this study KI, TTI, PW, equivalent potential temperature (θ_E), LI, CAPE, CIN, EHI and SREH are calculated and analysed as per the following definition:

(i) K Index is defined (George 1960) as:

$$KI = T_{850\,hPa} - T_{500\,hPa} + Td_{850\,hPa} - (T_{700\,hPa} - Td_{700\,hPa})$$

where T is temperature and Td is dew point temperature. $T_{850\,hPa} - T_{500\,hPa}$ indicates the lapse rate of temperature between the lower layer and middle layer. $Td_{850\,hPa}$ indicates the moisture content in the lower layer. $T_{700\,hPa} - Td_{700\,hPa}$ is a measure of the reduction of negative buoyancy through entrainment of dry air. The KI exceeding 28 K indicates the likelihood of convection (Fuelbarg and Biggar 1994).

(ii) Total Total Index (TTI) is defined (Sadowski and Rieck 1977) as:

$$TT = (T_{850\,hPa} + Td_{850\,hPa}) - 2T_{500\,hPa}$$

where T is the dry bulb temperature and Td is the dew point temperature.

(iii) Lifted Index (LI) is defined (Galway 1956) as:

$$LI = T_{500\,hPa} - Tp^*_{500\,hPa}$$

where $Tp^*_{500\,hPa}$ is the temperature of a parcel with the mean temperature and dew point temperature in the lowest 100 hPa lifted adiabatically until saturated, and then moist adiabatically to 500 hPa. $T_{500\,hPa}$ is the temperature at 500 hPa. Negative LI indicates the likelihood of convective activity. The LI explicitly reflects the condition in the boundary layer compared with the SSI.

(iv) Precipitable Water (PW) is defined (Huschke 1959) as:

$$PW = \int_{Ps}^{0} q\,dp = \int_{Ps}^{100hPa} q\,dp$$

where Ps is the surface pressure and q is the specific humidity. In this study, the top pressure is defined as 100 hPa because of the scarce amount of water vapour above 100 hPa.

(v) Convective Available Potential Energy (CAPE) is defined (Moncrieff and Miller 1976) as:

$$CAPE = g \int z_{LFC} z\, EL\,(T_{vp} - T_v)/T_v\,dz$$

where g is the gravitational acceleration, z_{LFC} is the level of LFC (level of free convection), and z_{EL} is the equilibrium level, where the temperature excess of a parcel lifted from the LFC first becomes zero above the LFC, and T_{vp} and T_v are the virtual temperatures of the air parcel and the environment, respectively. The CAPE is referred as the net work of the environment on a parcel per unit mass lifted from z_{LFC} to z_{EL} and the measurement of the development of convection.

(vi) Convective Inhibition (CIN) is defined (Colby 1984) as:

$$CIN = g \int z_i z LFC\,(T_{vp} - T_v)/T_v\,dz$$

where z_i is the initial level of the parcel lifted for calculating the CAPE, and T_{vp} and T_v are the virtual temperatures of the air parcel and the environment, respectively. The CIN is referred as the required net work to lift a negatively buoyant parcel per unit mass to the LFC. In general, the CIN is used for the measurement of stability of the atmosphere.

(vii) Storm Relative Environmental Helicity (SREH) index is defined (Davies-Johnes 1984) as:

$$SREH = -\int_0^h \kappa \cdot (V - C) \times (\partial V/\partial z)\,dz$$

where V is the horizontal velocity vector, C is the storm motion vector, k is the unit vector in the vertical and h is the depth over which the integration is performed (3 km herein).

(viii) Energy Helicity Index (EHI) is defined (Hart and Korotky 1991; Davies 1993) as:

$$EHI = (SREH \times CAPE)/1.6 \times 10^5$$

The EHI is the combination of the SREH and CAPE, and a measure of tornadic supercell. Rasmussen and Blanchard (1998) showed the EHI as highly correlated with the generation of supercells in the United States of America.

(ix) Equivalent potential temperature (θ_E) is characterised (Davies-Jones 2009) as:

$$\theta e = \theta \exp\left(L_0^* r / C_{pd} T_L\right)$$

where $L_0^* = 2.690 \times 10^6$ J kg^{-1}.

L_0 is the latent heat of condensation at temp T_L, θ is partial potential temperature, C_{pd} is specific heat of dry air, and r is mixing ratio in gram of water vapour per gram of dry air.

3 Results and Discussion

Simulated mean sea level pressure, surface and upper air relative humidity, wind at 10 m and upper levels vorticity ($\times 10^{-5}$ s^{-1}), divergence, vorticity, wind at 10 m for different MPs with CPs, recorded maximum wind at 10 m and rainfall of BMD on 19 April 2013 are plotted, presented and discussed in the following sub-sections.

3.1 Surface Pressure, Wind and Relative Humidity

Simulation depicted that a low pressure area formed over West Bengal of India and adjoining western part of Bangladesh at 0600 UTC which then became marked over the same area during next successive hours and extended to Bangladesh and its surrounding areas (Fig. 5). In response to this situation, the surface pressure of Bangladesh reduced and reached to its lowest level during 0900–1100 UTC and the lowest simulated minimum pressure of 1003.0 hPa was found at 0900 UTC for NGFR. Similar situation was observed at Chuadanga, Ishurdi, Khulna and Faridpur where STSs with rainfall were recorded (Fig. 6a). But in support of this situation, temperature at 2 m increased sharply over Bangladesh from morning to 0700 UTC (Fig. 6b). Due to decrease in SLP and increase in temperature, the RH decreases in surface levels but with the advection of moisture from the Bay of Bengal, the RH increases in the layer of 850–750 hPa. It decreases again in the layer of 700–500 hPa but increases in the layer above it due to the presence of positive vorticity and convergence. The vertical profile of area average RH for all combinations of MPs with CPs had similar pattern (Fig. 7). This feature was observed at all the places where STSs were recorded (Fig. 8).

3.2 Zonal and Meridional Winds

Area average zonal winds at 950 hPa over Bangladesh and its adjoining areas are negative in the morning but in the afternoon they become positive and strong (Fig. 9a). At 850 hPa it is positive in the night but decreases during morning and

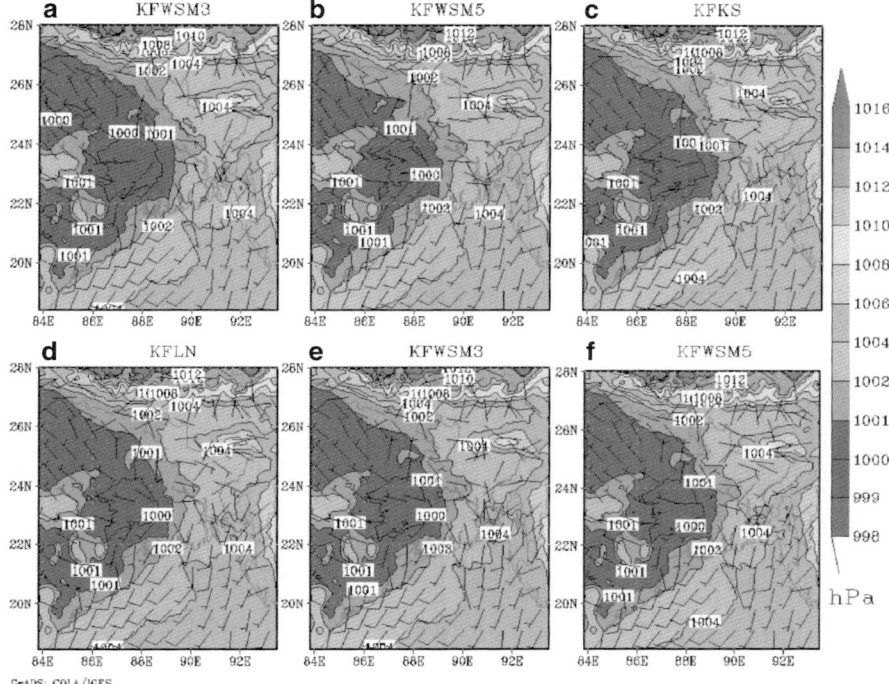

Fig. 5 Simulated MSLP and with (10 m) distribution for (**a**) KFFR, (**b**) KFKS, (**c**) KFLN, (**d**) KFTH, (**e**) KFWSM3 and (**f**) KFWSM6 over Bangladesh and adjoining areas at 1100 UTC of 19 April 2013

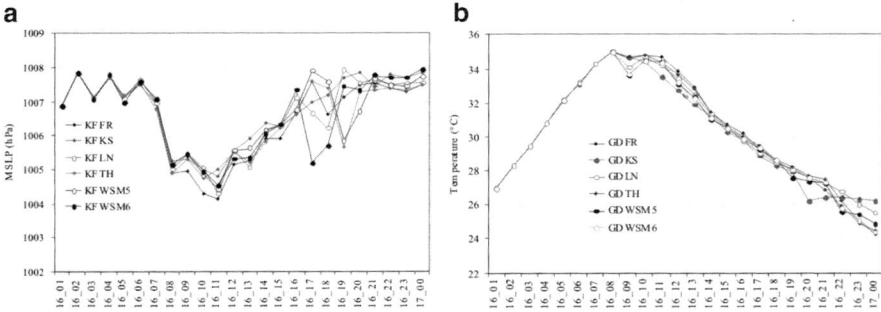

Fig. 6 Temporal variation of (**a**) SLP and (**b**) temperature (2 m) at Ishwardi for different combination of MPs and CPs on 19 April 2013

becomes negative temporarily in the early afternoon and then positive and increases till midnight (Fig. 9b). Above this layer, zonal wind components are absolutely positive but decrease in the afternoon. Similar situation was observed for different

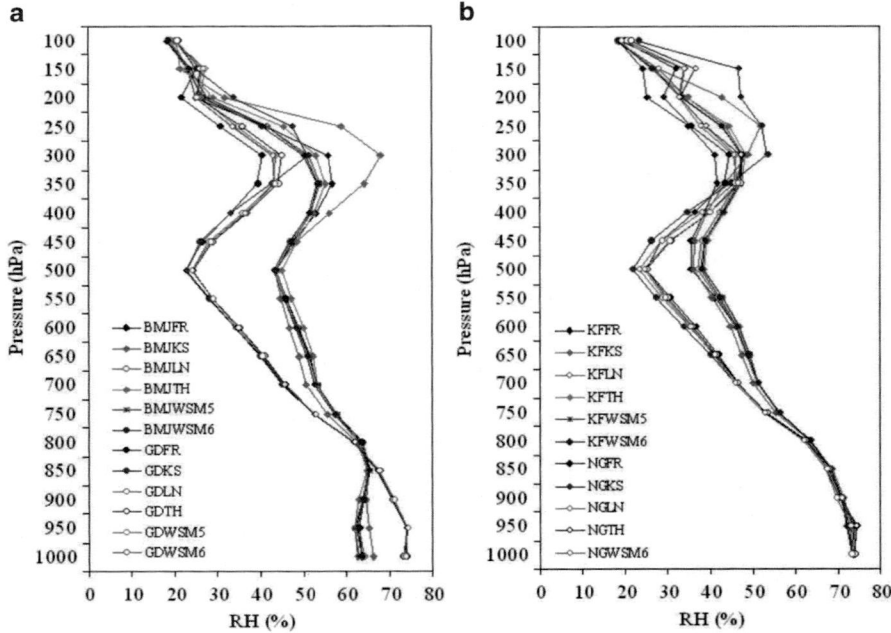

Fig. 7 Vertical profile area average RH: (**a**) for BMJFR, BMJKS, BMJLN, BMJTH, BMJWSM5, BMJWSM6, GDFR, GDKS, GDLN, GDTH, GDWSM5 and GDWSM6, and (**b**) for KFFR, KFKS, KFLN, KFTH, KFWSM5, KFWSM6, NGFR, NGKS, NGLN, NGTH, NGWSM5 and NGWSM6 over Bangladesh and adjoining areas on 19 April 2013

combinations of CPs and MPs at the places where the severe thunderstorms were recorded (Fig. 10).

Area average meridional winds over Bangladesh and its adjoining areas were absolutely positive in the layer 950–800 hPa but in the afternoon winds become strong (Fig. 11a). They were negative in the layer 750–500 hPa during morning but became positive and strong during the later period of the day (Fig. 11b). They were absolutely positive above this layer during the observed period. Accordingly, positive meridional winds in the afternoon period were stronger for different combinations of CPs and MPs at the places where the severe thunderstorms were recorded (Fig. 12).

3.3 Divergence and Vorticity

Following the surface wind field area, average divergence remained positive in the lower layer over Bangladesh during the early hours of the day and then convergence was established and became prominent as well as expanded to upper level.

Study of Severe Thunderstorms over Bangladesh... 13

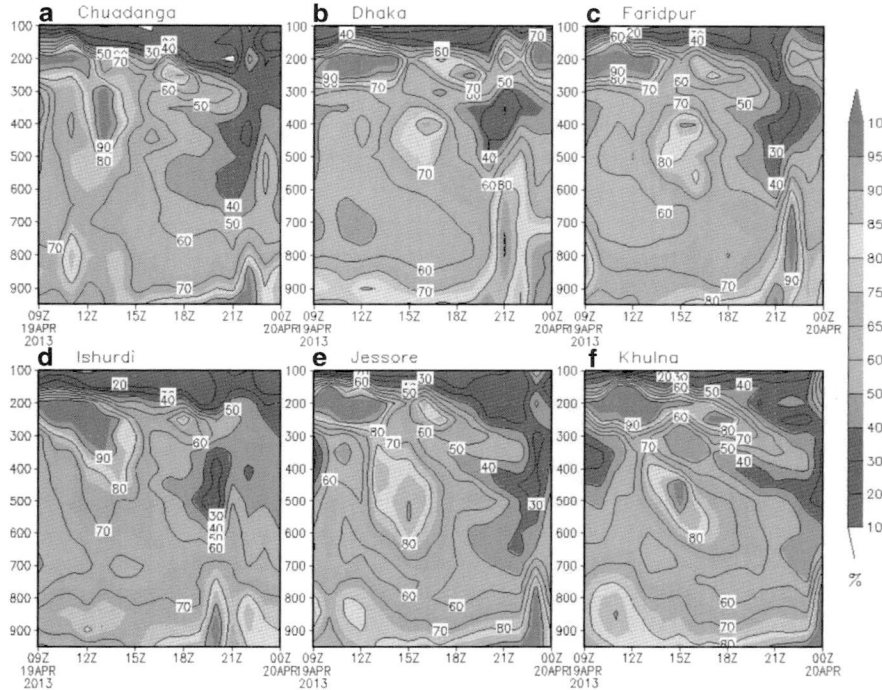

Fig. 8 Vertical profile RH for (**a**) Chuadanga, (**b**) Dhaka, (**c**) Faridpur, (**d**) Ishurdi, (**e**) Jessore, and (**f**) Khulna for KFWSM6 on 19 April 2013

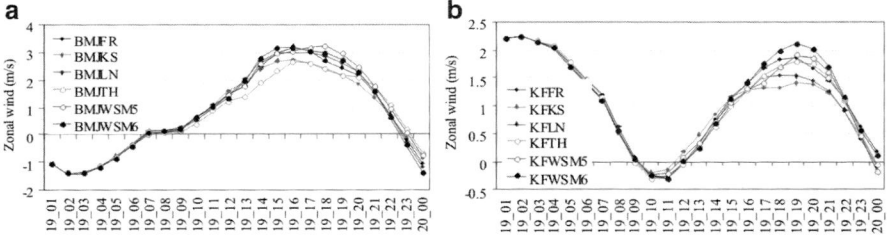

Fig. 9 Temporal variation of area average zonal wind for (**a**) 950 and (**b**) 850 hPa on 19 April 2013

Convergence persisted in the layer 400–150 hPa during the early hours of the day but it changed to divergence in the later period. Combination of this situation where convergence establishes in the lower layer and divergence in the upper layer during afternoon period and continues till the event happens was common over the places where the STS and rainfall occurred over Bangladesh and adjoining areas (Fig. 13).

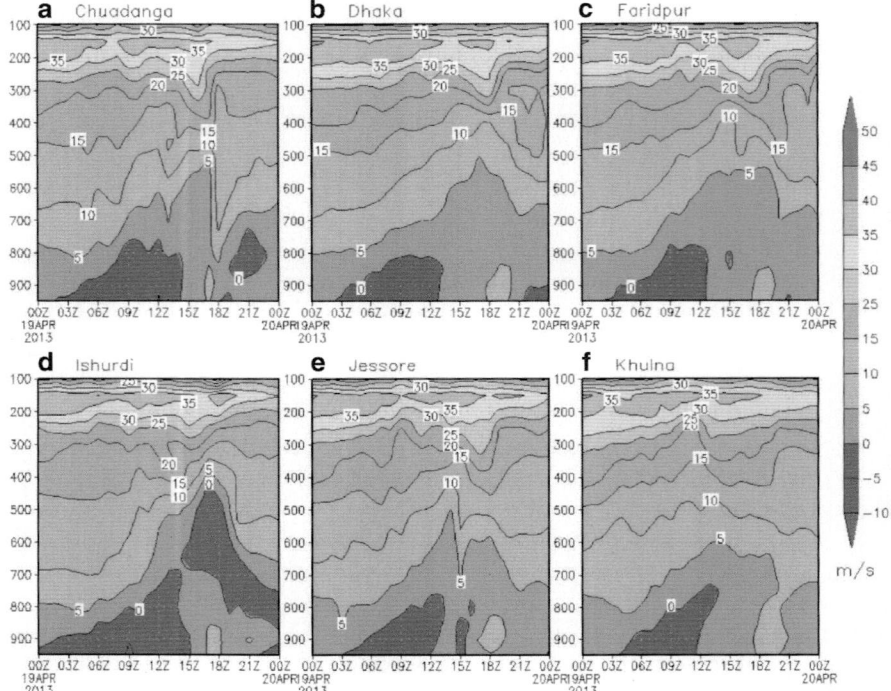

Fig. 10 Vertical profile zonal wind over (**a**) Chuadanga, (**b**) Dhaka, (**c**) Faridpur, (**d**) Ishurdi, (**e**) Jessore and (**f**) Khulna for KFTH on 19 April 2013

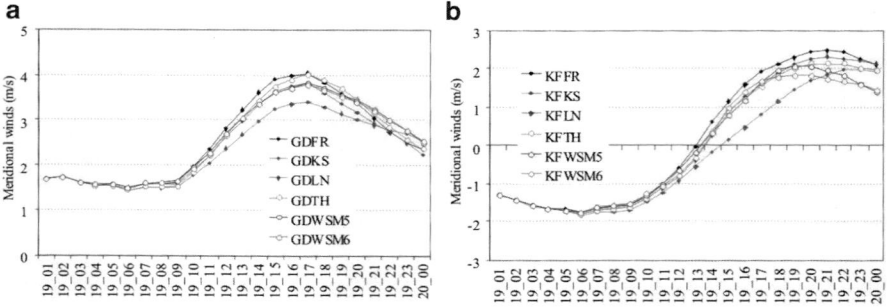

Fig. 11 Temporal variation of area average zonal wind for (**a**) 850 and (**b**) 700 hPa on 19 April 2013

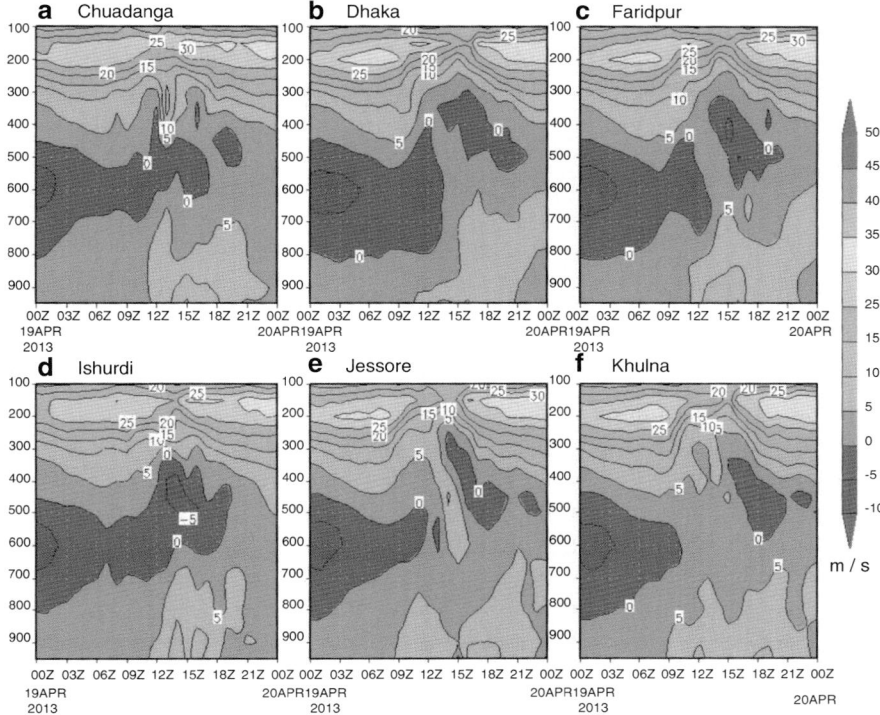

Fig. 12 Vertical profile meridional wind over (**a**) Chuadanga, (**b**) Dhaka, (**c**) Faridpur, (**d**) Ishurdi, (**e**) Jessore and (**f**) Khulna for NGFR on 19 April 2013

Area average vorticity remained positive during the early hours of the day and expanded to higher levels, but it changed to negative value during the later period of the day. It remained positive in the layer 600–200 hPa throughout the day. Accordingly, negative vorticity was established in the lower levels and positive vorticity in the upper levels. In other words, positive vorticity continued in the upper levels with extension in the lower levels. But the short-lived positive vorticity spikes were observed just before the event, at the places where STS was recorded (Fig. 14).

3.4 Convective Parameters

On the basis of the simulated lowest pressure for different MPs with CPs, the instability indices of K, TT, PW, θ_E, LI, CAPE, CIN, EHI and SREH were calculated. It was found that the convective thresholds were sufficiently strong and were found at

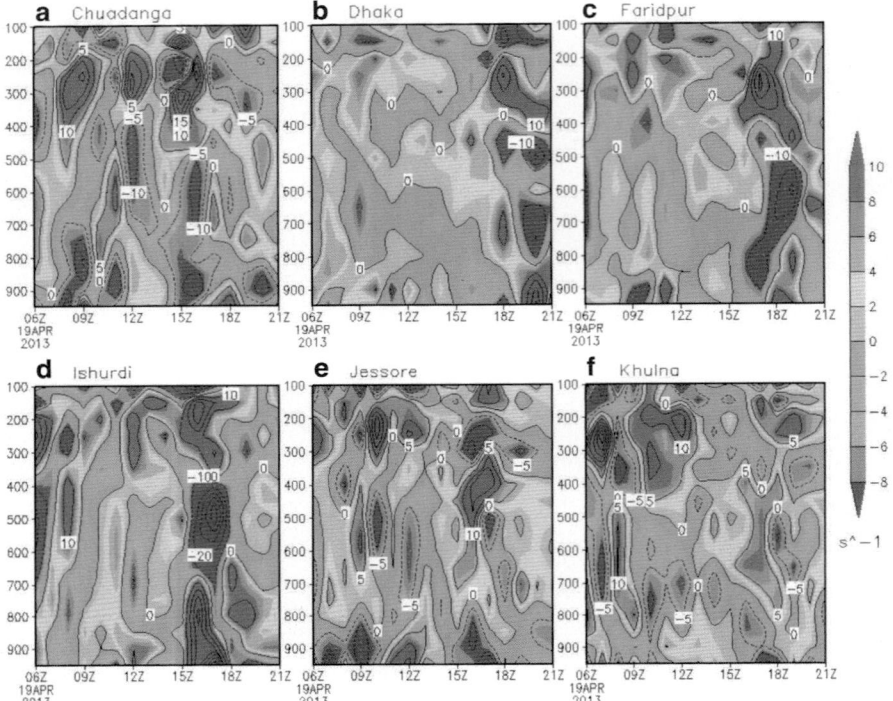

Fig. 13 Vertical profile divergence over (**a**) Chuadanga, (**b**) Dhaka, (**c**) Faridpur, (**d**) Ishurdi, (**e**) Jessore and (**f**) Khulna for KFWSM6 on 19 April 2013

Chuadanga, Ishurdi, Rajshahi and Satkhira where STSs were recorded. The brief description is summarized in Table 2.

The mean and median values for KI index on the day of STS over Bangladesh are 27.6 and 29.0 K (Yamane et al. 2010). The simulated KI indices at the places where STS was recorded were higher than these indices, indicating most unstable atmosphere. Simulated θ_E was much higher than its minimum thresholds of 320 K for severe thunderstorm. The mean and median values for LI are −0.2 and 0.1 K (Yamane et al. 2010) but simulated LI were within the range of −9 to −2 K. The mean and median values for the PW on STS are 38.2 and 38.5 g kg^{-1} (Yamane et al. 2010) but the simulated values were much higher than the minimum thresholds.

The mean and median values of CAPE for STS over Bangladesh are 1,363 and 1,170 J kg^{-1} (Yamane et al. 2010) but simulated CAPE was much higher than that. The mean and median values for CIN for STS over Bangladesh are 322 and 300 J kg^{-1} (Yamane et al. 2010) but the simulated CIN were much lower than that. The mean and median values of EHI for STS over Bangladesh are

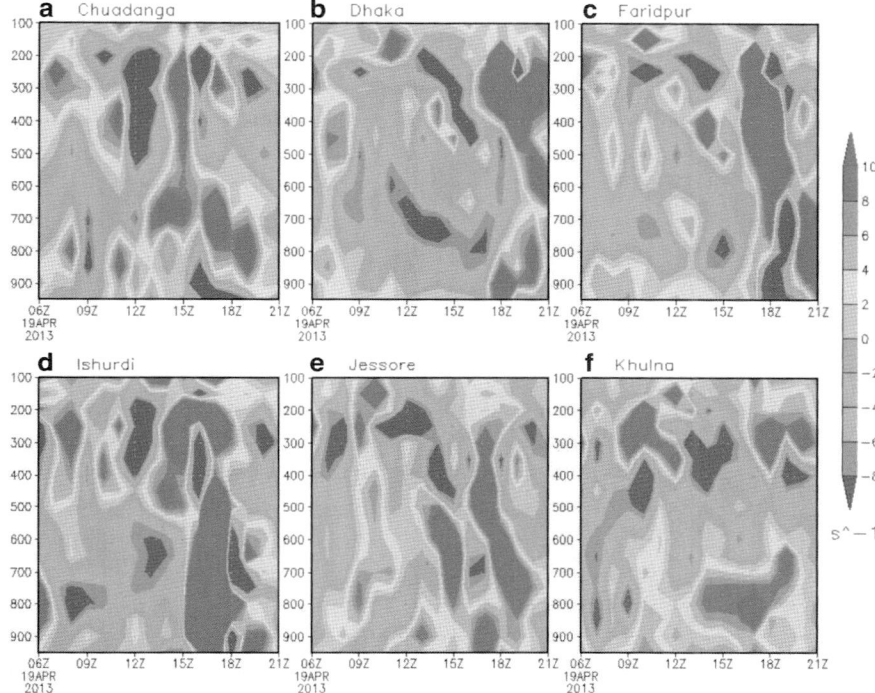

Fig. 14 Vertical profile vorticity over (**a**) Chuadanga, (**b**) Dhaka, (**c**) Faridpur, (**d**) Ishurdi, (**e**) Jessore and (**f**) Khulna for KFWSM6 on 19 April 2013

1.32 and 0.43 cm (Yamane et al. 2010), however the simulated thresholds were much higher than that. Similarly, simulated SREH were significantly higher than the threshold over Bangladesh. Therefore, the convective indices were found to be very strong indicator of very high instability at the places where STSs were recorded.

3.5 Rainfall

Simulated rainfall distribution and rainfall amounts for different combinations were different. Simulated rainfalls for BMJ group gave the signature of rainfall over southwestern part with the maximum over Satkhira region but it could not capture rainfall over other parts of Bangladesh. Simulated rainfalls for GD set showed rainfall over western and northern parts but the amounts are lower than the

Table 2 Simulated instability indices (thresholds) in Bangladesh on 19 April 2013

Combination	KI	TTI	PW	θ_E (K)	LI	CAPE	CIN	EHI	SREH
BMJFR	30	52	2.96	342	−3	816	12	66	12
BMJKS	33	50	3.45	340	−2	529	5	33	16
BMJLN	30	52	2.9	342	−3	775	24	25	−19
BMJTH	33	53	3.04	339	−4	608	44	79	50
MBJWSM5	29	51	3.09	341	−5	866	40	68	51
BMJWSM6	40	52	4.32	348	−5	1,355	0	40	70
GDFR	37	56	3.37	348	−7	1,791	4	24	33
GDKS	35	56	3.94	359	−9	3,325	0	31	2
GDLN	36	56	3.16	348	−7	1,808	23	35	33
GDTH	35	55	3.26	347	−6	1,733	22	52	56
GDWSM5	37	58	3.89	352	−9	2,297	8	18	34
GDWSM6	36	58	3.88	352	−8	2,291	7	19	33
KFFR	36	55	3.2	345	−6	1,481	63	18	14
KFKS	39	56	3.96	350	−7	2,064	9	35	57
KFLN	36	56	3.49	349	−7	1,952	27	27	20
KFTH	35	55	3.25	345	−6	1,480	54	38	35
KFWSM5	36	55	3.21	345	−6	1,602	71	36	33
KFWSM6	34	55	3.43	349	−7	1,938	25	36	30
NGFR	40	56	3.32	347	−6	1,640	0	16	23
NGKS	42	56	3.53	348	−6	1,721	2	48	63
NGLN	41	56	3.37	347	−5	1,627	0	16	25
NGTH	42	55	3.52	348	−6	1,693	1	48	61
NGWSM5	−	−	−	−	−	−	−	−	−
NGWSM6	42	56	3.47	348	−6	1,706	1	50	61

observed. KF cluster showed over western and central parts with the maximum over Chuadanga-Jessore-Ishurdi region for KFTH, KFWSM5 and KFWSM6 where maximum amounts of rainfall were recorded (Fig. 15). But there was no signature of rainfall over northern part of Bangladesh for this cluster. KF groups gave good coverage of rainfall and the amounts of rainfall were higher at some areas but the amounts of rainfall were comparable over Chuadanga-Jessore-Ishurdi region where high amounts of rainfall was recorded (Fig. 16).

3.6 Maximum Winds

Simulation depicted the maximum winds of 47.0 (Jessore), 43.9 (Khulna), 47.0 (Hatiya), 40.0 (Jessore), 48.8 (Jessore) and 48.1 (Jeesure) km/h respectively for BMJFR, BMJKS, BMJLN, BMJTH, BMJWSM5 and BMJWSM6. It was lower

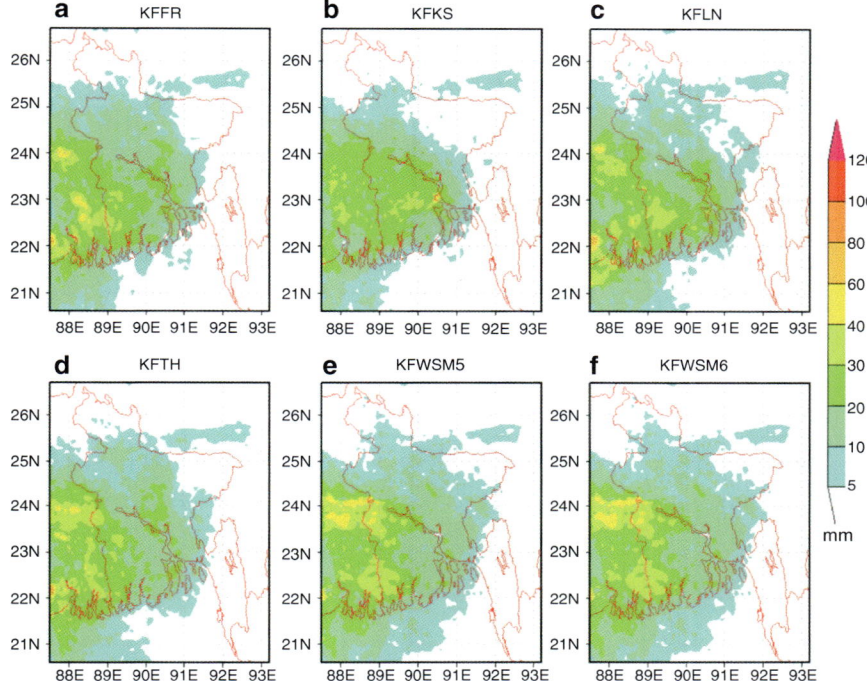

Fig. 15 Simulated rainfall for (**a**) KFFR, (**b**) KFKS, (**c**) KFLN, (**d**) KFTH, (**e**) KFWSM5, and (**f**) KFWSM6 over Bangladesh on 19 April 2013

for other combinations of CPs and MPs. Simulated maximum wind for all the combinations of CPs and MPs was much lower than observed at Chuadanga and Ishurdi. It was lower than observation at Dhaka and Faridpur. But it was comparable and close to observation at Khulna for BMJFR, BMJKS, BMJLN, BMJTH, BMJWSM5 and BMJWSM6 and lower than observed for all other combinations of CPs and MPs (Fig. 17a).

Analysis depicts that there was sudden increase of wind speed at 10 m height during 1200–1500 UTC over Bangladesh. The maximum winds simulated at the stations located over western parts of Bangladesh were recorded earlier than that at the stations located over southern parts, indicating southeastward motion of STS system (Fig. 17b). Spatial distribution of simulated maximum winds reveals that west-central and southwestern parts, including central part of Bangladesh are vulnerable to high wind and accordingly maximum winds were recorded over these parts (Fig. 18).

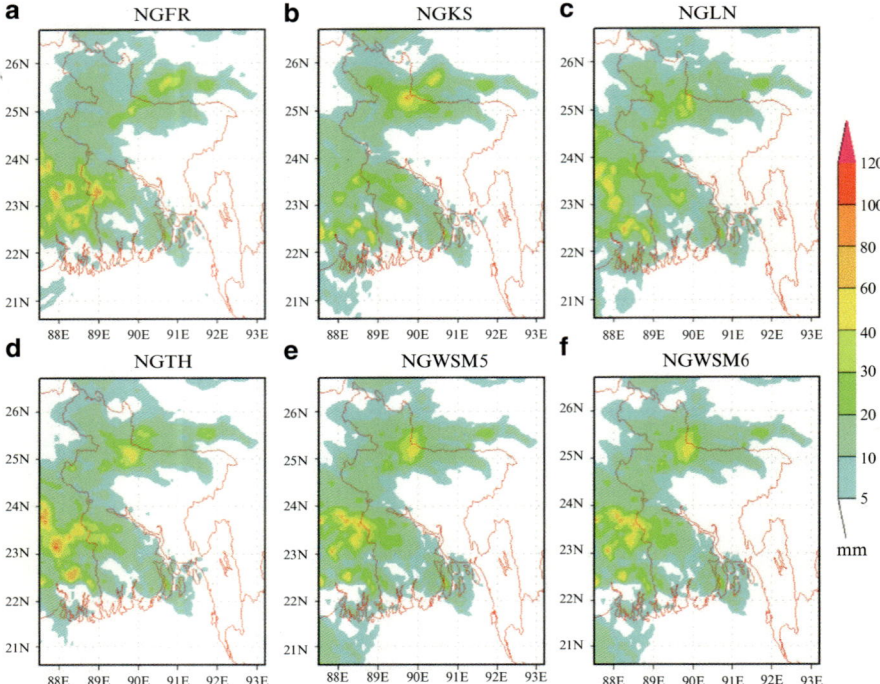

Fig. 16 Simulated rainfall for (**a**) NGFR, (**b**) NGKS, (**c**) NGLN, (**d**) NGTH, (**e**) NGWSM5, and (**f**) NGWSM6 over Bangladesh on 19 April 2013

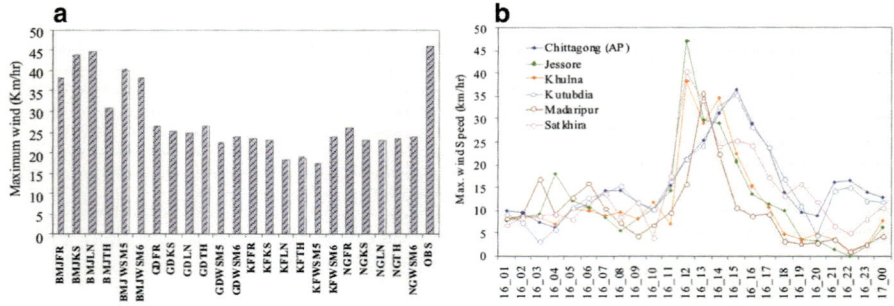

Fig. 17 (**a**) Simulated maximum wind at Khulna for different combinations of CPs and MPs and (**b**) temporal variation of maximum wind for BMJFR on 19 April 2013

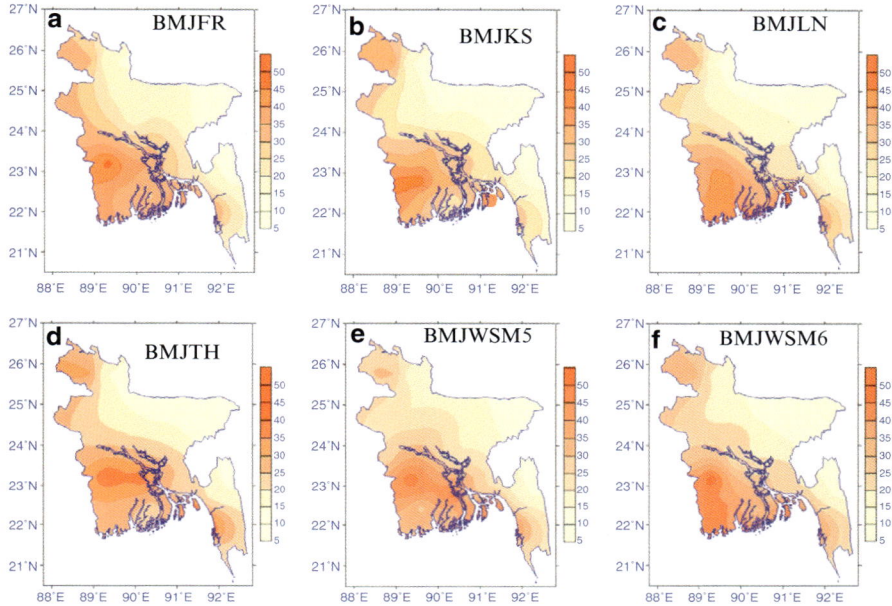

Fig. 18 Spatial distribution of simulated maximum wind for (**a**) BMJFR, (**b**) BMJKS, (**c**) BMJLN, (**d**) BMJTH, (**e**) BMJESM5 and (**f**) BMJWSM6 on 19 April 2013

Conclusion

Model efficiently simulates the genesis properties like divergence, vorticity and instability indices related to STS that occurred over Bangladesh on 19 April 2013. Simulated rainfall for BMJ and GD combinations do not match well with the observed and TRMM rainfall. Simulated rainfall for KF combination captures the rainfall well over western part but it could not capture the observed rainfall over northern part of Bangladesh. NG combination has good coverage for rainfall but there are limitations in calculating location specific rainfall. Model simulates the maximum wind, well over western and southern parts including central part but the magnitudes of maximum winds deviates from the observed maximum wind.

References

Brooks HE, Lee JW, Craven JP (2003) The spatial distribution of severe thunderstorm and tornado environments from global reanalysis data. Atmos Res 67–68:73–94

Chowdhury MHK, Karmakar S (1986) Pre-monsoon nor'westers in Bangladesh with case studies. In: Proceedings of the seminar on local severe storms, Dhaka. Bangladesh Meteorological Department, pp 147–166

Colby FP (1984) Convective inhibition as a predictor of convection during AVE-SESAM-2. Mon Weather Rev 112:2239–2252

Davies-Johnes RP (1984) Streamwise vorticity: the origin of updraft rotation in supercell storms. J Atmos Sci 41:2991–3006

Davies-Jones R (2009) On the computation of equivalent potential temperature. Mon Weather Rev 137:3137–3148. doi:10.1175/2009MWR2774.1

Fuelbarg HE, Biggar DG (1994) The pre-convective environment of summer thunderstorms over the Florida panhandle. Weather Forecast 9:316–326

George JJ (1960) Weather forecasting for aeronautics. Academic, New York

Hart JA, Korotky W (1991) The SHARP workstation v1.50 users guide. National Weather Service, NOAA, U.S. Department of Commerce, 30 pp

Houze RA (1993) Cloud dynamics. Elsevier, New York

Huschke RE (1959) Glossary of meteorology. American Meteorological Society, Boston

Karmakar S, Alam MM (2006) Instability of the troposphere associated with thunderstorms/ nor'westers over Bangladesh during the pre-monsoon season. Mausam 57(4):629–638

Moncrieff M, Miller MJ (1976) The dynamics and simulation of tropical cumulonimbus and squall lines. Q J R Meteorol Soc 102:373–394

Rasmussen EN, Blanchard DO (1998) A baseline climatology of sounding derived supercell and tornado forecast parameters. Weather Forecast 13:1148–1164

Sadowski AF, Rieck RE (1977) Technical procedures bulletin no. 207: stability indices. National Weather Service, Silver Spring, p 8

Schemetis MJ, Kok KJ, Vogelezang DHP, Van Western RM (2008) Probabilistic forecasts of (severe) thunderstorms for the purpose of issuing a weather alarm in the Netherlands. American Meteorological Society, pp 1253–1267. doi:10.1175/2008WAF2007102.1

Weisman ML, Klemp JB (1982) The dependence of numerically simulated convective storms on vertical wind shear and buoyancy. Mon Weather Rev 110:504–520

Wilks DS (1995) Statistical methods in the atmospheric sciences. Academic, New York

Yamane Y, Hayashi T (2006) Evaluation of environmental conditions for the formation of severe local storms across the Indian subcontinent. Geophys Res Lett 33:L17806. doi:10.1029/2006GL026823

Yamane Y, Hayashi T, Dewan AM, Akter F (2010) Severe local convective storms in Bangladesh: Part II. Atmos Res 95:407–418

Assimilation of Doppler Weather Radar Data Through Rapid Intermittent Cyclic (RIC) for Simulation of Squall Line Event over India and Adjoining Bangladesh

Kuldeep Srivastava, Vivek Sinha, and Rashmi Bharadwaj

1 Introduction

A squall line is a cluster of severe thunderstorms or storm cells that have formed into a line. Squall lines are hundreds of kilometres in length and having a life span of several hours, which is considerably longer than embedded thunderstorms. Squall lines generate gusty winds, sudden changes in the wind direction with an abrupt increase in wind speed and heavy rains with thunder which are more intense and extensive than individual thunderstorms. The severe gust associated with squall lines can exceed 100 km per hour. Some of them even carry hails and tornadoes. Area under influence of squall lines is extremely unstable and severely turbulent. Over Indian subcontinent, squall lines are often observed during the late pre-monsoon and early summer southwest monsoon over north-eastern states of India and adjoining Bangladesh. Squall line in Doppler Weather Radar (DWR) image will have solid line of heavy rainfall followed by a large area of light rainfall.

Most of the radar data assimilation (DA) studies based on other 3DVAR systems used only one or a few analysis at longer intervals (Xiao et al. 2007; Lin et al. 2011). However, to take full advantage of the high frequency of radar observations, Zhao and Xue (2009) assimilated radar data over a 6 h period at 30 min intervals using the ARPS 3DVAR and cloud analysis package. The results show that the assimilation of radial velocity (Vr) data help to improve the track and intensity forecast more while reflectivity data help to improve precipitation structure forecast. Over Indian region, Srivastava and Bhardwaj (2013a, b) have assimilated DWR data for cyclone

---

K. Srivastava • V. Sinha
India Meteorological Department, Lodhi Road, New Delhi, India

R. Bharadwaj (✉)
Guru Gobind Singh Indraprastha University, Dwarka, New Delhi, India
e-mail: rashmib22@gmail.com

and cloud burst events. These studies have demonstrated that nowcast of cloud burst event and very short-range forecast of cyclone is significantly improved with DWR data assimilation.

In the current study, ARPS3DVAR and cloud analysis package of ARPS model is used for radar data assimilation and WRF model for forecast of weather. Squall line event that occurred over India and Bangladesh on 09 May 2013 is considered to evaluate the impact of assimilation of DWR data through rapid intermittent assimilation cycle on squall-line structure and subsequent very short-range forecast, as compared to the one time assimilation. Doppler Weather Radar observation taken by Agartala DWR is assimilated in NWP model. Other aim of this study is to demonstrate successful coupling of ARPS3DVAR and WRF model to be used for real time operational implementation in near future.

In Sect. 2, assimilation procedure and forecast model are described. In Sect. 3, design of experiment is discussed. Squall line event occurred over India and Bangladesh on 09 May 2013 is described in Sect. 4. Result and discussion are presented in Sect. 5. Conclusions are summarized in Sect. 6.

2 Assimilation Procedure and WRF Model

Indian DWR data is assimilated into the coupled ARPS3DVAR-WRF model through ARPS3DVAR and cloud analysis. This coupled system is used to carry out various experiments and generate wind, pressure and precipitation forecast. WRF model and ARPS3DVAR assimilation system is briefly described here.

2.1 WRF Model

WRF model is a new generation meso scale numerical weather prediction system. It serves both operational forecasting and atmospheric research requirements. The WRF model has multiple dynamic cores. This study uses Advanced Research WRF model (ARW-WRF). ARW-WRF is based on an Eulerian solver for fully compressible non-hydrostatic equations, cast in flux conservation form and mass (hydrostatic pressure) vertical coordinates. It uses a third-order Runge–Kutta time integration coupled with split-explicit second-order time integration scheme for the acoustic and gravity wave modes. ARW-WRF carries multiple physical options for cumulus, microphysics, planetary boundary layer (PBL) and radiation physical processes. Details of the model are provided in Shamrock et al. (2005). Fifth-order upwind biased advection operations are used in the fully conservative flux divergence integration.

2.2 Data Assimilation System

2.2.1 Three-Dimensional Variational (3DVAR) Technique

ARPS3DVAR data assimilation technique is an advance technique of the ARPS model developed by CAPS, Oklahoma University, USA. This technique uses dynamic constraints appropriate for storm-scale analysis, as documented by Gao et al. (2004). The analysis variables in this procedure are three wind components (u, v and w), potential temperature (θ), pressure (p) and water vapour mixing ratio (q_v). In this system, the cross correlations between variables are not included in the background error covariance. The background error correlations for single control variables are modelled by a recursive spatial filter. The observation errors are assumed to be uncorrelated; hence, observation error covariance is a diagonal matrix, and its diagonal elements are specified according to the estimated observation errors.

One unique feature of the ARPS3DVAR is that multiple analysis passes can be used to analyze different data types with different filter scales to account for the variations in the observation spacing among different data sources. The main advantage of variational method (3DVAR) over classical data assimilation methods is the feasibility of assimilating directly the observed parameters (radial wind and reflectivity) at any given time and space location.

2.2.2 Cloud Analysis

The cloud analysis component of ADAS is derived from that of Local Analysis and Prediction System (LAPS) with a number of modifications (Bratseth 1986; Albers et al. 1996; Brewster 1996; Zhang et al. 1998). The radar reflectivity and thermodynamics are used to solve for the model's precipitating hydrometeors (rain, snow and hail). Because the radar reflectivity is generally a function of drop diameter raised to the sixth power and the water content is a function of diameter cubed, some assumptions must be made about drop size distribution (DSD). The present system uses relationships based on a Marshall-Palmer DSD. The analyzed temperature is used in the scheme in the diagnosis of precipitation species. Direct replacement of the background hydrometeors is done in areas where observed reflectivity is greater than a prescribed threshold (typically 10–20 dBZ). Precipitation is removed from the background in areas within the radar volume coverage and having reflectivity less than the precipitation threshold.

An important aspect of building and maintaining thunderstorm updrafts in a non-hydrostatic model is the inclusion of the effect of latent heat release due to condensation processes in the updraft regions. A moist adiabatic ascent from the analyzed cloud base with entrainment is calculated, and any excess in this temperature over the analyzed temperature is then added to the analyzed value.

3 Design of Experiment

Model domain for the squall line event on 09 May 2013 approximately covers the area between 21° to 27°N and 89° to 95°E with horizontal resolution of 9 km with 80×80 grid points in X and Y directions. This domain covers an area of 720×720 km with DWR Agartala at the centre of domain. The vertical grid stretched from surface to model top is located at about 20 km height at a vertical resolution of 500 m. IMD global forecast system (GFS) data at $1° \times 1°$ resolution are used to provide initial and boundary conditions for both the data assimilation experiments (i.e. rapid intermittent cycles and only one time) for both the events.

Two sets of experiments are carried out using coupled ARPS3DVAR-WRF model to investigate the impact of radar data assimilation on simulation of squall line event. In the first experiment data is assimilated through rapid intermittent cycles (cyclic) and in other experiment only one time (non-cyclic) assimilation is done. Both the reflectivity and wind data have been assimilated simultaneously through ARPS3DVAR and cloud analysis procedure. For 09 May 2013 event, first experiment (RIC) assimilates Agartala DWR data through rapid intermittent cycles every 20 min within 1 h assimilation window from 0300 UTC to 0400 UTC and second experiment (NORIC) assimilates DWR data only one time at 0400 UTC (Table 1).

4 Squall Line Event on 09 May 2013

4.1 Description of Event

On 9 May 2013 squall line occurred over northeast Indian region. Figure 1 shows domain of study which covers northeast India and Bangladesh. This study focusses on northeast to southwest oriented squall line initiated around 0300 UTC 09 May 2013 over northeast Bangladesh. At 0500 UTC it was hook shaped. Hook of the squall line is enclosed by the dotted red circle in Fig. 1. At 0700 UTC squall line is well developed and is seen as bow shaped in reflectivity field. This caused flash flood and nearly 200 houses were damaged in Mizoram and Tripura states of India. Squall line moved southeast ward throughout its life span and dissipated after 0900 UTC.

Table 1 Design of experiments

Experiments	Type of radar data assimilation	Time of data assimilation
RIC (Rapid intermittent cycle)	Reflectivity and radial velocity	0300, 0320, 0340 and 0400 UTC 09 May 2013
NORIC (No cycle, one time assimilation)	Reflectivity and radial velocity	0400 UTC 09 May 2013

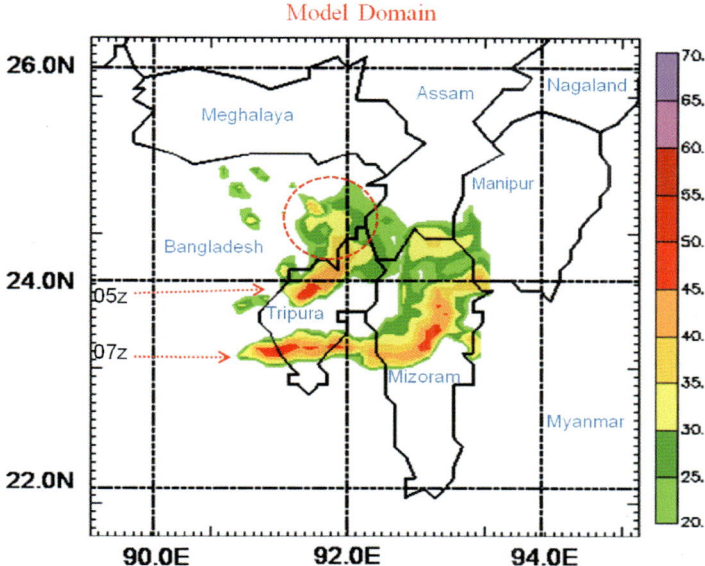

Fig. 1 Model domain and composite radar reflectivity observations (*colour*) at 0500 UTC 09 May and 0700 UTC on 09 May 2013

4.1.1 Radar Observation

Figure 2 shows hourly movement of squall line. The squall line from its initiation (around 0300 UTC on 9 May 2013) to dissipation (around 10 UTC 09 May 2013) was well observed by Agartala radar. Quality control of radial velocity and reflectivity data during assimilation cycle (03–04 UTC) is done by using WDSSII software. The method of DWR data quality control is same as described by Roy Bhowmik et al. (2011). Data assimilation experiments are performed using quality control data from Agartala to compare the relative importance of cyclic verses no cyclic data assimilation.

Figure 2 also shows that squall line formed around 03 UTC over northeast Bangladesh. It started moving southeast ward and lay over north Tripura and adjoining Bangladesh at 0500 UTC. At this time comma (hook) shaped echo was seen over northern end of squall line. Throughout its life period squall line moved southeast ward. At 0600 UTC it was well developed, bow shaped and lay over central part of Tripura and northwest part of Mizoram. During 0600–0800 UTC squall line was very intense, bow shaped and covered remaining southern parts of Tripura and northern parts of Mizoram. During 0800–1100 UTC system continued to move southeast ward and covered southern parts of Mizoram. The squall line started weakening after 1100 UTC and dissipated thereafter.

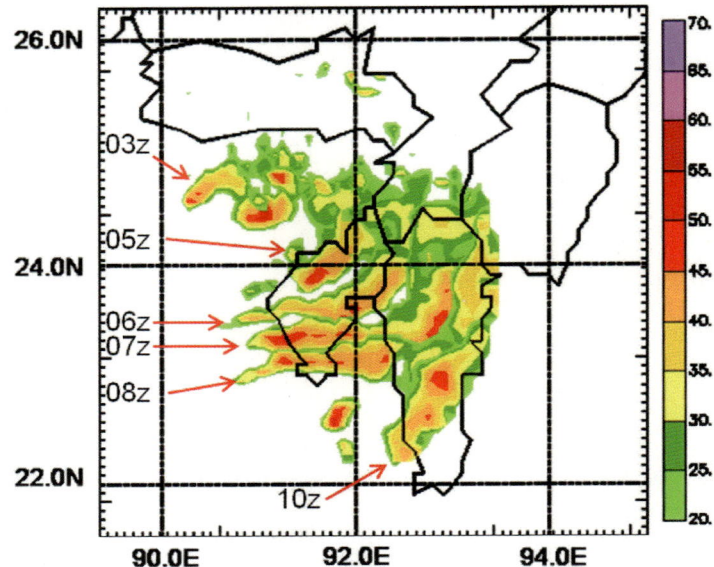

Fig. 2 Movement of squall line as observed by the Agartala radar during 03 UTC to 10 UTC on 09 May 2013

4.1.2 Observed Rainfall

In Fig. 3, shaded area shows the 24 h accumulated rainfall (in mm) during 03 UTC of 09 May 2013 to 03 UTC of 10 May 2013 estimated by the TRMM satellite. Figure 3 shows that the rainfall due to the squall line was mainly confined over northeast Bangladesh, Tripura and Mizoram and was in the range of 6–9 cm. This rainfall area is covered by the squall line during its southeast movement. Numbers shown in the figure are the amount of 24 h accumulated rainfall (in cm) observed by the synoptic observatories (Kailashahr – 8, Lengpui – 7, Silchar – 4, Cherrapunji – 6) located over the area covered by the squall line.

4.1.3 Synoptic Conditions

Figure 4 shows the wind pattern at 850, 500 and 200 hPa levels. Figure 4a shows that northsouth trough runs from north Bihar to Orissa coast through West Bengal at 850 hPa level. At the same level cyclonic circulation lies over Assam, Meghalaya and adjoining area. Figure 4a also shows that there is southwesterly wind flow over Bangladesh and Tripura in lower levels. Strong southwesterly wind flow of the order of 20–25 knots lead to the moisture incursion over the region, necessary for the development of thunderstorm/squall line. Upper level divergence at 200 hPa was also favourable for the development of TS/squall line.

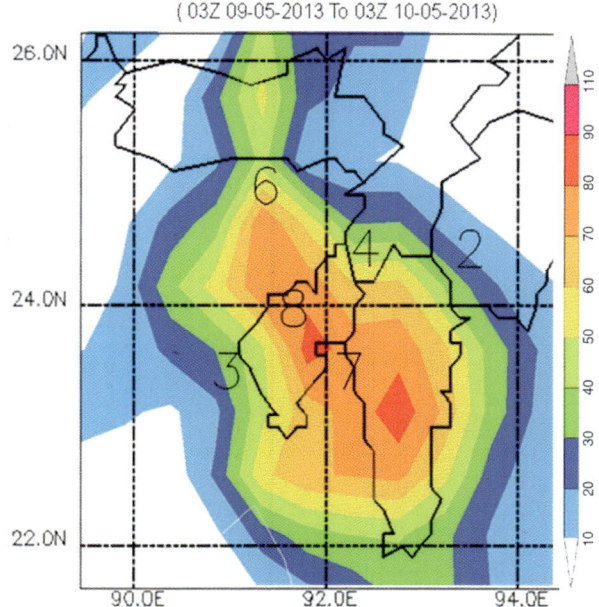

Fig. 3 24-h accumulated rainfall in cm (*shaded*) estimated by the TRMM satellite (*Number*) observed by the synoptic observatories

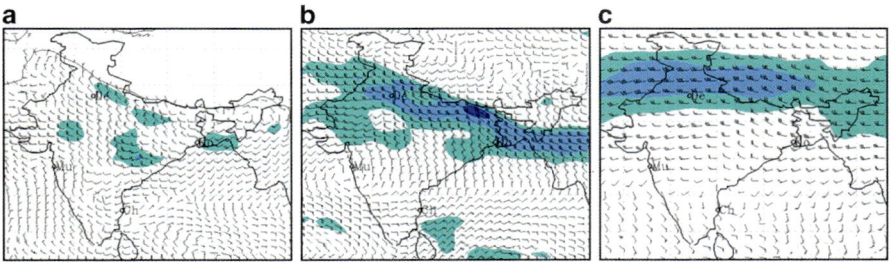

Fig. 4 Wind pattern at (**a**) 850 hPa, (**b**) 500 hPa and (**c**) 200 hPa levels at 00 UTC on 09 May 2013

5 Result and Discussion

5.1 Impact on Reflectivity Forecast

In this section we will evaluate whether the initialization through rapid intermittent assimilation cycle results in a better squall line structure in analysis and the subsequent very short range forecast, as compared to the one time assimilation. For this result from RIC experiment is analyzed and compared with NORIC experiment

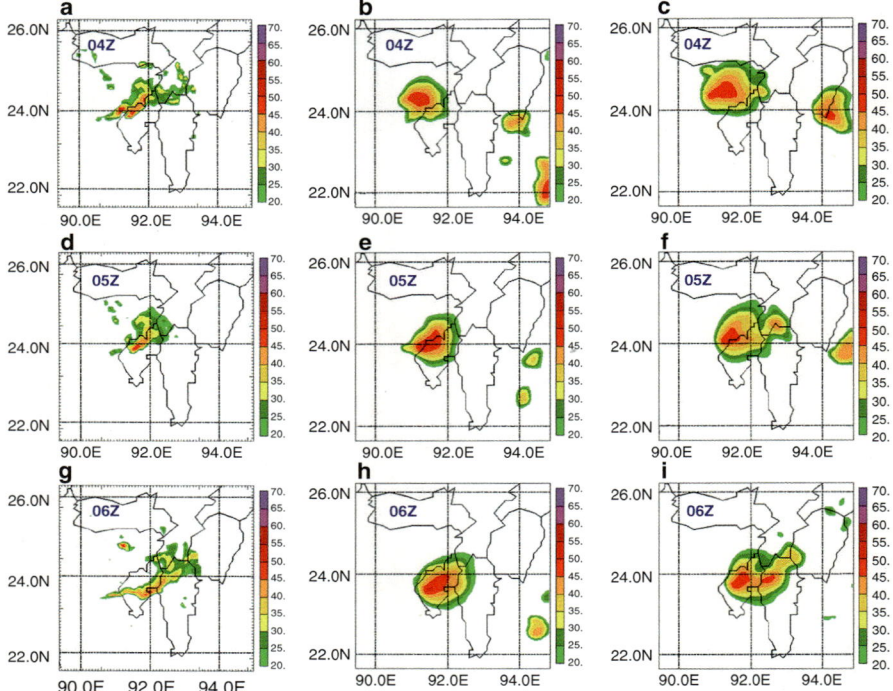

Fig. 5 Reflectivity fields at 0400, 0500 and 0600 UTC 09 May 2013: (**a, d, g**) observed by the Agartala radar, (**b, e, h**) simulated by NORIC experiment and (**c, f, i**) simulated by RIC experiment

(i.e. cycling versus no cycling). At the same time results from these experiments are compared with the radar observation. Figure 5 shows the simulated reflectivity patterns at 0400, 0500 and 0600 UTC 10 May from NORIC and RIC experiments, as well as the observations from the Agartala radar. Similarly Fig. 6 shows the reflectivity fields at 0700, 0800 and 0900 UTC.

It is noticed that throughout the life period of squall line (04–09 UTC), maximum reflectivity observed by DWR Agartala as well as simulated by both the experiments was of the order of 50 dBZ. Observed reflectivity plot at 0400 UTC depicts that squall line lies over northwest boundary of Tripura and have maximum value of 50 dBZ (Fig. 5a). Hook shaped echo is also seen at the northern end of squall line. Weak echoes are also seen over southern tip of Assam. NORIC experiment analyzes an oval shaped convective cell over the area (Fig. 5b). However, the location of reflectivity maxima (~50 dBZ) is slightly northwestward to the observed location. RIC experiment also analyzes an oval shaped convective cell but its horizontal extension is more than NORIC experiment and it has covered northeastern part of Bangladesh (Fig. 5c). In this experiment convective cell was slightly extended towards east and covered southern tip of Assam. Both the experiments also predict convective cell at the southern tip of Manipur, which moved eastward during subsequent forecast hours. Though both experiments analyze squall line as oval shaped convective cell, but RIC experiment analysis is better as it is extended eastward and covers southern tip of Assam. Reflectivity observation at 0500 UTC shows that squall line had moved slightly southeastward and covered northwest Tripura with maximum reflectivity of the order of 50 dBZ (Fig. 5d). Hook shaped echo at the northern end of squall line was slightly intensified. Echoes over Assam and Bangladesh at 0400 UTC had weakened. Figure 5e shows 1-h forecast by NORIC experiment; it depicts that simulated reflectivity pattern has shifted southeastward and covers north Tripura and adjoining Bangladesh. Similar to NORIC experiment, simulated reflectivity pattern by RIC experiment is also shifted southeastward and covers north Tripura and adjoining Bangladesh (Fig. 5f). Reflectivity field in this experiment is further extended eastward and eastern end is intensified.

Figure 5g shows that squall line continued to move southeastward and hook shaped echo lay over southern tip of Assam at 0600 UTC. At the same time squall line had covered central parts of Tripura, northwest part of Mizoram and southwest Manipur. Two-hour forecast by NORIC experiment at 0600 UTC depicted that simulated reflectivity pattern had shifted southeastward and covered almost entire Tripura and northwest boundary of Mizoram (Fig. 5h). Reflectivity field predicted by RIC experiment at 0600 UTC was east–west oriented and it further moved southeastward and covered north Tripura, northwest parts of Mizoram and southwest Manipur (Fig. 5i). At this time reflectivity pattern predicted by RIC experiment over Mizoram and Manipur is very close to the observation. Over Tripura, predicted reflectivity field has covered more area as compared to the observation.

Figure 6a shows that squall line continues to move southeastward and hook shaped echo is now disappeared at 0700 UTC. Squall line now (is now bow shaped echo) lies over southern parts of Tripura and extending up to northern part of Mizoram. Three-hour forecast by NORIC experiment at 0700 UTC depicts that simulated reflectivity pattern has further shifted southeastward and covers almost entire Tripura and also covers west-northwest area of Mizoram (Fig. 6b). East–west oriented reflectivity field predicted by RIC experiment at 0700 UTC is extended towards north; entire pattern has further moved southeastward and covers east Tripura and north Mizoram (Fig. 6c). At this time reflectivity pattern predicted by RIC experiment over Mizoram and Manipur is very close to the observation and reflectivity pattern also appears like bow shaped.

Figure 6d shows that squall line has further moved southeastward. Squall line now lies over extreme southern parts of Tripura, central Mizoram and extending up to eastern parts of Mizoram. Four-hour forecast by NORIC experiment at 0800 UTC depicts that simulated reflectivity pattern has further moved southeastward and covers western parts of Mizoram (Fig. 6e). Reflectivity field predicted by RIC experiment at 0800 UTC has further moved southeastward and covers central Mizoram (Fig. 6f). Reflectivity pattern predicted by RIC experiment over Mizoram is close to the observation. However reflectivity pattern is broader than observation.

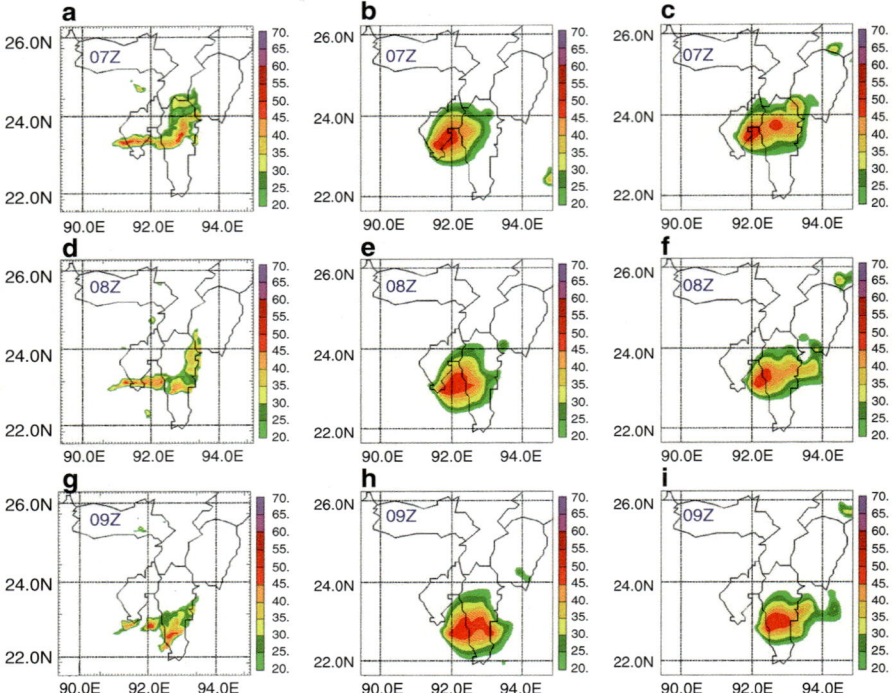

Fig. 6 Reflectivity fields at 0700, 0800 and 0900 UTC 09 May 2013: (**a, d, g**) observed by the Agartala radar, (**b, e, h**) simulated by NORIC experiment and (**c, f, i**) simulated by RIC experiment

0900 UTC radar observation in Fig. 6g depicts that squall line continues to move southeastward and now lies over southern parts of Mizoram. Five-hour forecast by NORIC experiment at 0900 UTC depicts that simulated reflectivity pattern has further moved southeastward and covers southern parts of Mizoram (Fig. 6h). Similar reflectivity field is predicted by RIC experiment at this time. However smaller area is covered in RIC experiment as compared to NORIC experiment. Location of predicted reflectivity field in both the experiments is very close to the observation. Above discussion indicates that RIC experiment demonstrates that squall line is much better simulated than NORIC experiment.

5.2 Impact on Precipitation

Figure 7 shows three-hourly rainfall estimated by TRMM satellite and simulated by NORIC and RIC experiment during 03–06 UTC and 06–09 UTC on 09 May 2013. Comparison of Fig. 7a–c shows that rainfall during 03–06 UTC in RIC experiment

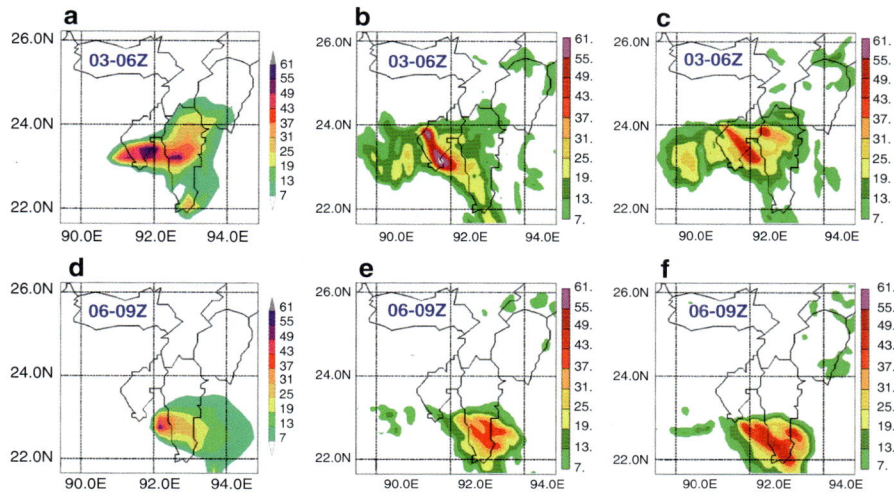

Fig. 7 Three-hourly precipitation during 03–06 UTC and 06–09 UTC on 09 May 2013: (**a, d**) Estimated by TRMM satellite, (**b, e**) simulated by NORIC experiment and (**c, f**) simulated by RIC experiment

is better predicted than the NORIC experiment as rainfall over north Mizoram is well captured when data is assimilated through rapid intermittent cycle. However rainfall pattern during 06–09 UTC in both the experiments is similar and close to the observation.

Conclusions

Squall line event occurred over India and Bangladesh on 09 May 2013 is investigated to see the impact of assimilation of DWR data through rapid intermittent assimilation cycle on squall-line structure and subsequent very short range forecast, as compared to the one time assimilation. Based on the experiments carried out, following conclusions can be drawn:

- Coupling of data assimilation package ARPS3DVAR and cloud analysis with WRF model has been successfully accomplished for Indian DWR data.
- Horizontal extension, merger of cells, intensity and direction of movement of squall line are better predicted when data is assimilated through rapid intermittent cycles (RIC).
- Amount and precipitation structure are significantly improved in cyclic assimilation.

Acknowledgement Authors are thankful to Guru Gobind Singh Indraprastha University, New Delhi for their support to carry out this study. The authors are also grateful to the Director General of Meteorology, IMD, New Delhi for encouraging to do present work. Authors also thankfully acknowledge the support of DWR Delhi; NWP Division, IMD, New Delhi; NWP system (ARPS) of CAPS, University of Oklahoma, USA and HPCS system for this study.

References

Albers SC, McGinley JA, Birkenhuer DL, Smart JR (1996) The Local Analysis and Prediction System (LAPS): analysis of clouds, precipitation and temperature. Weather Forecast 11:273–287

Bratseth AM (1986) Statistical interpolation by means of successive corrections. Tellus 38A:439–447

Brewster K (1996) Application of a Bratseth analysis scheme including Doppler radar data. In: Preprints 16th conference on weather analysis and forecasting. American Meteorological Society, pp 92–95

Gao JD, Xue M, Brewster K, Droegemeier K (2004) A three dimensional data analysis method with recursive filter for Doppler radars. J Atmos Ocean Technol 21:457–469

Lin HH, Lin PL, Xiao QN, Kuo YH (2011) Effect of Doppler radial velocity data assimilation on the simulation of a typhoon approaching Taiwan: a case study of Typhoon Aere (2004). Terr Atmos Ocean Sci 22(3):325–345. doi:10.3319/TAO.2010.10.08.01(A)

Roy Bhowmik SK, Sen Roy S, Srivastava K et al (2011) Processing of Indian Doppler Weather Radar data for mesoscale applications. Meteorol Atmos Phys 111:133–147

Shamrock WC, Klemp JB, Dudhia J, Gill DO, Barker DM, Wang W, Powers JG (2005) A description of the advanced research WRF version 2 NCAR. Technical note NCAR TN-468 1STR

Srivastava K, Bhardwaj R (2013a) Assimilation of Doppler Weather Radar data in WRF model for simulation of tropical cyclone Aila. Pure Appl Geophys. doi:10.1007/s00024-013-0723-5

Srivastava K, Bhardwaj R (2013b) Real time nowcast of a cloudburst and a thunderstorm event with assimilation of Doppler Weather Radar data. Nat Hazards. doi:10.1007/s11069-013-0878-5

Xiao Q, Kuo YH, Sun J, Lee WC, Barker DM, Lim E (2007) An approach of radar reflectivity data assimilation and its assessment with the inland QPE of Typhoon Rusa (2002) at landfall. J Appl Meteorol Clim 46:14–22

Zhang J, Carr F, Brewster K (1998) ADAS cloud analysis. In: Preprints 12th conference on numerical weather prediction, Phoenix, AZ. American Meteorological Society, Boston

Zhao K, Xue M (2009) Assimilation of coastal Doppler radar data with the ARPS 3DVAR and cloud analysis for the prediction of Hurricane Ike (2008). Geophys Res Lett 36:L12803, 6 pp

Impact of Data Assimilation in Simulation of Thunderstorm Event over Bangladesh Using WRF Model

Nazlee Ferdousi, Sujit K. Debsarma, Md. Abdul Mannan, and Md. Majajul Alam Sarker

1 Introduction

Local Severe Storm (here after referred as LSS) is a mesoscale meteorological phenomenon of strong vortex over land. Nor'westers or tornadoes are land-based local severe thunderstorms. If there is a sudden fall of pressure over a small area then, due to differential heating, supported by deep layer of atmospheric instability, there are chances of formation of local thunderstorms (Debsarma 2004). In Bangladesh, thunderstorms occur in all seasons of pre-monsoon, monsoon and post-monsoon and are characterized by strong surface wind generally from 40 to 100 miles per hour and accompanied by heavy rain and often destructive hail with lightning. The frequently occurring thunderstorms are observed during pre-monsoon months; these are known as nor'westers, because they come mostly from northwesterly direction (Afroze et al. 1981).

In India and Bangladesh, some works have been done on the space and time variations of thunderstorms, the physical characteristics of the atmosphere for their formation and frequency distribution of the days of thunderstorm (Chowdhury and Karmakar 1986; Karmakar 2000, 2001; Karmakar and Alam 2005).

Koteswaram and Srinivasan (1958) expressed that the Bay of Bengal could provide warm humid air masses from the south. The Himalayan range could spill cold dry air masses from the north. Warm, dry air masses could arrive from central India. It is likely that these different air masses form a dry line, much like that occurs in the southern Plains of the United States. Hossain and Karmakar (1998) emphasized the significance of jet stream and the intersecting moist and dry air masses associated with the local severe storm formation.

N. Ferdousi (✉) • S.K. Debsarma • Md.A. Mannan • Md.M.A. Sarker
SAARC Meteorological Research Centre (SMRC), Dhaka 1207, Bangladesh
e-mail: nferdousi@yahoo.com

Nor'westers or LSSs are short lived mesoscale severe weather phenomena which sometimes turn into natural hazard causing huge damage and play a vital impact on the socio-economic condition of Bangladesh. Improvement of understanding and proper prediction using Numerical Weather Prediction (NWP) models assimilating local observations of Automatic Weather Stations (AWS), Doppler Weather Radars (DWR) and satellites data is the objective of our study.

Almost every model has the limitation to predict the time and location of the occurrences of LSS due to lack of observations at mesoscale resolution or deficiency lying in the initial and boundary conditions. The data assimilation process may be one of the best suitable techniques which have the facility to assimilate heterogeneous mesoscale observations directly to improve the estimation of the model's initial condition. The present study is an attempt to simulate the LSS that occurred over Ishurdi, Dhaka and Chittagong regions of Bangladesh and Shillong of India on 26 April 2010 and produced gusty winds followed by squalls using the Weather Research and Forecasting (WRF) model. An effort is also made to simulate the event using WRF 3DVAR Data Assimilation (3DVAR DA) technique in WRF model with synoptic and upper air data.

2 Data Sources and Methodology

2.1 Data Sources

Monthly thunderstorm records of all available stations for the period of 1981 to 2010 were collected from Bangladesh Meteorological Department (BMD) to prepare thunderstorm climatology. Daily rainfall data of 34 stations of BMD (Fig. 3) and synoptic and upper air data of 26 April are also collected from BMD. Hourly cloud imagery and associated derived parameters, Synoptic, AWS and upper air data of 26 April 2010 are also collected from India Meteorological Department (IMD). To run the model six hourly FNL Data are collected from NCEP and used as initial and boundary condition. Upper air data is also collected from Wyoming University website and used to assimilate in the WRF model using WRF 3DVAR DA technique.

2.2 Methodology

In this study, Weather Research and Forecasting (WRF) Model version 3.2.1 has been used to simulate the LSS developed on 26 April 2010. The special observations such as Synoptic, AWS and upper air data of 0000 UTC of 26 April 2010 collected during STORM Field Experiment have been assimilated using WRF 3DVAR DA technique.

2.3 WRF Model Configuration

For the present study, the WRF-ARW (version 3.2.1) has been used over the SAARC STORM domain (Fig. 1). In this experiment WRF Model is set up as per the following configuration:

Model features	Configurations
Version	3.2.1
Horizontal resolution	9 km
Vertical levels	27
Topography	USGS
Dynamics	
Time integration	Semi implicit
Time steps	50 s
Vertical differencing	Arakawa's energy conserving scheme
Time filtering	Robert's method
Horizontal diffusion	2nd order over Quasi-pressure, surface, scale selective
Physics	
Convection (cumulous paraleterization scheme)	Kain-Fritsch (KF)
PBL surface layer	YSU (Yonsei University) Scheme Monin-Obukhov
Cloud microphysics	WSM 3-Class simple ice
Radiation for long wave	RRTM (Rapid Radiative Transfer Model)
Radiation for short wave	Dudhia
Land surface processes	Unified NOAA land surface model

3 Climatology of Thunderstorm over Bangladesh

Thunderstorms are frequent in the months of pre-monsoon, monsoon and post-monsoon season in Bangladesh. But the frequencies of thunderstorms are the highest in May followed by April during pre-monsoon months. It is the highest in September during post-monsoon months. In winter months thunderstorm frequencies are very less (Fig. 2a). The frequency of thunderstorm in pre-monsoon, monsoon, post-monsoon and winter are 57, 80, 12 and 5 respectively but the annual frequency is 154 (Fig. 2b). Spatial distribution reveals that thunderstorm frequencies are maximum in the northeastern parts of Bangladesh decreasing as we move south-westwards. Thunderstorm frequencies are less in the southeastern and northwestern parts of Bangladesh (Fig. 3).

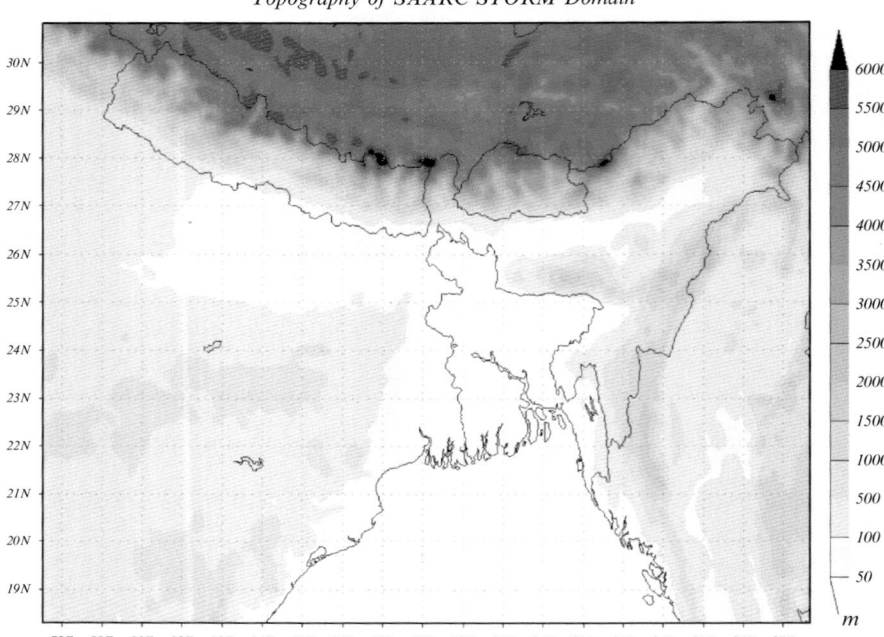

Fig. 1 Selected domain for the experiment with topography (*shaded*) in metres used for simulation and data assimilation

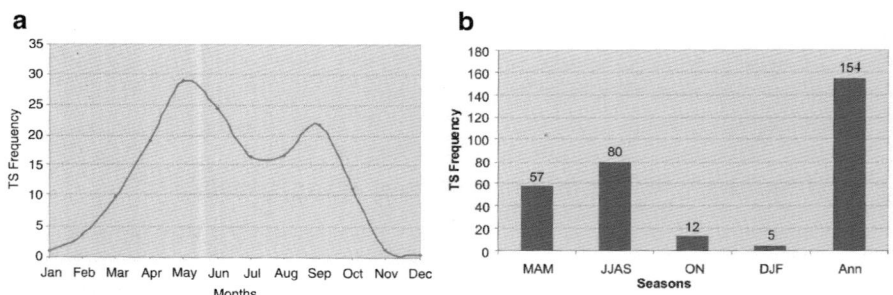

Fig. 2 Variation of (**a**) monthly and (**b**) seasonal thunderstorm frequency during 1981–2010 over Bangladesh

4 Results and Discussion

4.1 Thunderstorm Event of 26 April 2010

A moderate LSS occurred over Ishurdi, Dhaka and Chittagong regions of Bangladesh and Shillong of India on 26 April 2010 and produced gusty winds followed by squalls. It was a non-Intensive Observation Period (IOP) day according to SAARC

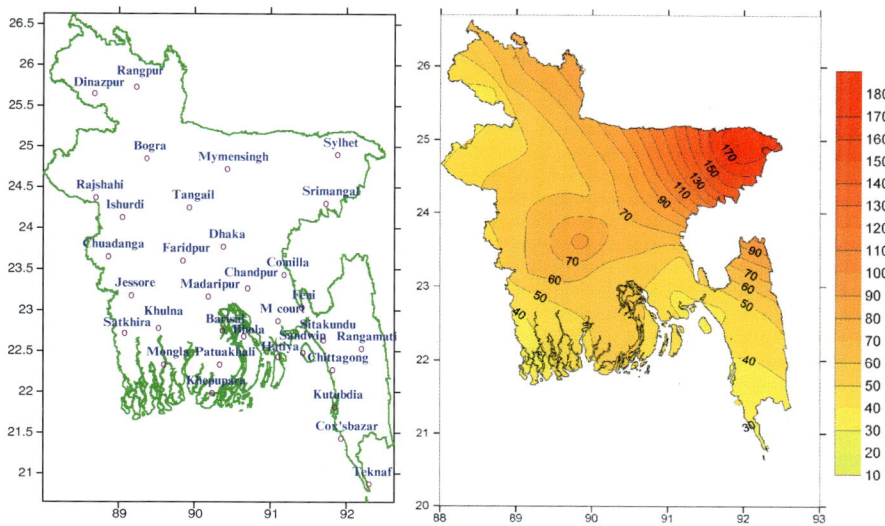

Fig. 3 Spatial distribution of thunderstorm frequency in pre-monsoon season over Bangladesh (1981–2010)

Table 1 Observed weather event at different places of Bangladesh and India

Stations	Time/duration of thunderstorm (UTC)	Gusty/squally wind	Maximum wind (km/h)
Shillong, India	0500	Squally wind	40.8
Ishurdi, Bangladesh	1120–1150	Squally wind	74.0
Dhaka, Bangladesh	1320	Squally wind	74.0
Chittagong, Bangladesh	1651–1702	Squally wind	65.0

STORM Pilot Experiment which was conducted during 15 April–31 May 2010. The observed weather at different places is shown in Table 1.

In addition, the recorded rainfall over India was Canning 15 mm, Hashimara 1 mm, Kolkata 10 mm, Naharlagun 4 mm, Tawang 11 mm, Guwahati 19 mm, Dimapur 12 mm, Kohima 15 mm, and Gangtok 45 mm. In Bangladesh, thunderstorms were observed and the recorded rainfall was Ishurdi 22 mm, Dhaka 11 mm and Chittagong 1 mm.

4.1.1 Observed Features

Observed precipitation from TRMM and BMD rain-gauge data, DWR echoes from Kolkata, Cox's Bazar and Dhaka radars and Kalpana Satellite imageries are shown in Figs. 4, 5, 6, and 7.

Figure 4a represents the TRMM rainfall accumulated on 26 April 2010. TRMM product has the signature of rainfall with the maximum amounts of 80–90 mm over northeastern, central and southwestern parts of Bangladesh. According to BMD, the

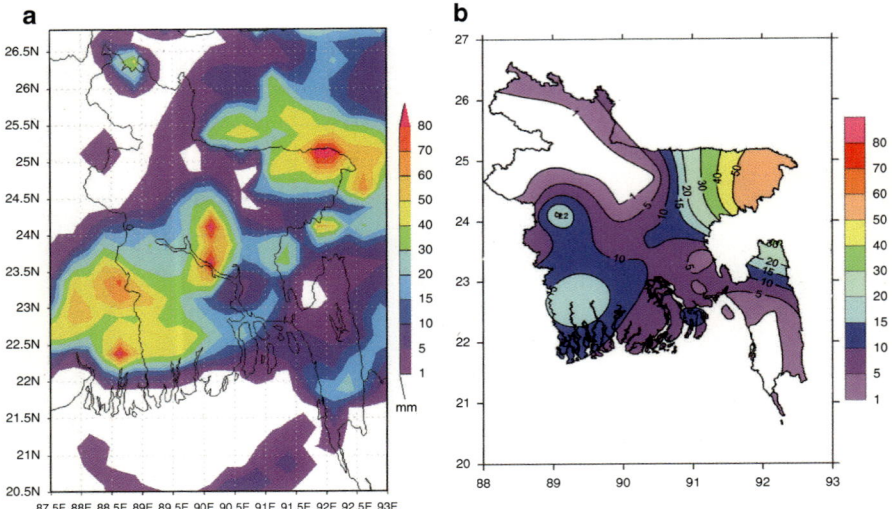

Fig. 4 (**a**) TRMM precipitation accumulated for 24 h during 00 Z of 26 April to 00 Z of 27 April 2010 and (**b**) recorded rainfall (mm) in Bangladesh

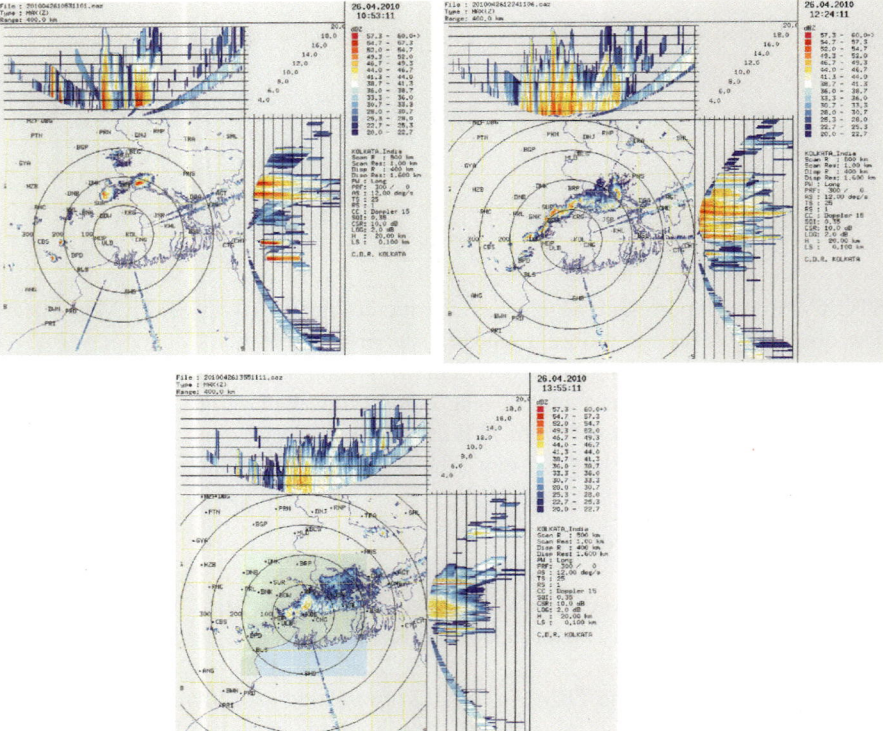

Fig. 5 Reflectivity observed by Kolkata Doppler Radar on 26 April 2010 at different time steps

Impact of Data Assimilation in Simulation of Thunderstorm Event over Bangladesh 41

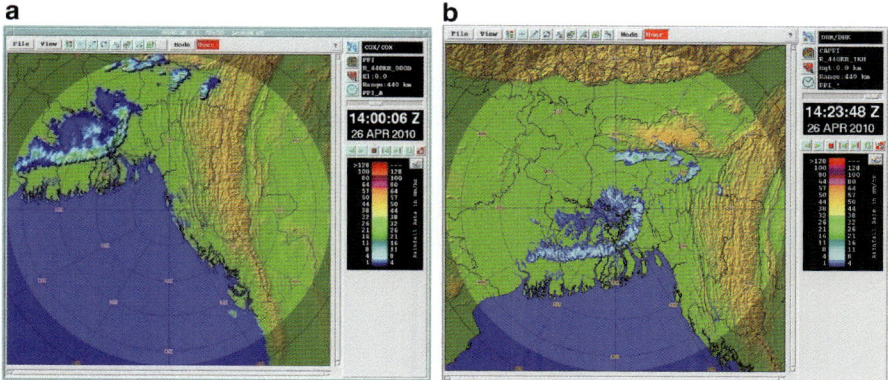

Fig. 6 Reflectivity observed by (**a**) Cox's Bazar and (**b**) Dhaka Doppler Radar on 26 April 2010 at two time steps

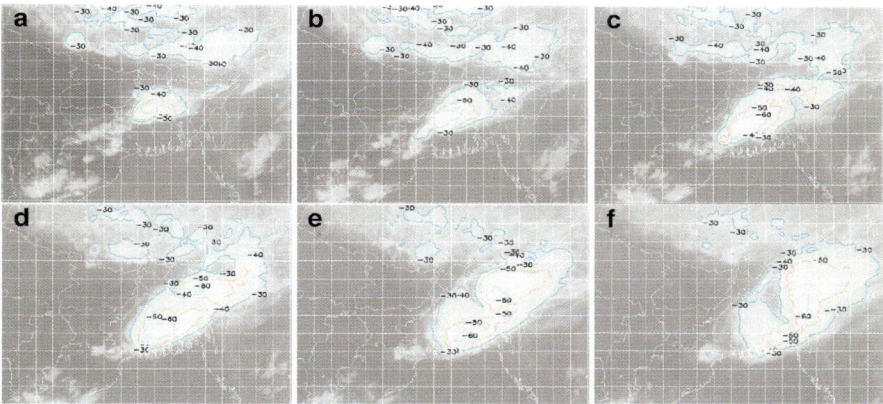

Fig. 7 Cloud top temperature (CTT) over Bangladesh at (**a**) 1100, (**b**) 1200, (**c**) 1300, (**d**) 1400, (**e**) 1500 and (**f**) 1600 UTC of 26 April 2010 as observed from Kalpana-I satellite

maximum amount of rainfall of 57 mm occurred over northeastern part of Bangladesh (Fig. 4b). The Doppler Radar of Kolkata depicted two strong echoes over West Bengal and mid-western part of Bangladesh at 1053 UTC (Fig. 5). It then intensified into a NE-SW oriented squall line by 1224 UTC and moved south-southeastwards and then dissipated over the North Bay of Bengal by 1600 UTC. Similar features were found from the imageries of Cox's Bazar and Dhaka Doppler Radars (Fig. 6). The length of squall line was about 300 km. Kalpana-I satellite imageries of 1100, 1200, 1300, 1400, 1500 and 1600 UTC are shown in Fig. 7. According to Fig. 7, convection developed over northeastern part of West Bengal, India and mid-western part of Bangladesh at 1100 UTC. At this time Cloud Top Temperature (CTT) was −40 °C over that area. Then, it moved eastward, intensified and expanded over Bangladesh. Isolated convective cells also developed over northeastern extreme parts of Bangladesh, adjoining areas of Meghalaya and Assam of India at 1200 UTC. These cells then merged together and CTT lowered to

−60 °C over the same area at 1400 UTC. After that it moved eastward and finally dissipated over eastern part of India at 2330 UTC as observed.

4.1.2 Assimilation of STORM Field Observations

Simulated products of rainfall (mm/h), accumulated rainfall for 24 h, wind field at 925 hPa level and wind speed at 10 m height, reflectivity, vorticity, CAPE, CINE, vertical wind shear between 950 and 500 hPa levels, wind fields at 850, 700, 500 and 200 hPa levels and vertical velocity are shown in Figs. 8, 9, 10, 11, 12, 13, 14, and 15.

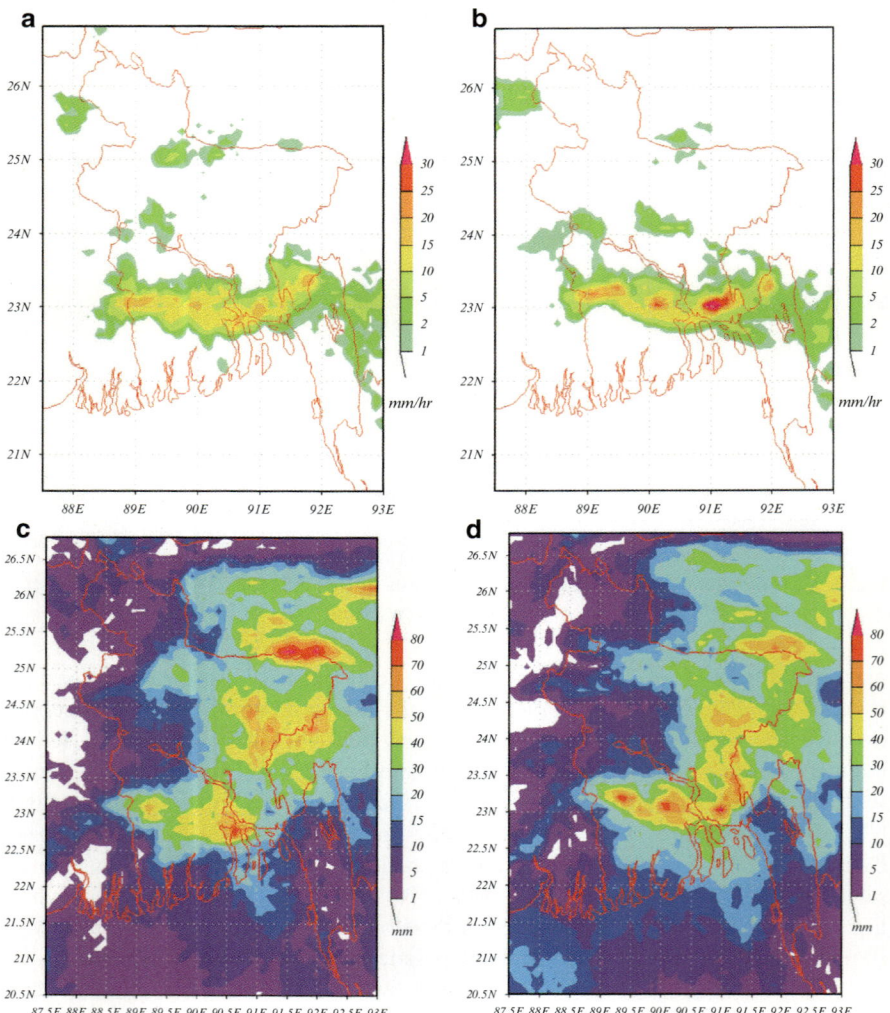

Fig. 8 Rainfall (mm/h) for (**a**) CTRL run and (**b**) with DA valid at 1200 UTC; and 24 h accumulated rainfall for (**c**) CTRL run and (**d**) with DA of 26 April 2010

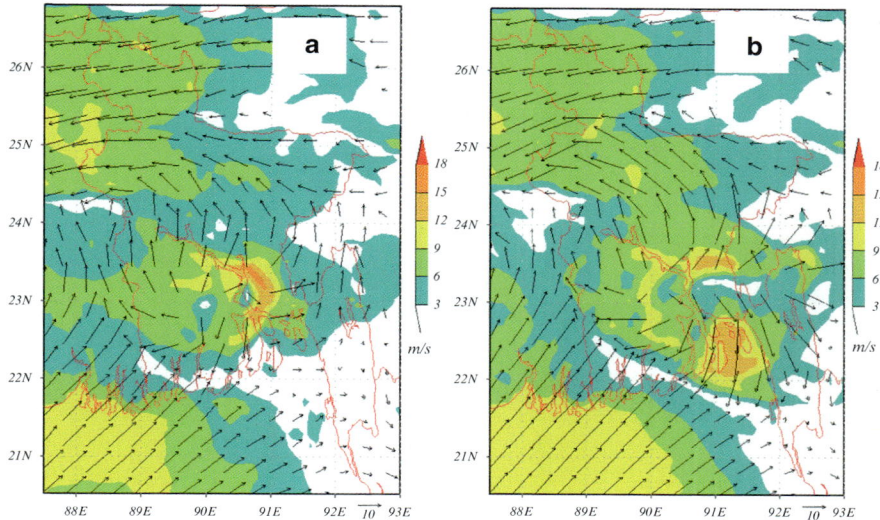

Fig. 9 Wind field at 950 hPa level and wind speed (m/s) 10 m height for (**a**) CTRL run and (**b**) with DA at 1400 UTC of 26 April 2010

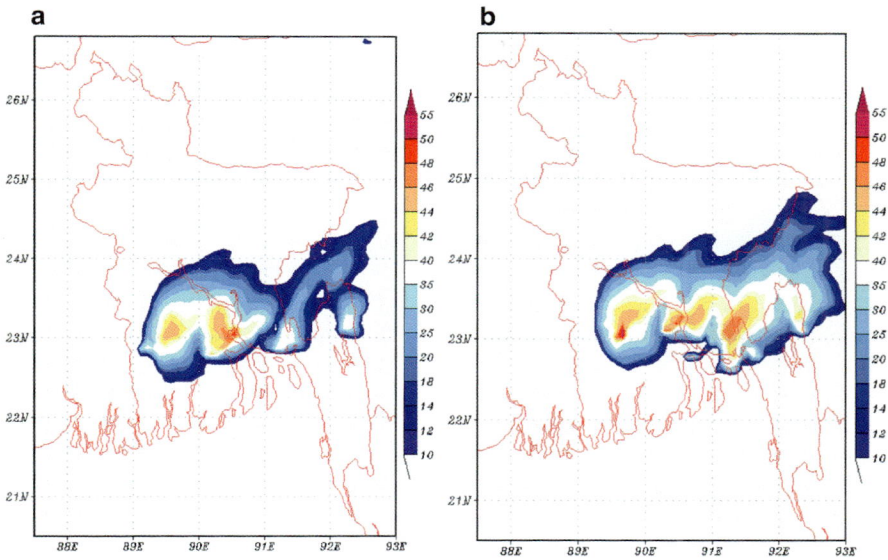

Fig. 10 Reflectivity (dBZ) for (**a**) CTRL run and (**b**) with DA at 1300 UTC of 26 April 2010

Figure 8a, b represent the simulated rainfall with and without data assimilation at 1200 UTC of 26 April 2010. Figure 8a depicts that there is a signature of simulated rainfall over south-central part of Bangladesh but after assimilation the intensity of rainfall increases within the rainfall band as observed over the same area (Fig. 8b). Model simulated 24 h accumulated rainfall for CTRL run is shown in Fig. 8c and it

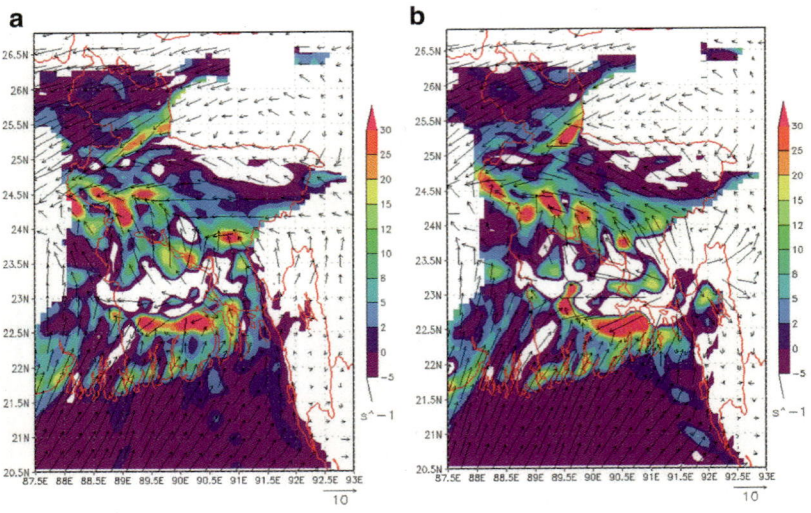

Fig. 11 Surface wind and vorticity for (**a**) CTRL run and (**b**) with DA at 1300 UTC of 26 April 2010

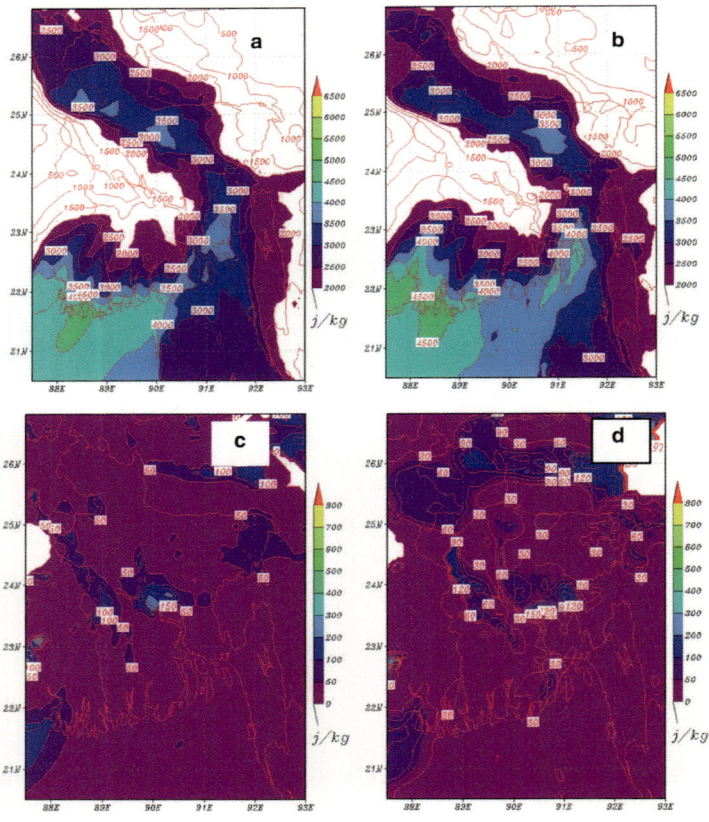

Fig. 12 CAPE for (**a**) CTRL run and (**b**) with DA run; Mean CINE (**c**) CTRL run and (**d**) with DA run valid at 0900 UTC of 26 April 2010

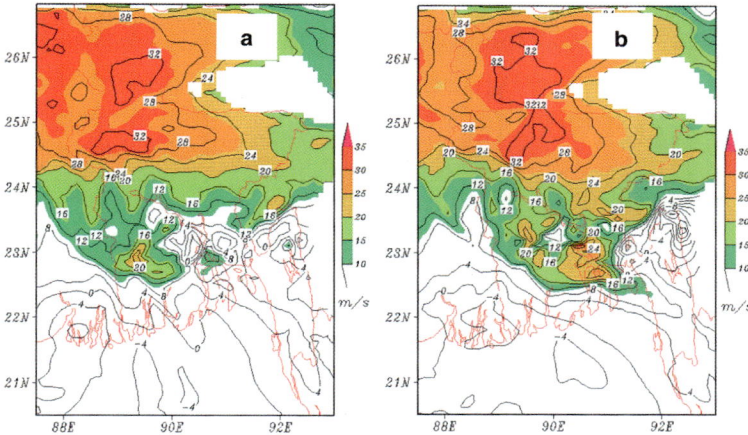

Fig. 13 Wind shear between 500 and 950 hPa for (**a**) CTRL run and (**b**) with DA run valid at 1300 UTC of 26 April 2010

captured maximum rainfall over northeastern part of Bangladesh and also a signature of rainfall over mid-eastern and south-central parts of Bangladesh. But model with data assimilation (Fig. 8d) has underestimated the rainfall amount over northeastern part of Bangladesh as compared to the observation (Fig. 4).

Figure 9 represents the simulated wind field at 925 hPa level and wind speed at 10 m height at 1400 UTC of 26 April 2010. It is seen that the control runs of the model simulated wind speed of 15 m/s over the region of 22–23.5°N and 90.5–91.5°E. But after data assimilation model simulated wind speed of 13–15 m/s at the same level over the same area. Simulated reflectivity position and patterns are similar for both the cases but there is an area enlargement observed for data assimilation in WRF model at 1300 UTC of 26 April 2010 as shown in Fig. 10.

Figure 11a, b represent the model simulated vorticity in the midst of surface wind with and without data assimilation at 1300 UTC of 26 April 2010. Both the figures clearly depict the lower level moisture incursion with the vorticity of 25–30×10^{-5} s^{-1} which is very much favourable for convection to intensify the system over the same area as observed in Figs. 6 and 7.

Figure 12 shows the simulated mean CAPE (J/kg) and mean CIN (J/kg) valid at 0900 UTC of 26 April. Both CTRL and DA run extract the maximum CAPE and minimum CINE around the region of strong convection as observed in Figs. 6 and 7. Figure 13 shows the wind shear between 500 and 950 hPa levels for both CTRL and DA in WRF model valid at 1300 UTC of 26 April 2010. The regions indicated in Fig. 13 are the shear values above 10 m/s. Both the simulations are able to produce maximum wind shear of >20 m/s over the region where maximum reflectivity is simulated. Higher shear zones are widely spread out in the north of the observed convective area in CTRL run as well as DA run.

Simulated result shows that both the CTRL and DA run are able to capture the wind convergence over the thunderstorm formation area. But the convergence extracted by the model with DA is stronger than that generated by CTRL run as

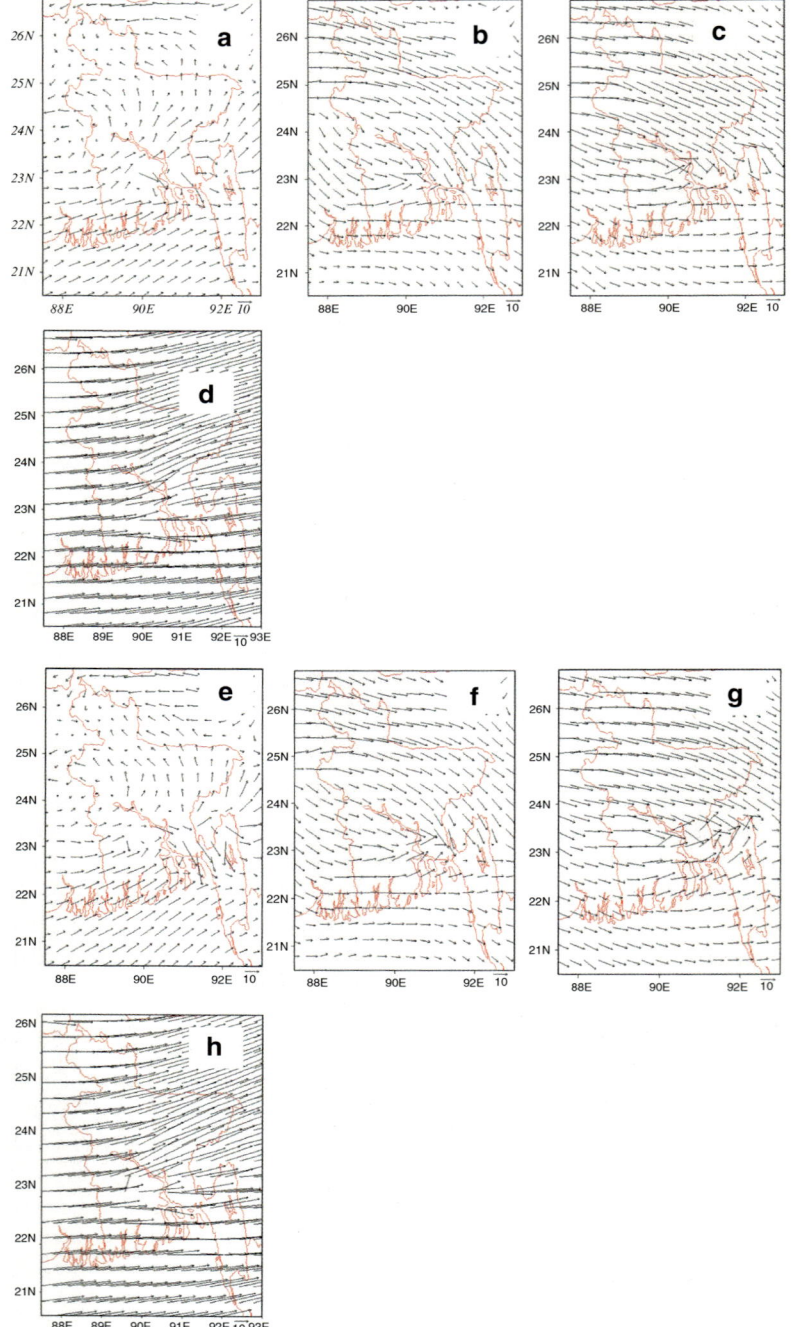

Fig. 14 Wind field at different levels (**a**) 850, (**b**) 700, (**c**) 500 and (**d**) 200 hPa for CTRL run and (**e**) 850, (**f**) 700, (**g**) 500 and (**h**) 200 hPa with DA at 1300 UTC of 26 April 2010

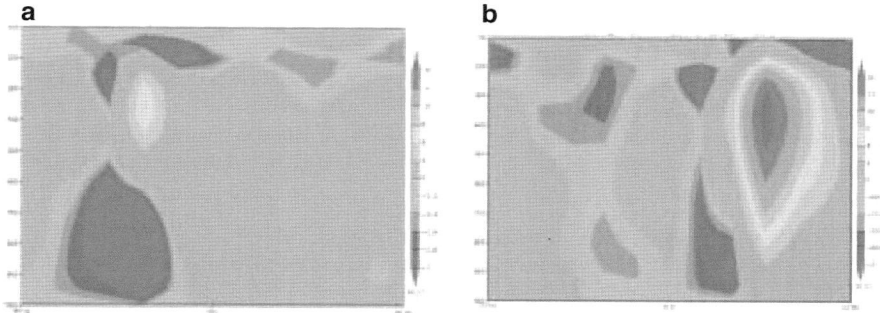

Fig. 15 East-West cross-section of the vertical velocity (m/s) for (**a**) CTRL and (**b**) with DA run along the 23.2°N associated with LSS of 26 April 2010

observed at 850 hPa level at 1300 UTC of 26 April 2010 (Fig. 14a, e). Similarly, the existence of westerly flow at 700 hPa and 500 hPa levels is observed by both the CTRL and DA run which are noticed in Fig. 14b, f and 14c, g valid at 1300 Z of 26 April 2013. Very strong westerly jet stream is captured by both control and DA run of WRF model as depicted in Fig. 14d, h.

Figure 15 shows that the model simulated vertical cross-section of the vertical velocity (m/s) both for CTRL and DA run valid at 1200 UTC. It is seen that the CTRL run simulates the vertical velocity of 6–14 m/s whereas the DA run simulates the vertical velocity of more than 18 m/s over same region of observed convection (Fig. 15).

Conclusions

Data assimilation in WRF model facilitates better simulation of Local Severe Storm (LSS) of 26 April 2010. Simulated wind fields at surface and lower levels with and without data assimilation (DA) are close to each other which are also favourable for thunderstorm generation. Simulated reflectivity, CAPE and Vertical Wind Shear are clearly defined by the model but clearer picture has been obtained from model with DA technique. Simulated rainfall pattern obtained from the model with and without DA are close to observation but the maximum amount of rainfall has decreased with DA facility in WRF model. The model does not capture the exact location and time of occurrence of the thunderstorms for both experiments. Further studies are required to make DA technique useful to the operational meteorologist.

References

Afroze S et al (1981) Investigation of nor'westers and tornadoes in Bangladesh by weather satellite pictures. Nucl Sci Appl 12, 13(B):100–104

Chowdhury MHK, Karmakar S (1986) Pre-monsoon nor'westers in Bangladesh with case studies. In: Proceedings of the seminar on local severe storm, Bangladesh Meteorological Department, held on 17–21 January 1985

Debsarma SK (2004) Post mortem (analysis) of the nor'wester that hit Dhaka on 14 April 2003. SAARC Meteorological Research Centre (SMRC), Dhaka

Hossain A, Karmakar S (1998) Some meteorological aspects of the Saturia tornado, 1989—a case study. J Bangladesh Acad Sci 22(1):109–122

Karmakar S (2000) Probabilistic extremes of thunderstorm days over Bangladesh during pre-monsoon season. J Bangladesh Acad Sci 2(2):225–238

Karmakar S (2001) Climatology of thunderstorm days over Bangladesh during pre-monsoon season. J Bangladesh Acad Sci 3(1):103–112

Karmakar S, Alam M (2005) On the probabilistic extremes of thunderstorm frequency over Bangladesh during pre-monsoon season. J Hydrol Meteorol (SOHAM-Nepal) 2(1):41–47

Koteswaram P, Srinivasan V (1958) Thunderstorms over Gangetic West Bengal in the pre-monsoon season and the synoptic factors favourable for their formation. Indian J Meteorol Geophys 9(4):301–312

SAARC STORM Pilot Field Experiment (2010) SAARC Meteorological Research Centre (SMRC), 2011

Numerical Simulation of a Hailstorm Event over Delhi, India on 28 Mar 2013

A. Chevuturi and A.P. Dimri

1 Introduction

Over north Indian region, pre-monsoon storm events bring precipitation over the region during the months of March-April-May. Such storm events occur during convectively unstable atmospheric conditions culminating due to transient disturbances observed in the air mass. These may be categorized as severe storms, if the storm is associated with heavy precipitation, hail and high winds. The severe storms occur during strong vertical wind shear, which are ideal conditions for hail formation (Orville and Kopp 1977; Houze Jr. 1981).

This paper focuses on the severe precipitation events and hailstorms over Indian region. These severe storms usually depict a multi-cellular structure, where each cell continuously evolves and desiccates in the cloud itself (Chalon et al. 1976). The cloud formation during the hailstorm occurrence has a larger vertical extent (cumulonimbus), with an anvil structure formation at the top of the cloud which causes formation of the frozen precipitation particles from supercooled liquid (Fankhauser 1976). Hail formation is usually associated with simultaneous cells depicting updrafts and downdrafts of wind movement caused by severe convective activity. These generate regions of strong turbulence at the edges of the cells, enhancing the instability within the storm. Hail formation is promoted by the cycling of the hail particles through the cloud (Musil et al. 1976). During the formation, the updrafts cause freezing of the ice over the nuclei and shedding during the downdrafts. Hailstones are uncommon over the Indian region, with the hailstorms occurrence usually seen around the hot summer or pre-monsoon seasons. Despite their rare occurrences, hailstorms can cause heavy damage to the surroundings (Chatterjee et al. 2008). Studies such as Browning et al. (1976) have even focused on the topic

---

A. Chevuturi • A.P. Dimri (✉)
School of Environmental Sciences, Jawaharlal Nehru University, New Delhi, India
e-mail: apdimri@hotmail.com

of hail suppression as a way to avoid hailstorms. Specifically, in the context of urban cities like National Capital Region (NCR) of India, such storms cause damage to infrastructure and also cause the life to come to a standstill by disrupting transport and other amenities. Thus, there is a need to study these severe storms in the context of Indian region to understand the atmospheric processes involved, for better development of a early warning and prediction system.

A case of pre-monsoon heavy precipitation event with hail occurrence is the focus of this study on the 28 Mar 2013 as reported in NNDC-CDO (2013). The storm evolution started at 1330 IST and lasted till 1930 IST over the NCR region. This storm vertically extended upto 12 km in height and developed the associated cumulonimbus clouds with strong wind speeds. The storm was associated with heavy rains, thunder and lightning with some hail formation. Figure 1 shows the reported Doppler weather radar (DWR) reflectivity of the storm, which shows a peak around 1242–1252 UTC. The peak reflectivity of the storm reached around 60 dBZ.

The aim of the paper is to use numerical weather prediction model (Weather Research and Forecasting, WRF) for simulation of the above discussed storm event. The focus of the study is the simulation at finer horizontal resolution to understand the atmospheric processes involved in the storm. To understand in-depth the variability inherent in the microphysical properties of the storm, with the WRF model microphysical parameterization varied to optimize maximum focus to hail formation in the simulation. The objective of the paper is the detailed analysis of the various forcings causing the hailstorm formation and propagation.

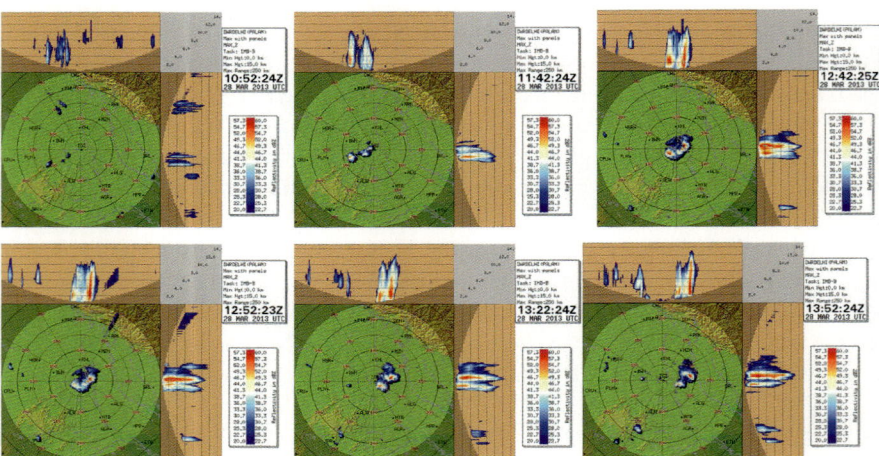

Fig. 1 Observed reflectivity (dBZ) by IMD New Delhi (Palam) Doppler Weather Radar for 28 Mar 2013, 1052–1352 UTC (*Source*: IMD Radar output archived at http://ddgmui.imd.gov.in/storm2012/)

The following document is divided in various sections, with the data and methodology discussed in Sect. 2. The results and discussion pertaining to the study are described in Sect. 3. Finally, a conclusion is presented in Sect. 4.

2 Data and Methodology

In the present study, the WRF model (v3.0) with the Advanced Research WRF (ARW) dynamic solver is used to simulate the severe precipitation event described in the section above. The model domain is set up over the NCR region with a central domain point at Delhi (28.6°N 77.2°E), where the heavy precipitation was reported and observed in the DWR output. The extent of these domains with the topography of the region is shown in Fig. 2 with the model detail description seen in Table 1.

WRF model is initialized with the global forecasting system's final run analysis (FNL) dataset available at every 6 h (NCEP et al. 2000). Model simulations for the experiment begin at 27 Mar 2013 0000 UTC to account for the spinoff due to the coarser resolution of the FNL initial and boundary condition (1°×1° spatial

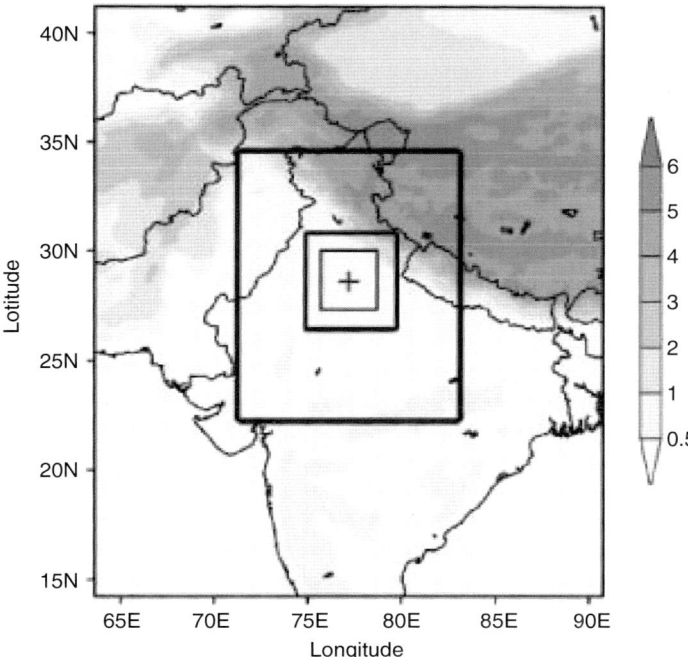

Fig. 2 Model domain and topography (×10³ m; *shaded*). *Shaded* region in figure corresponds to model domain 1 (27 km horizontal model resolution), and *boxes* with *solid black lines* indicate model domain 2 (9 km), domain 3 (3 km), domain 4 (1 km) are marked with corresponding *boxes*. *Plus sign* indicates location of NCR (28.6°N 77.2°E)

Table 1 Details of model configuration and experimental design

Model	WRF Version 3.0
Map projection	Mercator
Horizontal resolution	Nest: 27 km, 9 km 3 km and 1 km
Simulation temporal extent	27 Mar 2013 0000 UTC to 30 Mar 2013 0000 UTC
Central point of domain	28.6°N 77.2°E
Horizontal grid scheme	Arakawa C-grid
Time step	162 s
Microphysics	GCE scheme
	1. Hail option
	2. Graupel option
Land surface model	Noah land surface model
Surface layer model	MM5 similarity model
Radiation scheme	Shortwave – Dudhia scheme
	Longwave – RRTM
Planetary boundary layer	Yonsei University scheme
Cumulus parametrization	Kain-Fritsch scheme

resolution). The ARW model in the study is configured with the Goddard Cumulus Ensemble (GCE) microphysics scheme (Tao and Simpson 1993). It generates six hydrometeor outputs in the simulation including: vapour, cloudwater, rainwater, snow, ice (hail) and graupel. The varied option for the hail or graupel options suggest variation in the ice formation microphysics and how the model simulates parameters like hydrometeor population etc. In the following experiment, two simulations have been made with both the hail and graupel option of the GCE microphysics to understand the variation in the output and the microphysical processes involved in hail formation.

For model verification, NASA's Modern Era Retrospective-analysis for Research and Applications (MERRA) (Rienecker et al. 2011) six hourly analysis dataset is used, which is at a horizontal spatial resolution of 0.5°×0.7° available from 1979 onwards. For precipitation verification, Tropical Rainfall Measuring Mission (TRMM) (Huffman et al. 2007) 3B42 V7 daily derived precipitation dataset is used, available at 0.25°×0.25° spatial resolution available from 1998 onwards. The daily observation data used for outgoing longwave radiation (OLR) (Liebmann and Smith 1996) is data provided by Earth Systems Research Laboratory (ESRL), NOAA, USA, available at 2.5°×2.5° spatial resolution from 1974 onwards.

3 Results and Discussion

3.1 Model Simulated Synoptic Situation

Model simulated 3-h accumulated precipitation for 28 Mar 2013 0900–1800 UTC is depicted in Fig. 3 with the corresponding observed precipitation from TRMM. The model simulates heavy rainfall between 1200 and 1500 UTC. This time period can

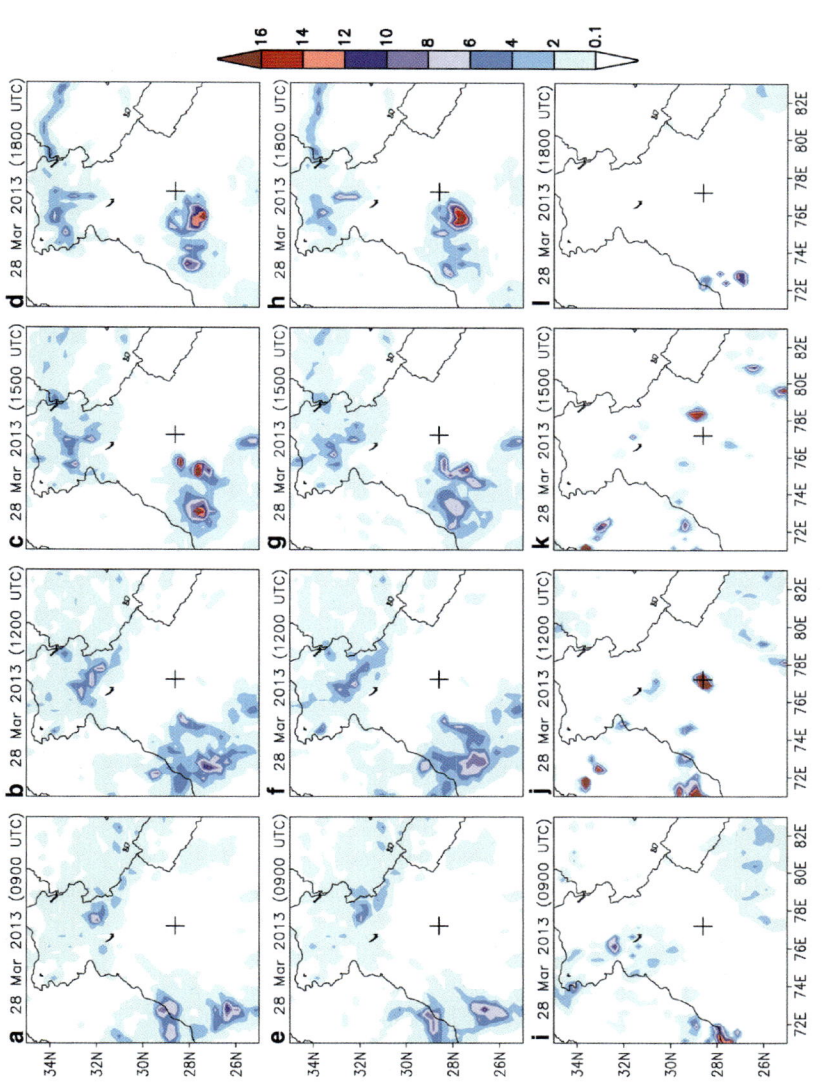

Fig. 3 3-h cumulative precipitation (mm) from 28 Mar 2013 0900–1800 UTC for model-simulated domain at 27 km horizontal resolution with (**a–d**) hail option, (**e–h**) graupel option, and (**i–l**) with TRMM observational analysis

be corroborated from the DWR reflectivity images from Fig. 1 which depict the peak of the storm around 1242–1252 UTC. But when compared with the TRMM observational dataset, it is observed that there is a slight temporal shift in the rainfall estimates for the precipitation maxima reproduced over the NCR. In model simulation precipitation maxima shows a spatial shift towards the west by approximately 150 km. Despite the spatial shift, when the precipitation amounts are compared between the model simulation and TRMM output, it is observed that the model simulation with hail option and TRMM observation data matches well by the model output. Especially with the hail option and observation data, precipitation maxima intensity is ~20 mm whereas graupel option simulation shows lesser intensity with precipitation peaking at ~15 mm only. With this analysis it can be concluded that the model simulation is able to capture the heavy precipitation event, but is showing displacement in the location of the precipitation intensity. With the shift in the model simulation the localized precipitation pattern near NCR will be considered in detail for this study.

When large scale flow is analyzed (figure not depicted), it shows the 500 hPa wind circulation and speed along with the geopotential height for 28 Mar 2013 0900–1800 UTC with model simulation and corresponding MERRA observational data. There is deep trough observed over the western Indian and Pakistan region, seen through the curved contours of the geopotential height and wind vectors. The wind speed shown in shaded grey region depicts the flow intensifying along the curved circulation. This depicts the formation of cyclonic circulation in the mid-tropospheric level, indicating the approaching western disturbance (WD) over the region. WDs are extratropical synoptic scale weather systems which move eastward from Mediterranean to Indian region as disturbances in the subtropical westerly jet and cause precipitation majorly in the winter time over northern India through large scale convection (Pisharoty and Desai 1956; Ramaswamy 1956; Dutta and Gupta 1967; Singh et al. 1981). Even in March, such weather system and the associated anomalous circulation can be considered the cause of storm formation by the generation of instability. From the figure it can be seen that the model simulation with both options shows a very good match with the observational analysis. The model simulation and observation show similar pattern of trough development and decrease in the values of the geopotential height. The development of a strong cyclonic circulation is conducive for the generation of convection for storm evolution over the NCR.

The hail and graupel precipitation amounts for both the experimental simulations are detailed in Fig. 4. Here it is noted that there is no graupel precipitation at all (figures not presented), whereas the sedimentation of hail is clearly captured by the hail option at 28 Mar 2013. Thus, the ice precipitation is clearly simulated in the hail option rather than the graupel option. Though the precipitation amounts or intensity of the hail is very low which is similar to the observed hail during the event. On 1500 UTC there is a spatial shift in the hail formation, coinciding with the displacement in rainfall patterns. It can be concluded that, the hail option is better representative of conditions during the hailstorm specifically where ice precipitation is concerned.

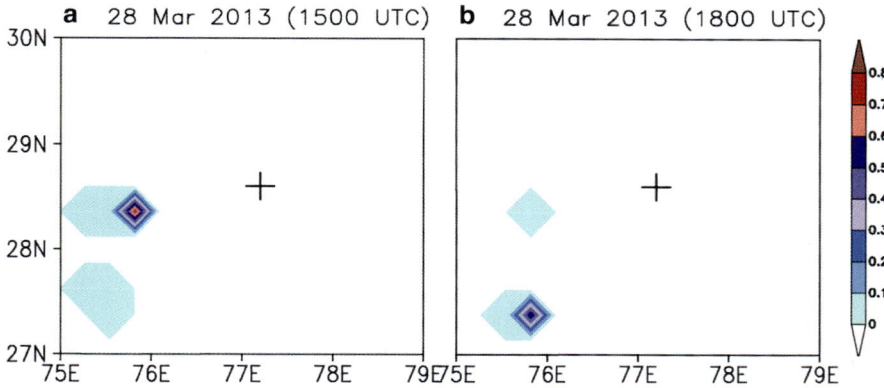

Fig. 4 (**a, b**) 3-h cumulative hail precipitation ($\times 10^{-3}$ mm) from 28 Mar 2013 1500–1800 UTC for model-simulated domain at 27 km horizontal resolution with hail option

The comparative analysis of precipitation (PREC), maximum reflectivity (REFL) and convective available potential energy (CAPE) and convective inhibitive energy (CINE) with both options of model output is presented in this paragraph. It is observed that the precipitation at 1 km resolution shows higher precipitation intensity than the coarser domains. This can be attributed to the fact that in the fine resolution model simulation the terrain representation is also at a much finer scale, which shows higher values than interpolation of observed parameters. The hail option shows a much heavier precipitation output than the graupel option. As observed in the 27 km resolution, there is a shift in the precipitation spatial distribution towards the west of the NCR. Thus, the study focuses on the precipitation localized in that region. The precipitation pattern shows similar localized nature as seen in the observation output showing the generation of a local storm. When considering the REFL, it shows a small defined region of peak intensity, which is indicative of the localized storm observed in the precipitation patterns. In both the model output options the reflectivity patterns match the high precipitation regions. High REFL values describe the region of concentration of the hydrometeors which precipitate during the storm. Increase in these hydrometeor concentrations indicate the enhancement of microphysical processes which ultimately lead to the sedimentation of these particles. The thermodynamic forcing of the storm is described by the energy dynamics. As per the figure we note that there is an increase in the CAPE around the region of heavy precipitation, whereas, in the hail option, lower values of CINE is observed as compared to the graupel option. The CAPE is the energy available due to the release of instability in the atmospheric column. The development of instability is discussed in detail in the next section of the paper. When the CAPE is released, it gets converted to kinetic energy which drives the storm propagation. So, higher values of CAPE promote the storm. On the other hand, the CINE is an inhibitive energy, which works against the storm evolution. There is increase in the values of

CINE for the graupel option, showing development of a weaker storm than in the hail option due to this energy working against convection. Thus, the hail option provides higher precipitation intensity when compared with the graupel option.

3.2 Various Forcings Related to the Storm

Considering the spatial shift in the precipitation patterns in the model simulated outputs, this section describes in details the various forcings in the atmosphere related to the storm development at the displaced location of precipitation maxima. From the above section it is derived that this location would coincide with 28.4°N, 75.6°E. Further, analysis in this section is carried out with the focus on this location.

Figure 5 depicts the various hydrometeors mixing ratios in a temporal-pressure cross section over the region of focus. Here it is noted that all the hydrometeors including cloudwater (CW), rainwater (RW), cloud ice (CI), hail and graupel show higher values of mixing ratios around 1200 UTC. This increased mixing ratios cause the increased precipitation of the hydrometeors during the peak precipitation time of 1500 UTC during which the mixing ratios reduce due to the sedimentation. In case of CW, the small droplets of cloud get formed and remain suspended in mid-tropospheric layer, as these get accreted to form the larger RW droplets, and then start descending. Thus, the CW droplets are suspended higher than the RW. Higher values of the CW and RW correspond to the increased precipitation in liquid form or rainfall. Hail option shows higher CI values, as this hydrometeor on accretion onto the hail particle form the hailstone. It is also noted that the hail option shows lower ice particle formation, as hail formation is much more complex when compared with graupel. There is variation in their formations due to the continuous process of conversion in between the various hydrometeors. So much so that even with higher graupel mixing ratios, there is no ice precipitation in graupel option as detailed in the above section. As discussed in the introduction, the hailstone formation is associated with cycling of hail throughout the cloud with continuous melting and re-freezing. Thus, an increase in the mixing ratio concentrations of not only ice particles but also liquid particles is seen. When considering the hail option the model simulates lower mixing ratios as compared to graupel option. Hail processes associated with the hail formation include the accretion of ice along the hail nuclei in concentric circles (Pruppacher and Klett 1980). Higher concentrations of ice mixing ratios indicate availability of ice for deposition on the hail particle as the particle ascends in the atmosphere during the updraft. The mixing ratios reach the maxima at the top of the troposphere. When the hailstone descends in the downdrafts, there is shedding and melting of the ice. The melting of the hail also releases latent energy during the storm process. In comparison the graupel particle is caused by the rimming (or clumping) of snow particles over a nucleus in non-uniform clumps (Pruppacher and Klett 1980). This formation does not exceed the middle layers of troposphere, but in higher reaches hail processes are enhanced due to colder conditions.

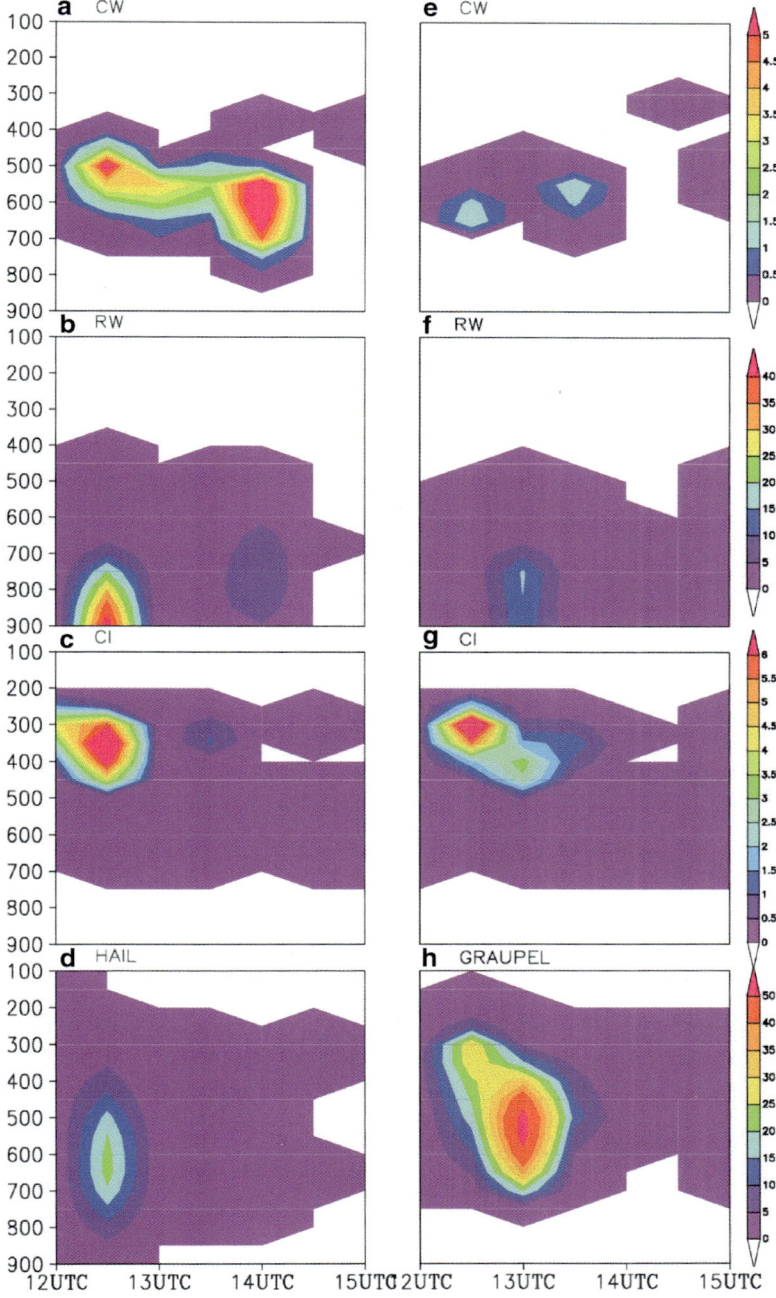

Fig. 5 Temporal-pressure cross section of mixing ratios for (**a**) CW ($\times 10^{-4}$), (**b**) RW ($\times 10^{-4}$), (**c**) hail ($\times 10^{-4}$) and (**d**) graupel ($\times 10^{-4}$) with hail option at 28 Mar 2013 1200 UTC for model-simulated domain 4 at 28.4°N, 75.6°E. Similarly, (**e–h**) for graupel option

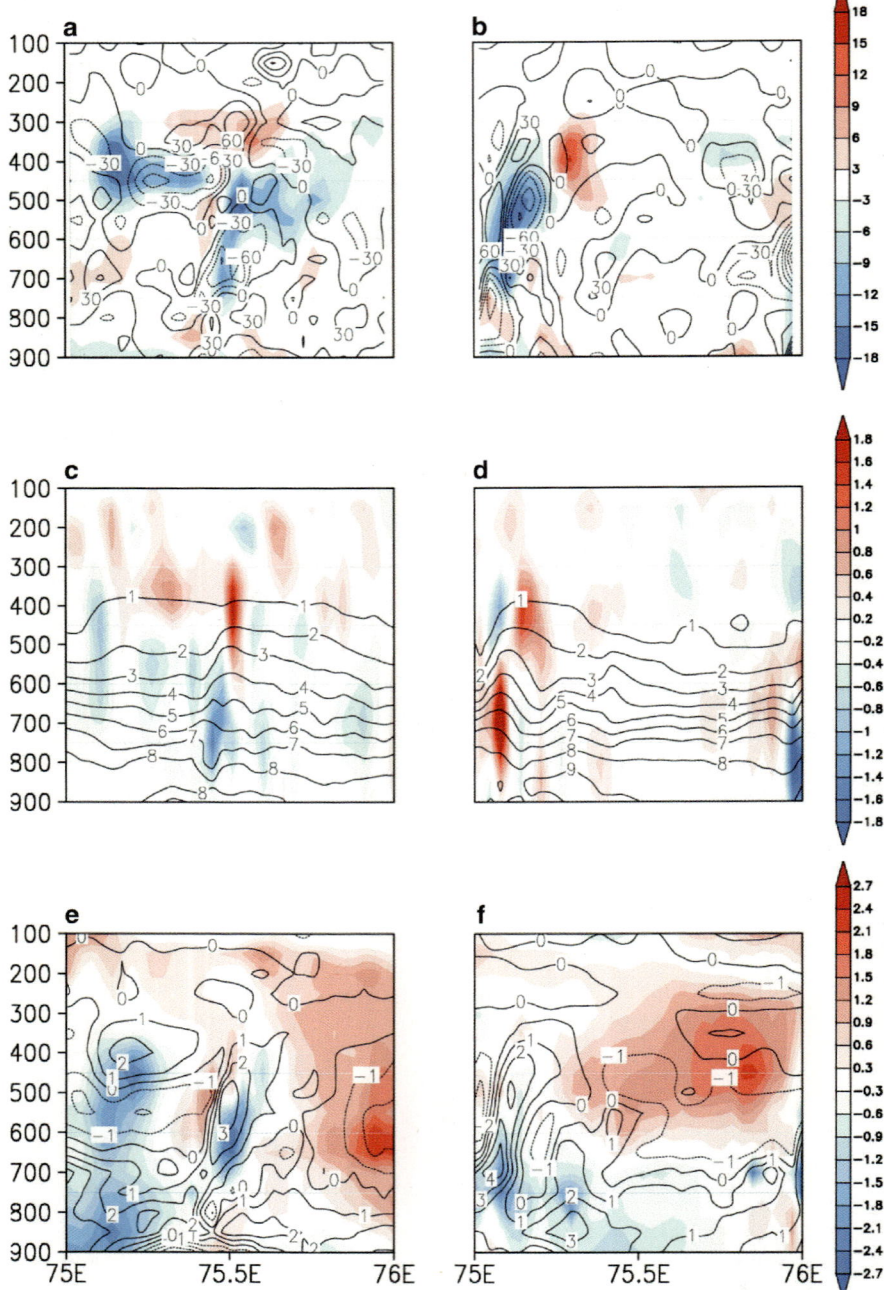

Fig. 6 Longitude-pressure distribution of model simulated (**a**) vorticity (×10^{-4} s^{-1}; *shaded*) and divergence (×10^{-5} mm; *contour*) (**b**) vertical wind (m/s) and specific humidity (×10^{-3} g/g) and (**c**) standardized anomaly for geopotential height and EPT with hail option at 28 Mar 2013 1200 UTC for model-simulated domain 4 at longitude 28.4°N. Similarly, (**d–f**) for graupel option

Thus, hail option in the microphysics focusing on the hail formation microphysical processes will provide a better representation of this storm event.

The longitude-pressure cross sections of various parameters like vorticity, divergence, specific humidity, vertical wind and anomaly of geopotential height and equivalent potential temperature (EPT) is shown in Fig. 6. The storm development is described in terms of instability and convection. The hail option shows the various forcings at the location of precipitation maxima whereas graupel option does not represent them well. The hail option captures the positive values of vorticity and increased specific humidity over the region of interest. The moisture convergence is observed coinciding with the region of heavy precipitation and this region shows upward flowing and downward flowing wind cells. The anomaly of geopotential height shows lower values around the mid-troposphere levels, whereas the EPT values show an increase. The decrease in the geopotential height shows a region of lower pressure developed due to the incoming WD. This causes the development of cyclonic circulation and is seen in the form of positive values of the vorticity along 300 hPa. Along with this, there is an incursion of moisture which converges near the cyclonic circulation and increases the specific humidity in the atmospheric column. The higher values of EPT from the surrounding regions show the development of instability along with moisture incursion signifying the convection. This convective activity generates the vertical wind movement, which is observed in the form of updrafts and downdrafts clearly seen in the hail option of model simulation. The region between these two cells show a region of vertical wind shear which is another source of released energy. The development of these updrafts and downdrafts are commonly observed in hailstorms, which are conducive for the development of hail due to cycling of the hailstones through the cloud continuously.

Acknowledgements The authors would like to thank India Meteorological Department (IMD) for the DWR output used as observation in the study. This study is benefitted in part by the Council of Scientific and Industrial Research (CSIR, India) fellowship to a. Chevuturi.

Conclusions

This study aims to understand the severe storm event and the associated hail occurrence over NCR on 28 Mar 2013. The WRF model simulation is used for analyzing this event based on GCE microphysics with hail and graupel options. When compared with the respective observations, the model well represents the weather event, though there is some spatial shift in the precipitation output. The model simulation shows the passage of a WD over the region, along with the moisture incursion from Arabian Sea which enhances convection and causes heavy precipitation during the storm.

When the graupel and hail experiments are compared, it is noted that the model simulation promoting hail processes show a storm reaching the upper

(continued)

tropospheric layers, with the associated hail precipitation which was not seen in the graupel option. With higher concentration of the hydrometeors, the microphysical processes of ice formation in colder cloud tops are enhanced. This storm pattern is also associated with cells of updrafts and downdrafts which enhances the cycling of hailstones through the cloud.

Winter months of December-January-February show low temperatures over the northern Indian region. Due to such lower temperatures strong convective storms are not common during this time period. But the pre-monsoon months of March-April-May show a steady increase in surface temperatures and with the approach of a WD there is enough instability generated for a storm to reach upper tropospheric layers, which is in turn favourable for hail formation.

There is a need to study such severe weather events with better methods of numerical weather prediction techniques like data assimilation to reduce the gap found in this study. Future studies should include very fine resolutions to understand the localized nature of such hailstorms which in severe conditions may cause major damage. Representation of air pollutant loading, urban heat island effects, etc. should be incorporated in the model outputs for a focussed study on storm patterns, specifically in an urbanized environment.

References

Browning KA, Fankhauser JC, Chalon JP, Eccles PJ, Strauch RG, Merrem FH, Musil DJ, May EL, Sand WR (1976) Structure of an evolving hailstorm. Part V: synthesis and implications for hail growth and hail suppression. Mon Weather Rev 104:603–610

Chalon JP, Fankhauser JC, Eccles PJ (1976) Structure of an evolving hailstorm. Part I: general characteristics and cellular structure. Mon Weather Rev 104:564–575

Chatterjee P, Pradhan D, De UK (2008) Simulation of hailstorm event using mesoscale model MM5 with modified cloud microphysics scheme. Ann Geophys 26(11):3545–3555

Dutta RK, Gupta MG (1967) Synoptic study of the formation and movement of western depression. Indian J Meteorol Geophys 18(1):45

Fankhauser JC (1976) Structure of an evolving hailstorm. Part II: thermodynamic structure and airflow in the near environment. Mon Weather Rev 104:576–587

Houze RA Jr (1981) Structures of atmospheric systems: a global survey. Radio Sci 16:671–689

Huffman GJ, Adler RF, Bolvin DT, Gum G, Nelkin EJ, Bowman KP, Hong Y, Stocker EF, Wolff DB (2007) The TRMM multi-satellite precipitation analysis: quasi-global, multi-year, combined-sensor precipitation estimates at fine scale. J Hydrometeorol 8(1):38–55

Liebmann B, Smith CA (1996) Description of a complete (interpolated) outgoing longwave radiation dataset. Bull Am Meteorol Soc 77:1275–1277

Musil DJ, May EL, Smith PL Jr, Sand WR (1976) Structure of an evolving hailstorm. Part IV: structure from penetrating aircraft. Mon Weather Rev 104:596–602

NCEP, NWS, NOAA, U.S. Department of Commerce (2000) NCEP FNL operational model global tropospheric analyses, continuing from July 1999. Research Data Archive at the National Center for Atmospheric Research, Computational and Information Systems Laboratory. http://rda.ucar.edu/datasets/ds083.2. Accessed 18 Apr 2013

NNDC-CDO (2013) NOAA national data center climate data online. http://www7.ncdc.noaa.gov/CDO/cdo. Accessed 11 May 2013

Orville HD, Kopp FJ (1977) Numerical simulation of the life history of a hailstorm. J Atmos Sci 34:1596–1618

Pisharoty P, Desai BN (1956) Western disturbances and Indian weather. Indian J Meteorol Geophys 7:333–338

Pruppacher HR, Klett JD (1980) Microphysics of clouds and precipitation. D. Reidel Publishing Company, Dordrecht

Ramaswamy C (1956) On the sub-tropical jet stream and its role in the development of large-scale convection. Tellus 8:26–60

Rienecker MM, Suarez MJ, Gelaro R, Todling R, Bacmeister J, Liu E, Bosilovich MG, Schubert SD, Takacs L, Kim GK, Bloom S, Chen J, Collins D, Conaty A, da Silva A, Gu W, Joiner J, Koster RD, Lucchesi R, Molod A, Owens T, Pawson S, Pegion P, Redder CR, Reichle R, Robertson FR, Ruddick AG, Sienkiewicz M, Woollen J (2011) MERRA: NASA's modern-era retrospective analysis for research and applications. J Clim 24:3624–3648

Singh MS, Rao AVRK, Gupta SC (1981) Development and movement of a mid tropospheric cyclone in the westerlies over India. Mausam 32(1):45–50

Strauch RG, Merrem FH (1976) Structure of an evolving hailstorm. Part III: internal structure from Doppler radar. Mon Weather Rev 104:588–595

Tao WK, Simpson J (1993) The Goddard Cumulus Ensemble model. Part I: model description. Terres Atmos Ocean Sci 4:35–72

Simulation of Mesoscale Convective Systems Associated with Squalls Using 3DVAR Data Assimilation over Bangladesh

Mohan K. Das, Someshwar Das, and Md. Mizanur Rahman

1 Introduction

Thunderstorms of pre-monsoon season (March–May), locally known as "Kal Boishakhi" or nor'westers, develop from a variety of mesoscale convective structures as they mature to meso α scale (200–1,000 km) systems. These systems develop mainly due to merging of cold dry northwesterly winds aloft and southerly low level warm moist winds from the Bay of Bengal.

Finch (http://bangladeshtornadoes.org) has made a detailed documentation of the severe storms of Bangladesh. Several factors lead to the active thunderstorm season across Bengal (Bangladesh and its adjoining regions in India). North and central India heats up and dries out in late March or early April. A deep, dry mixed layer develops. Low level flow from the Bay of Bengal increases markedly during this time. Westerly mid-level flow around the Tibetan Plateau advects the Indian mixed layer over the Bengal moist tongue. This leads to the Elevated Mixed Layer (EML). It may be noted that parts of the East Indian plateau are 'elevated' (1–3,000 ft) compared to Bangladesh which is near sea level. The mid-level flow is fairly strong in April with 30–50 kt (~15–25 m sec^{-1}) speed at 700 hPa and 35–50 kt (~18–25 m sec^{-1}) at 500 hPa. The high-level jet is usually over or just north of Bengal in April. The southern branch of the polar jet often retreats north of the Tibetan Plateau by May, leaving light, mid to high level flow across the Bengal region. All these factors result in severe storms maximum in early to mid April. In short, vertical wind shear and instability are maximized and the jet is in a favourable position during this time.

M.K. Das (✉) • Md.M. Rahman
SAARC Meteorological Research Centre (SMRC), Dhaka, Bangladesh
e-mail: mohan28feb@yahoo.com

S. Das
National Centre for Medium Range Weather Forecasting, Noida, India

Squall is defined as a sudden onset of strong winds with speeds increasing to at least 16 knots (18 miles per hour) and sustained at 22 or more knots (25 miles per hour) for at least 1 min. Several episodes of squalls occurred during April and May 2010. Widespread outbreaks of intense squalls occurred on the subsequent days affecting some parts of Bangladesh on 26 April 2010.

2 Observed Characteristics by Radar

The Bangladesh Meteorological Department (BMD) has S-band weather radar (~10 cm wave length) at Dhaka (90.4°N, 23.7°E) since the year 2000. It has a maximum radius about 400 km horizontally and effective radius about 250 km. It is operated at frequency of 2,700–2,900 MHz, and beam width of 1.7°. The radar is operated at zero elevation angle to collect hourly PPI scan data (pixel size 2.5 km) from 2300 (5 AM) to 1700 UTC (11 PM). The radar is operated for 1 h followed by a gap of 2 h during the period. The precipitation rates observed by radar on 26 April 2010 is depicted in Fig. 1. The frame-by-frame analysis of the radar observations indicate the three convective cells developed around West Bengal and west of Bangladesh area at about 11 UTC.

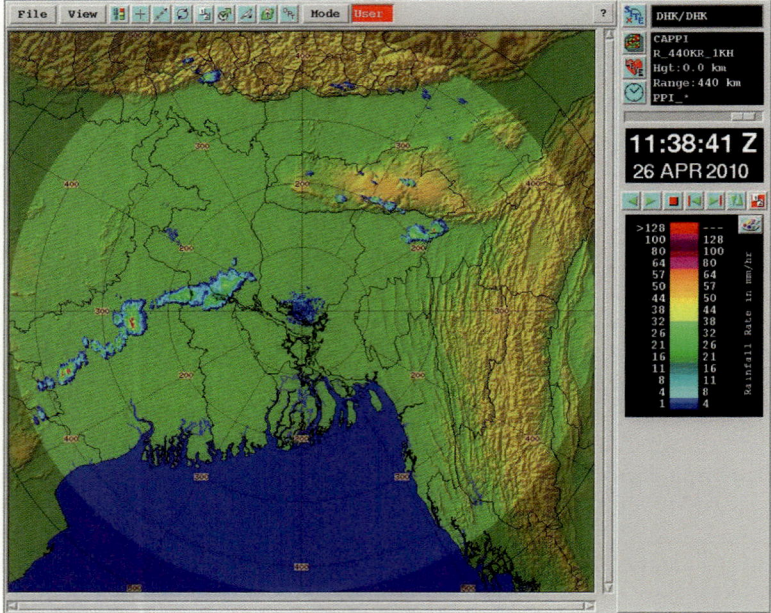

Fig. 1 Rain rate observed by Dhaka weather radar on April 26, 2010 at 1138 UTC

3 Numerical Model

The Weather Research and Forecasting (WRF) model (version 3.4.1) has been used for simulation of the tornadic storms in this study. The WRF model is a new-generation mesoscale Numerical Weather Prediction (NWP) system designed to serve both operational forecasting and atmospheric research needs. It features multiple dynamical cores, a 3-dimensional variational (3DVAR) data assimilation system, and a software architecture allowing for computational parallelism and system extensibility. WRF is suitable for a broad spectrum of applications across scales ranging from metres to thousands of kilometres. Applications of WRF include research and operational numerical weather prediction (NWP), data assimilation and parameterized-physics research, downscaling climate simulations, driving air quality models, atmosphere-ocean coupling, and idealized simulations (i.e., boundary-layer eddies, convection, baroclinic waves). There are two dynamics solvers in the WRF system: the Advanced Research WRF (ARW) solver (originally referred to as the Eulerian mass or "em") developed primarily at NCAR, and the NMM (Nonhydrostatic Mesoscale Model) solver developed at NCEP. The ARW system consists of the ARW dynamics solver with other components of the WRF system needed to produce a simulation. For the purpose of simulating the squall the model was run at 9 km resolutions with 27 vertical levels using initial and boundary conditions data obtained from NCEP FNL (Final) Operational Global Analysis, which is at about $1° \times 1°$ horizontal resolution (Table 1). Das et al. (2013) conducted several sensitivity experiments with different combinations of physical parameterization schemes of the model and found that the best skill scores were obtained by the combinations of no-cumulus, Milbrandt and YSU schemes for cumulus convection, cloud microphysics and planetary boundary layer respectively for the simulation of nor'westers over Indian and Bangladesh region.

This combination of physical processes provided least RMSE values for rainfall, wind speed at surface and time of occurrences of storms in the model simulations. Figure 2 shows the locations of the squall affected area Doppler Weather Radar stations in Bangladesh.

3.1 The 3DVAR System

In recent years efforts have been initiated towards the development of variational data assimilation systems to replace previously used schemes such as the Cressman, Newtonian nudging, optimum interpolation, and analysis algorithms. The 3DVAR is designed for a community data assimilation system flexible enough to allow a variety of research studies apart from its operational utilization. The basic goal of

Table 1 WRF ARW mode configuration

Model features	Configurations
Horizontal resolution	9 km
Vertical levels	Total 27 (1.000, 0.990, 0.978, 0.964, 0.946, 0.922, 0.894, 0.860, 0.817, 0.766, 0.707, 0.644, 0.576, 0.507, 0.444, 0.380, 0.324, 0.273, 0.228, 0.188, 0.152, 0.121, 0.093, 0.069, 0.048, 0.029, 0.014, 0.000)
Topography	USGS
Dynamics	
Time integration	Semi implicit
Time steps	50 s
Vertical differencing	Arakawa's energy conserving scheme
Time filtering	Robert's method
Horizontal diffusion	2nd order over quasi-pressure, surface, scale selective
Physics	
Convection	No-cumulus
PBL surface layer	YSU scheme Monin-Obukhov
Cloud microphysics	Milbrandt and Yau (2005)
Radiation	RRTM (LW), Dudhia (SW)
Gravity wave drag	No
Land surface processes	Unified NOAH land surface model

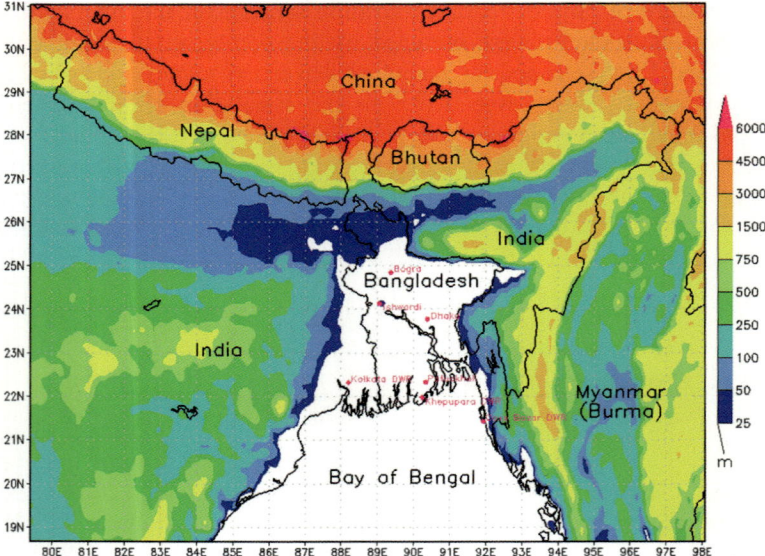

Fig. 2 The model domain under study with topography (*shaded*) in metres used for simulation and data assimilation

Table 2 Selected squalls and gusty wind events in Bangladesh

Date	2010-04-26
Station	Ishwardi
Occurrence time (UTC)	1120–1150
Wind speed (m/s)	20.577
Direction (ly)	NW
Rainfall 24 h (mm)	22

the 3DVAR system is to produce an "optimal" estimate of the true atmospheric state at any desired analysis time through iterative solution of a prescribed cost-function.

$$J(x) = J^b + J^0 = \frac{1}{2}(x-x^b)^T B^{-1}(x-x^b) + \frac{1}{2}(y-y^0)^T (E+F)^{-1}(y-y^0) \quad (1)$$

where x is the analysis state, xb is the background, y^o is the observation, B, E and F are the background, observation (instrumental) and representative error covariance matrices respectively. Error of Representativeness is an estimate of inaccuracies introduced in the observation operator. H is used to transform the gridded analysis x to observation space $y = Hx$.

In this study, Doppler Weather Radar (DWR) observations (radial winds and reflectivity) of Bangladesh Meteorological Department (BMD) are used to study the two episodes of squalls mentioned earlier in order to update the initial and boundary conditions (IC/BC) through 3DVAR technique within the Weather Research Forecasting (WRF) modelling system. It indicates that NWP models are very important for obtaining guidelines for the prediction of local severe storms.

Realized weather phenomena over Bangladesh and surroundings selected events of 26 April 2010 are shown in Table 2.

4 Results and Discussion

The WRF model was run for 24 h to simulate the squalls close to the time of their observations. The results are presented below for 9 km resolution. The idea is to diagnose the structure of the squalls by the model, and compare with available observations from ground based radar. The results presented below correspond to the mature stage of the squalls as simulated by the model.

4.1 Convective Available Potential Energy (CAPE), Storm-Relative Environment Helicity (SREH), Bulk Richardson Number Shear (BRNSHR) and Simulated Precipitation

4.1.1 CAPE

CAPE is the positive buoyancy of an air parcel. It is the amount of energy a parcel of air would have if lifted a certain distance vertically through the atmosphere. It is an indicator of atmospheric instability. It is defined as

$$\text{CAPE} = \int g \frac{(Tvp - Tve)}{Tve} dz \qquad (2)$$

where Zf and Zn are the levels of free convection and neutral buoyancy respectively. Tvp and Tve are the virtual temperatures of the air parcel and environment respectively. The threshold values of CAPE for different stability regimes are given below.

CAPE < 1,000: Instability is weak
CAPE > 1,000 < 2,500: Moderate instability
CAPE > 2,500: Strong instability

4.1.2 SREH

The helicity is a measure of the amount of rotation found in a storm's updraft air. If there is significant rotation in a storm's updraft air, the storm will more than likely become a supercell and possibly spawn one or more tornadoes. Helicity is a parameter that defines the amount of streamwise vorticity (i.e., directional shear) a steady storm updraft will ingest as a result of a given storm motion. In meteorology, helicity corresponds to the transfer of vorticity from the environment to an air parcel in convective motion. Here the definition of helicity is simplified to only use the horizontal component of wind and vorticity:

$$H = \int V_h \, \zeta h \, dZ = \int V_h \cdot \nabla \times V_h \, dZ \qquad (3)$$

where Z = altitude, V_h is the horizontal velocity and ζ_h is the horizontal vorticity. According to this formula, if the horizontal wind does not change direction with altitude, H will be zero as the product of V_h and $\nabla \times V_h$ are perpendicular one to the other making their scalar product nil. H is then positive if the wind turns (clockwise) with altitude and negative if it backs (counter-clockwise). Helicity has energy units per units of mass ($m^2 \, s^{-2}$) and thus is interpreted as a measure of energy transfer by the wind shear with altitude, including directional.

This notion is used to predict the possibility of tornadic development in a thundercloud. In this case, the vertical integration is limited below cloud tops (generally 3 km or 10,000 ft) and the horizontal wind is calculated to wind relative to the storm in subtracting its motion:

$$\text{SREH} = \int (V - C) \cdot \nabla \times V_h \, dZ \tag{4}$$

where C is the cloud motion to the ground. Critical values of SRH (Storm Relative Helicity) for tornadic development, as researched in North America, are:

SREH = 150–299: Supercells possible with weak tornadoes according to Fujita scale
SREH = 300–499: Very favourable to supercells development and strong tornadoes
SREH > 450: Violent tornadoes

When calculated only below 1 km (4,000 ft), the cut-off value is 100.

4.1.3 BRNSHR

The bulk Richardson number (BRN) is used to quantify the relationship between buoyant energy and vertical wind shear (Moncrieff and Green 1972), and is defined as

$$\text{BRN} = \frac{\text{CAPE}}{0.5\left(\overline{u}^2 + \overline{v}^2\right)} \tag{5}$$

where u and v are the wind components of the difference between the density-weighted mean winds over the lowest 6,000 m and the lowest 500 m above ground level. As discussed in Droegemeier et al. (1993), the BRN is only a gross estimate of the effects of vertical wind shear on convective storms, since it does not measure the turning of the wind profile with height. However, Weisman and Klemp (1984) showed using cloud-scale model simulations that the BRN can distinguish between supercell and multicell storms, with modelled supercells likely when $10 \leq \text{BRN} \leq 50$ and multicells storms likely when BRN > 35. It is important to note that there is no well-defined threshold value for BRN, since there is an overlap in these values used to specify storm type. Brooks et al. (1994a, b) hypothesized that the midlevel, storm-relative winds are important to the development of low-level rotation in thunderstorms.

In thunderstorm forecasting, CAPE is used to define the region in which convection is possible, SREH is used to define the region in which thunderstorms are likely to be supercells, and BRNSHR is used to define the region in which low-level mesocyclogenesis is more likely. These results highlight the potential value of analyzing various severe weather parameters in forecasting tornadic thunderstorms. By combining the storm characteristics suggested by these parameters, it is possible to use mesoscale model output to infer the dominant mode of severe convection. There are many other parameters that should be used in forecasting severe-weather threat (Johns and Doswell 1992; Thompson 1998). But we focus here on these three parameters CAPE, SREH and BRNSHR for simplicity. It should be recognized that there remains great uncertainty in, and debate about, the best parameters to use for forecasting tornadoes. Figure 3 depicts the combined graphic of the three fields of

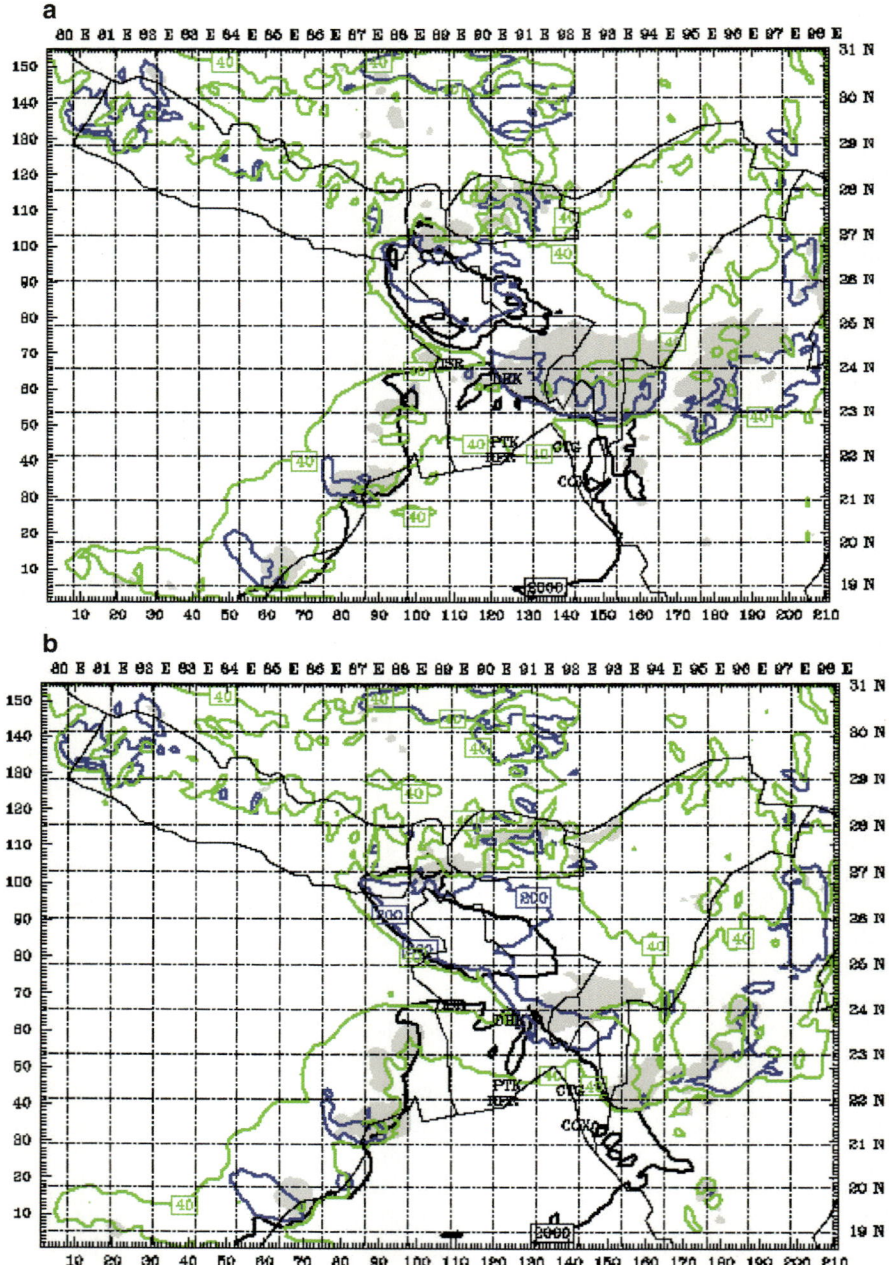

Fig. 3 CAPE (>2,000 J kg^{-1} in *black contours*), SREH (>200 m^2 s^{-2} in *blue contours*) and BRNSHR (>40 m^2 s^{-2} in *green contours*) as simulated by the model for 26 April 2010: (**a**) CTRL, and (**b**) 3DVAR. Rainfall (>0.25 mm) are *shaded*

CAPE (>2,000 J kg^{-1} in black contours), SREH (>200 m^2 s^{-2} in blue contours) and BRNSHR (>40 m^2 s^{-2} in green contours). It highlights areas in which all the three fields are in the ranges that are favourable for low-level mesocyclones in all the two cases of squalls (one case figure shown for brevity).

4.1.4 Precipitation

Rainfall values above 0.25 mm have been shaded in the diagrams to indicate areas of convection. Results show that on 26 April, the most favourable region for the development of squalls was near the northwest and east of Bangladesh. The border areas between Bangladesh-Assam-Meghalaya were favourable for the development of squalls. All the three parameters suggested the development of the storms.

4.2 850 hPa Wind Vector and 10 m Wind Speed Analysis

The impact of 3DVAR assimilation upon the first guess is studied here. It is seen that the experimental 3DVAR runs have simulated 10 m wind speed of about 18–20 m/s in small patches in the West Bengal region.

There is a strong trough at 850 hPa simulated by the 3DVAR; this is absent in the CTRL run (Fig. 4). The 850 hPa horizontal wind shows a trough and high wind velocity over the squalls location. The phenomena are stronger in the 3DVAR run.

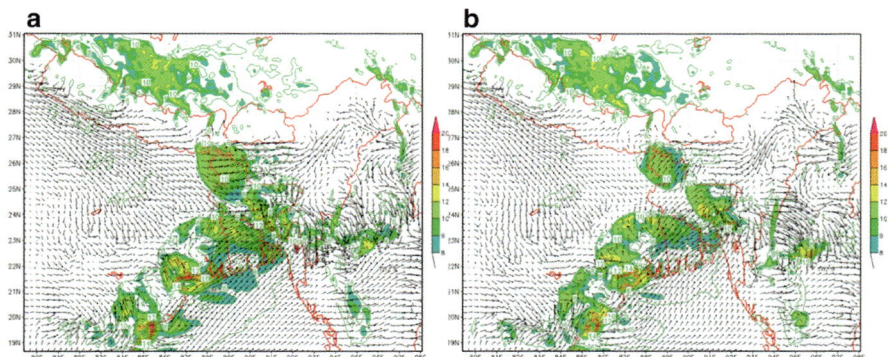

Fig. 4 Vector wind at 850 hPa and 10 m wind speed (m/s) forecasts valid on the days of the squalls on 26 April 2010 at 1200 UTC: (**a**) CTRL and (**b**) 3DVAR run

Conclusion

Bangladesh is prone to severe thunderstorms of tornadic intensity due to its geophysical location. The simulated results provided a basis to study the microphysical and dynamical characteristics of the squalls, which are generally not available from observations. The model could not simulate the bow type squall lines as observed by the radar. The model underestimated the strength of the squall lines in general. The simulated wind speed at surface was well simulated more than about 18 ms^{-1}.

This study is the first attempt to simulate squall event over Bangladesh by using the WRF model at 9 km horizontal resolution with new parameterization schemes. Results have provided many interesting findings, but also indicated many weaknesses in our present understanding and capability to forecast the squalls with sufficient lead time and accuracy. There is a severe scarcity of data to understand the observational characteristics of the severe storm in this region.

The impact of data assimilation is clearly visible as the experimental simulations from the WRF-3DVAR are able to capture the squalls closer to the observations compared to the CTRL run. The position and intensity of the simulated squalls in the experimental runs is close to the observed values, as compared to the DWR derived products. Assimilation of satellite and dense network of surface observations in the model may improve the accuracy of the forecasts.

Acknowledgements The authors would like to thank the National Center for Atmospheric Research (NCAR), USA for their excellent community service done by providing the WRF model. The initial and boundary conditions to run the model were obtained from NCEP. They would like to thank the Bangladesh Meteorological Department (BMD) and SMRC for providing the radar observations, and data.

References

Banerji BN (1938) Nature of 'Nor'westers' of Bengal and their similarity with others. Beitr Physik fr Atmos 24:231–233

Brooks HE, Doswell CA III, Wilhelmson RB (1994a) On the role of midtropospheric winds in the evolution and maintenance of low-level mesocyclones. Mon Weather Rev 122:126–136

Brooks HE, Doswell CA III, Cooper J (1994b) On the environments of tornadic and nontornadic mesocyclones. Weather Forecast 9:606–618

Das S, Sarkar A, Das MK, Rahman MM, Islam MN (2013) Composite characteristics of nor'westers observed by radar, TRMM and simulated by model. J Earth Syst Sci (Revised version submitted)

Droegemeier KK, Lazarus SM, Davies-Jones R (1993) The influence of helicity on numerically simulated convective storms. Mon Weather Rev 121:2005–2029

Thompson RL, Mead CM, Edwards R (2007) Effective storm-relative helicity and bulk shear in supercell thunderstorm environments. Weather Forecast 22:102–115

Simulation of Severe Convective Weather Events over Southern India Using WRF Model

S. Stella and Geeta Agnihotri

1 Introduction

The impacts of severe convective weather events like thunderstorm and hailstorm accompanied with squall during the pre-monsoon season are very high with devastating catastrophic effect. Nowcasting of such events are done with near 100 % accuracy with the help of Doppler Weather Radar products. But short-range forecasting of the same is a challenging task.

As per the electronic/print media, hailstorms with squall over several districts of Telangana region (Andhra Pradesh) from 1030Z to 1730Z on 1st April 2013 caused severe damages to mango, paddy, maize crops and 150 acres of banana plantations. More than 40 persons got injured due to hailstorms. Bangalore and its suburbs also experienced hailstorm with gusty winds that created havoc on 22nd May 2013. Since hailstorm events over the southern peninsular India have not been studied so much with NWP models, these two case studies have been simulated using WRF-ARW model and the output has been validated with the hail warning information product of DWR. The dynamical parameters vis-a-vis, winds at 10 m, constant pressure level, winds, relative humidity, water vapour mixing ratio, at 925, 850, 700, 600, 500, 400, 300 and 200 hPa levels, convective available potential energy (CAPE), convective inhibition energy (CINE), vertical velocity, vertical wind shear, maximum reflectivity and qgraupel simulated by the model were studied.

S. Stella (✉)
Regional Meteorological Centre, Chennai, India
e-mail: samuelstella@yahoo.com

G. Agnihotri
Meteorological Centre, Bangalore, India

2 Tools Used

2.1 Description of the WRF-ARW Model

- WRF-ARW version 3.0 model is a non-hydrostatic model designed to serve both operational forecasting and research needs.
- FNL data from GFS of NCEP archived on $1° \times 1°$ grids, available four times a day at six hourly interval has been used to initialise the model simulation.
- A single domain configuration 10°N–20°N/73°E–82°E with horizontal resolution of 9 km and vertical spacing of 27 pressure levels and the following combination of schemes of different physical processes have been used for the simulation of the hailstorms.
- Long-wave radiation scheme: RRTM
- Short-wave radiation scheme: Dudhia
- Surface layer parameterization: Monin-Obhikov scheme
- Cumulus parameterization: Grell and Devenyi
- Boundary layer: YSU
- Microphysics: Morrisson (2-moment scheme)

2.2 Doppler Weather Radar

DWR is an ultimate tool for nowcasting of severe convective weather events like thunderstorms, hailstorms and tornado and is being widely used worldwide. It helps the forecasters to understand the intensity and direction and speed of the movement of the convective event and nowcast with near 100 % accuracy. The products such as Maximum reflectivity, Max (Z), Plan Position Indicator (PPI), PPI (V) and PPI (Z), Volume Velocity Processing, VVP_2, Surface Rainfall Intensity (SRI) and Precipitation Accumulation (PAC) (24 h) are commonly used every day. Another DWR product namely Hail warning information HHW has been utilized for the study. The DWR products of Hyderabad have been utilized for the validation of the WRF model simulated output for hailstorm over Telangana region on 1st April 2013.

3 Methodology and Data

The WRF-ARW model was run with the resolution of 9 km. NCEP final analysis (fnl) data of $1° \times 1°$ resolution has been used as initial and boundary condition. Surface observation data over the domain were collected from India Meteorological Department (IMD). Doppler Radar products from the DWR, Hyderabad have been used.

The model simulated outputs for the dynamical parameters vis-a-vis wind at 10 m, constant pressure level winds, relative humidity, water vapour mixing ratio, at 925, 850, 700, 600, 500, 400, 300 and 200 hPa levels, CAPE, CINE, vertical velocity, vertical wind shear, maximum reflectivity and qgraupel were studied.

4 Hailstorm Cases

Two hailstorm events have been taken up for the study. First one is at Bangalore city (12.98 °N/77.58 °E) and nearby urban areas on 22nd May 2013. Hailstorm with gusty winds and rainfall occurred, between 16 UTC and 17 UTC and was reported in synoptic observation by Meteorological Centre, Bangalore and in electronic/print media.

The second hailstorm with squall at Telangana on 1st April 2013 between 1030 UTC and 1730 UTC was reported by the electronic/print media. Hailstorm with squall and rainfall occurred over Nizamabad (18.7°N/78.1°E), Medak (18.0°N/78.3°E), Adilabad (19.7°N/78.5°E), Karimnagar (18.4°N/79.1°E) and at Uravakonda (14.9°N/77.3°E) and Bhukaraya mandals of Anantapur district of Andhra Pradesh that caused severe damages to mango, paddy and maize crops ready for harvest and 150 acres of banana plantations. More than 40 persons got injured due to hailstorm and one person died of wall collapse.

5 Results and Discussions

5.1 Case (i): Hailstorm at Bangalore and Its Suburbs on 22nd May 2013

A trough in the lower levels was seen extending from Chattisgarh to south interior Tamilnadu across south interior Karnataka. Wind and moisture convergence from Bay of Bengal had taken place to initiate the convection over Bangalore and suburbs facilitated by Orographic lifting as Bangalore is 921 m above M.S.L and the Nandi hills are 1,478 m above M.S.L, located 60 km west of Bangalore. The Kalpana IR satellite imageries [Fig. 1a, b] with very high resolution (VHR) and cloud top temperature of −70 °C indicate very severe convection over Bangalore and neighbourhood.

Figure 2a, b show the model simulated absolute vorticity at 850 hPa at 00Z that is nil and very high at 14 UTC. Figure 2c, d show the model simulated very high absolute vorticity at 700 and 200 hPa at 16 UTC. Low level convergence and upper level divergence with high water vapour mixing ratio were noticed. Figure 2e, f show the model simulated watervapour mixing ratio at 700 hPa and 700 to 600 hPa at 15 UTC and 16 UTC respectively.

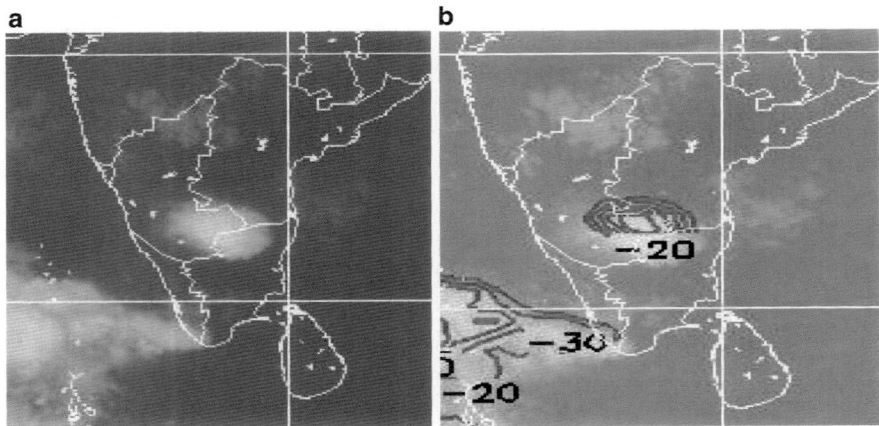

Fig. 1 (**a**) The Kalpana satellite imageries and (**b**) CTT (−70 °C) at 16 UTC on 22nd May 2013 over Bangalore and neighbourhood

Figure 3a shows high wind shear with maximum of 30 mps at 17 UTC. The trough at lower levels with low level convergence of wind and moisture, orographic lifting, high CAPE value, water vapour mixing ratio of 12 g/kg at 700 hPa and high vertical wind shear were all favourable for the severe convective weather over Bangalore city and suburbs on 22nd May 2013. Figure 3b shows the model simulated 24-h rainfall amount with the maximum of 2.5 cm over many areas in the study area. Actual rainfall recorded was Sompura (Bng-urban) 6 cm, Bangalore, GKVK, HAL airport, Goplanagara 4 cm each and Anekal (Bng-urban) 2 cm. Hence rainfall prediction was underestimated by the model.

Hailstorm occurred between 15 UTC and 16 UTC at Bangalore and its suburbs. The model has very well simulated the reflectivity of 45–60 dBZ at 15 UTC and 16 UTC over the area of study, shown in Fig. 3c, d predicting the definite occurrence of hailstorm. Figure 3e, f show the well simulated qgraupel of maximum 5 g/kg at 15 UTC and 16 UTC over Bangalore and suburbs which has indicated the hail.

5.2 Case (ii): Hailstorm over Telangana on 1st April 2013

On 1st April 2013, as per the electronic/print media, severe thunderstorm, hailstorm with squall and rainfall occurred at various places in several districts of Telangana region from 1030 UTC to 1730 UTC. Nizamabad (18.7°N/78.1°E), Medak (18.0°N/78.3°E), Adilabad (19.7°N/78.5°E), Karimnagar (18.4°N/79.1°E) and Uravakonda (14.9°N/77.3°E) and Bhukaraya mandals of Anantapur district of Andhra Pradesh are the areas that experienced the high impact of the severe convective weather. Figure 4a shows the maximum reflectivity of more than 55 dBZ upto

Simulation of Severe Convective Weather Events

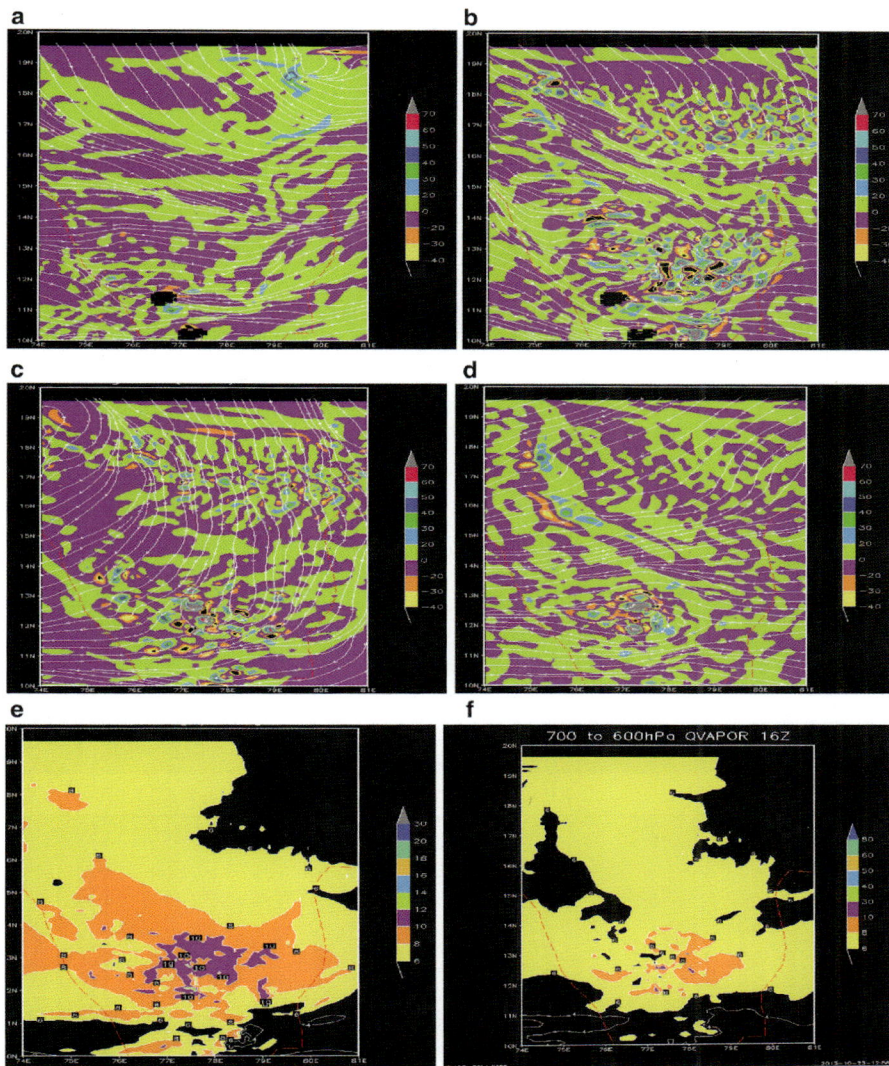

Fig. 2 Model simulated parameters on 22nd May 2013: (**a**) and (**b**) absolute vorticity at 850 hPa at 00Z and 14 UTC; (**c**) and (**d**) absolute vorticity at 700 and 200 hPa at 16 UTC; and (**e**) and (**f**) water vapour mixing ratio at 700 hPa and 700–600 hPa at 15 UTC and 16 UTC

the height of 9 km over Nizamabad district at 11:11 UTC. Figure 4b is the hail warning product (HHW) showing the maximum reflectivity of 56 dBZ over Nizamabad district at 11:11 UTC. As Adilabad and Anantapur districts of Telangana are beyond 200 km from DWR, Hyderabad, the Max(Z) and hail warning product could not be seen for those districts in Fig. 4a–i.

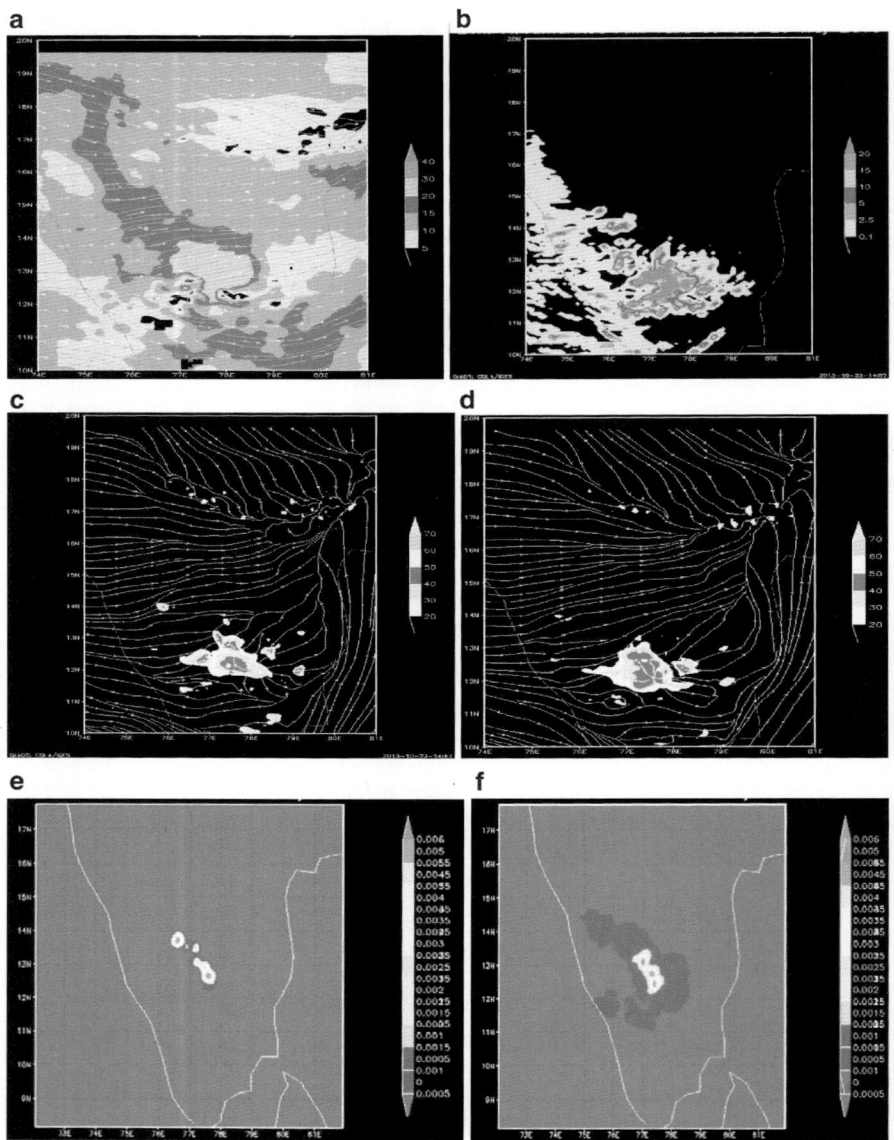

Fig. 3 Model simulated parameters on 22nd May 2013: (**a**) Wind shear at 17 UTC; (**b**) 24-h rainfall (03Z/22nd May 2013 to 03Z/23rd May 2013); (**c**) and (**d**) Reflectivity at 15 UTC and 16 UTC; and (**e**) and (**f**) Qgraupel at 15 UTC and 16 UTC

Simulation of Severe Convective Weather Events

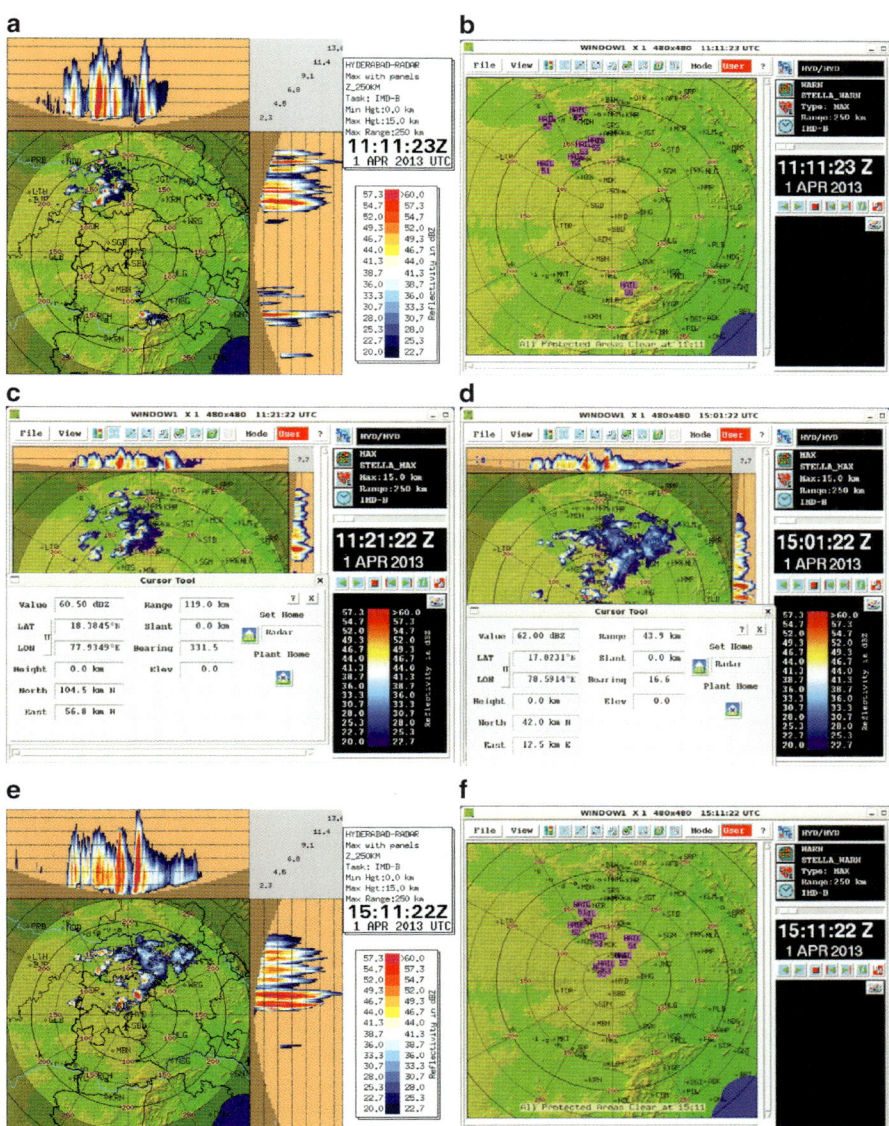

Fig. 4 (**a**) DWR derived reflectivity max(Z); (**b**) Doppler Radar hail warning product HHW at 11:11 UTC; (**c**) reflectivity Max(Z) showing 60.5 dBZ over Nizamabad at 11:21 UTC and (**d**) showing 62 dBZ over Medak at 15:01 UTC. DWR products (**e**) shows the Max(Z); (**f**) hail warning product HHW at 15:11 UTC; (**g**) shows the Max(Z) near Hyderabad area; (**h**) HHW at 16:11 UTC, 1st April'13; (**i**) reflectivity Max(Z) showing 64.5 dBZ over Nizamabad at 16:11 UTC and (**j**) showing the 24-h accumulated precipitation (PAC) on 2nd April 2013

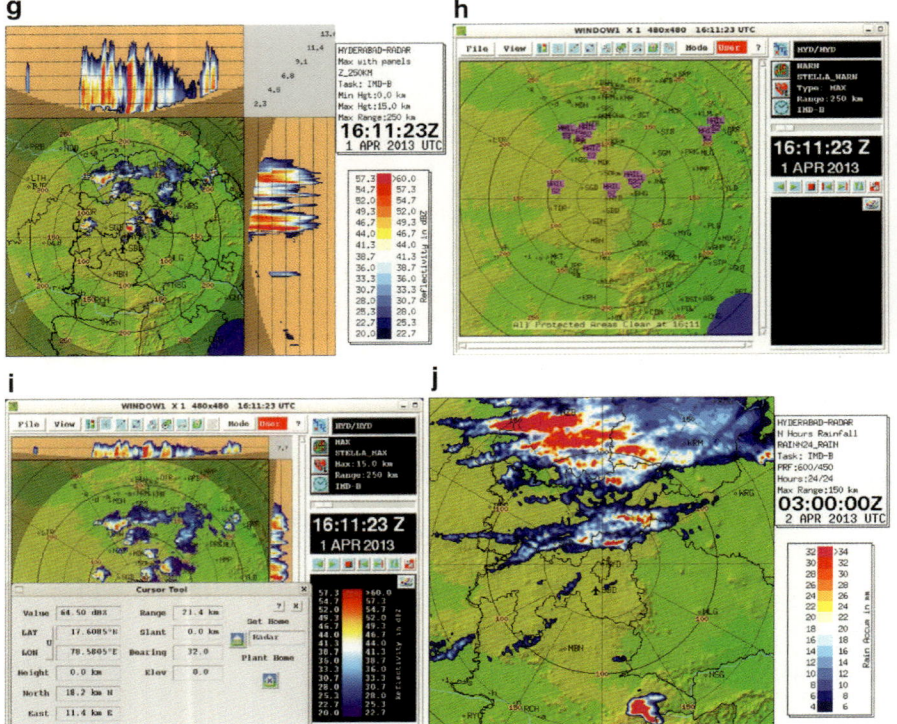

Fig. 4 (continued)

DWR products in Fig. 4c reflectivity Max(Z) show 60.5 dBZ over Nizamabad at 11:21 UTC and (d) show 62 dBZ over Medak at 15:01 UTC; (e) show the Max(Z) of more than 55 dBZ over Nizamabad at 15:11 UTC and (f) hail warning product (HHW) of 57 dBZ at 15:11 UTC; (g) shows the Max(Z) near NE of Hyderabad area at 16:11 UTC; (h) HHW of 57 dBZ near Hyderabad at 16:11 UTC; (i) shows the reflectivity Max(Z) of 64.5 dBZ over Nizamabad at 16:11 UTC and (j) shows the 24-h accumulated precipitation (PAC) with the maximum of 3.4 cm on 2nd April 2013.

As per the 00 UTC analysis, a cyclonic circulation was seen extending from surface upto 850 hPa and as a trough/wind discontinuity upto 700 hPa extending from Telangana to north interior Tamilnadu. Model simulated output for winds at 10 m, 925 and 850 hPa (Fig. 5a–f) show very good convergence of wind and moisture from both Arabian Sea and Bay of Bengal. Figure 5g, h show the model simulated CAPE of 1,600–2,000 J/kg at 00Z and 1,000–1,200 J/kg at 12 UTC respectively. Model simulated CINE as shown in Fig. 5i, j was 160–220 J/kg at

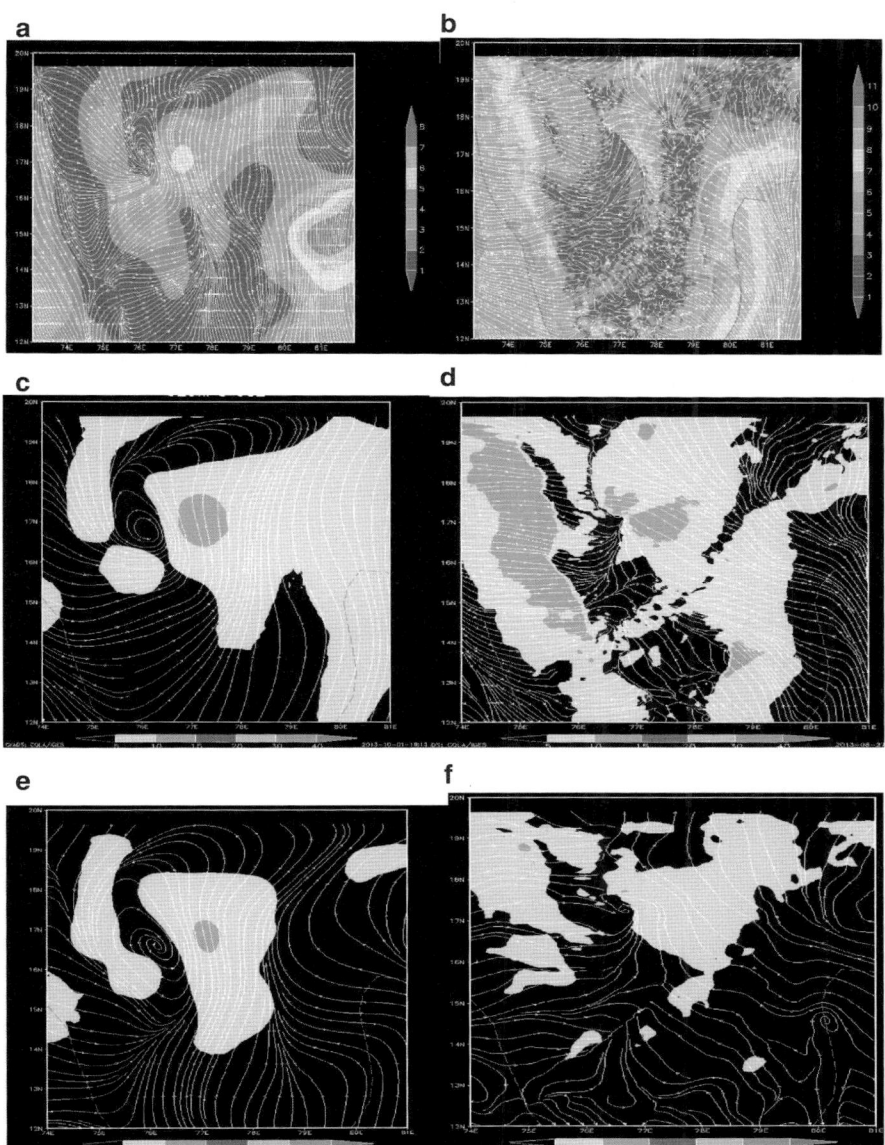

Fig. 5 Model derived winds. (**a**) and (**b**) Winds at 10 m at 00 UTC and 12 UTC respectively; (**c**) and (**d**) winds at 925 hPa at 00 UTC and 15 UTC respectively; (**e**) and (**f**) Winds at 850 hPa at 00 UTC and 15 UTC respectively; (**g**) and (**h**) Model derived CAPE valid at 00 UTC and 12 UTC respectively; (**i**) and (**j**) Model derived CINE valid at 00 UTC and 12 UTC respectively; and (**k**) and (**l**) Model derived water vapour mixing ratio valid at 00 UTC and 12 UTC respectively

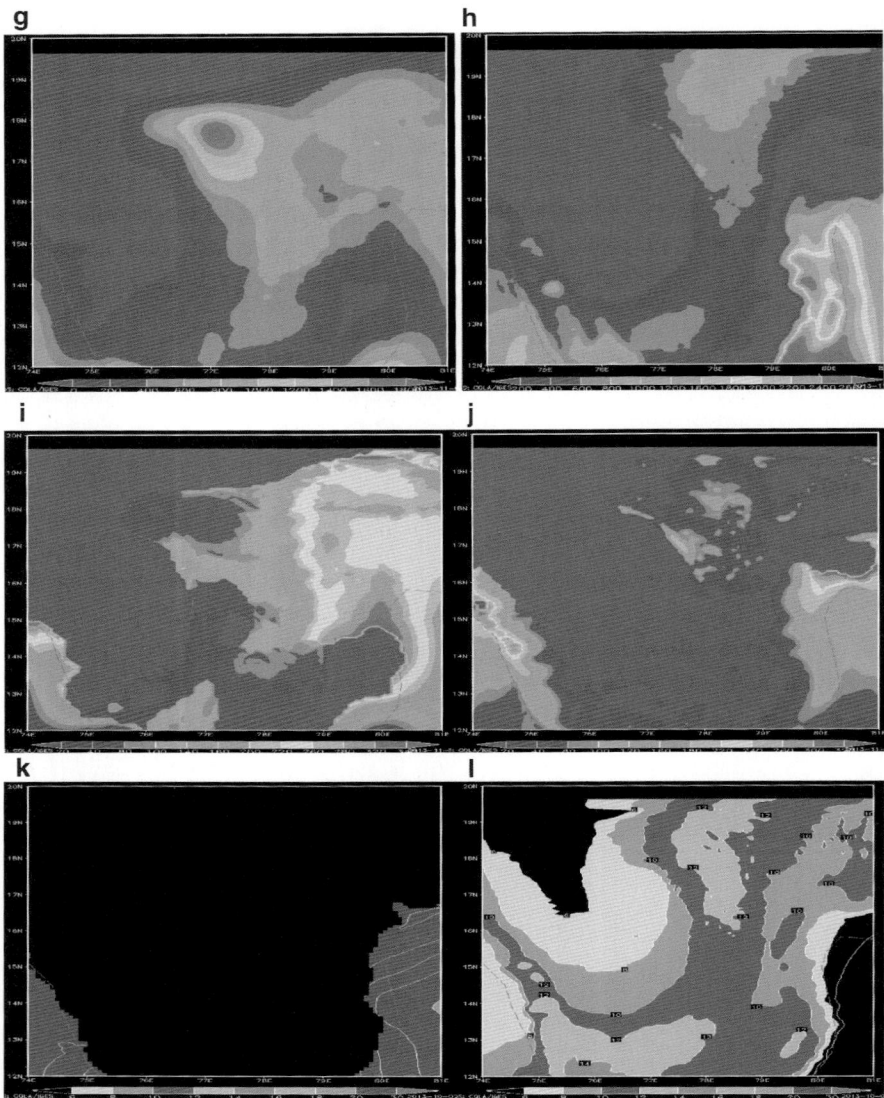

Fig. 5 (continued)

00 UTC and 80–140 J/kg at 12 UTC. The water vapour mixing ratio as simulated by the model (Fig. 5k, l) started increasing from 0 g/kg at 00 UTC and reached 14 g/kg at 12 UTC due to the moisture incursion from both Arabian Sea and Bay of Bengal. The rainfall was underestimated by the model over Telangana with the maximum of 1.5 cm. The actual rainfall amount recorded at various places of Nizamabad district

and Medak district: 5.2 mm to 2.7 cm, Karimnagar dist.: 4.0 mm to 3.6 cm, Adilabad dist.: 2.2 mm to 2.8 cm, Anantapur: 3.0 mm to 2.3 cm with maximum of 4.4 cm and mostly 1.1–2.2 cm. The model also could not simulate the reflectivity of more than 45 dBZ required for hailstorm.

Conclusion

WRF-ARW model could simulate the severe convective event over southern peninsular India, but with underestimated rainfall amount in both hailstorms at Bangalore and Telangana (AP). Maximum reflectivity and qgraupel were simulated well by the model to forecast the hailstorm with accurate space and time over Bangalore, but it underestimated reflectivity and qgraupel over Telangana.

More case studies on hailstorm should be taken up for southern peninsular India.

Acknowledgement The authors would like to thank the Director of DWR Hyderabad for the valuable guidance and supply of radar products in time.

References

Asnani GC (2005) Tropical meteorology
Chatterjee P, Pradhan D, De UK (2008) Simulation of hailstorm event using Mesoscale Model MM5 with modified cloud microphysics scheme. Ann Geophys 26:3545–3555
Misra PK, Prasad SK (1980) Forecasting of hailstorm over India. Mausam 31:385–396
Smitha J (2007) Numerical modeling of thunderstorm over Indian region. Vatavaran 31(1)
Tyagi A (2007) Thunderstorm climatology over Indian region. Mausam 58(2):189–212

Part II
Tropical Cyclones

Early Warning Services for Management of Cyclones over North Indian Ocean: Current Status and Future Scope

M. Mohapatra, B.K. Bandyopadhyay, Kamaljit Ray, and L.S. Rathore

1 Introduction

Tropical cyclones are the most devastating phenomena among all natural disasters, having taken more than half a million lives all over the world in the last five decades. Cyclones are accompanied by very strong winds, torrential rains and storm surges. The havoc caused by cyclones to shipping in the high seas and coastal habitats along the Indian coasts due to above mentioned adverse weather have been known since hundreds of years. The tropical warm north Indian Ocean (NIO), like the tropical North Atlantic, the South Pacific and the NW Pacific, is a breeding ground for the disastrous TC phenomenon. Historically, in terms of loss to human life, the Bay of Bengal TCs have accounted for deaths ranging from 1,000–300,000. The Bay of Bengal has experienced more than 75 % of the total world-wide TCs causing human death of 5,000 or more in last 300 years (Dube et al. 2013).

The low pressure systems over the NIO are classified (Table 1) based on the associated maximum sustained surface wind (MSW) at the surface level (IMD 2003). The systems with the intensity of depressions and above are considered as cyclonic disturbances. The 'cyclone' is a generic term associated with a low pressure system with MSW of 34 knots or more. It corresponds to the definition of tropical storms over other ocean basins like Pacific and Atlantic Oceans. Over these basins, it is called as TC/Typhoon when the MSW is 64 knots or more corresponding to very severe cyclonic storm (VSCS) over the NIO.

The reduction of cyclone disasters depends on several factors including hazard analysis, vulnerability analysis, and preparedness & planning, early warning, prevention and mitigation. The early warning is a major component for the south Asian region due to its socio-economic conditions. The early warning component includes skill

---

M. Mohapatra (✉) • B.K. Bandyopadhyay • K. Ray • L.S. Rathore
India Meteorological Department, Lodhi Road, New Delhi 110003, India
e-mail: mohapatraimd@gmail.com

Table 1 Classification of cyclonic disturbances over the NIO (since 1999)

Low pressure system	Maximum sustained surface winds
Low pressure area (L)	<17 knots
Depression (D)	17–27 kts
Deep depression (DD)	28–33 kts
Cyclonic storm (CS)	34–47 kts
Severe cyclonic storm (SCS)	48–63 kts
Very severe cyclonic storm (VSCS)	64–119 kts
Super cyclonic storm (SuCS)	120 kts and above

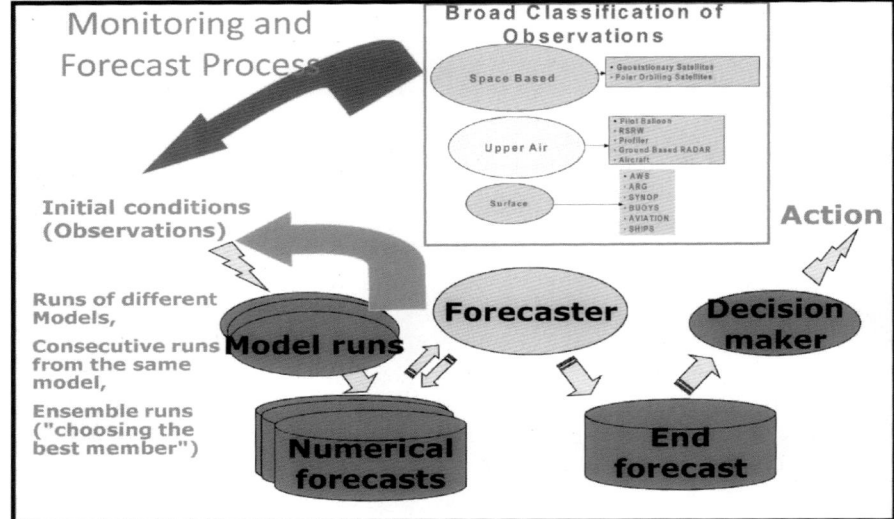

Fig. 1 Monitoring and forecasting process of tropical cyclone (After Mohapatra et al. 2013a)

in monitoring and prediction of cyclone, effective warning products generation and dissemination, coordination with emergency response units and the public perception about the credibility of the official predictions and warnings. The entire process of cyclone early warning system is shown in a schematic diagram (Fig. 1). It is important to continuously upgrade all the components of early warning based on latest technology for effective management of TCs.

Considering all these, the initiative has been taken recently to modernize early warning system for TC and hence to maximize relevance and effectiveness of the cyclone warning during emergent situations. The above objective is accomplished through:

(i) Modernization of observational system
(ii) Modernization of cyclone analysis and prediction system

(iii) Updating of Standard Operation Procedure (SOP) for cyclone monitoring and forecasting
(iv) Institutional mechanism with various NWP modelling centres and disaster management agencies
(v) Building forecast demonstration projects on land falling cyclones
(vi) Value added warning products generation, presentation and dissemination and
(vii) Measures for enhancing confidence of disaster managers and public in forecast and warning through forecast verification and preparation of reports
(viii) Capacity building through training

1.1 Climatological Characteristics of Cyclones over the NIO

It is now a well known fact of climatology (IMD 2008) that about 5–6 cyclones occur in a year over the NIO prominently during the pre-monsoon season (March-April-May) and the post-monsoon season (October-November-December). It accounts for about seven per cent of the global cyclones. The maximum frequency is in the 2 months of May and November. Cyclones generally move in a northwesterly direction. However, they may recurve sometimes depending upon the environmental conditions. The Bay of Bengal cyclones more often strike Odisha-West Bengal coast in October, Andhra coast in November and the Tamil Nadu coast in December. Over 50 % of the cyclones in the Bay of Bengal strike different parts of the east coast of India, 30 % strike coasts of Bangladesh, Myanmar and Sri Lanka and about 20 % dissipate over the sea itself. Percentage of cyclones dissipating over the Arabian Sea is higher (60 %) as the western Arabian Sea is cooler. Maximum landfall occurs over Gujarat coast (18 % of total cyclones in Arabian Sea) of India followed by Oman coast. Life period of a cyclone over the NIO is 5–6 days. It has VSCS (64 knots or more) intensity for 2–3 days as against 6 days of global average.

About 50 % of the depressions develop into cyclone intensity and only less than 25 % of cyclones further intensify into severe cyclones. While it takes about 2–4 days to develop a low pressure area into a depression, the intensification from a depression to severe or very severe cyclones can occur in 24–48 h.

Several efforts have been made in the IMD to update climatological records on cyclones of the NIO as it provides useful guidance on early warning and planning of coastal regions. Recently an electronic atlas has been published for tracks of TCs over the Bay of Bengal and Arabian Sea (IMD 2008). It is available in cyclone page of IMD's website with free access (www.imd.gov.in).

2 Observational Systems for Cyclones over the NIO

It is important to correctly determine the location and intensity of the cyclone, as initial error in location and intensity can lead to increase in error in forecast location and intensity (Mohanty et al. 2010). Hence, there is a need of dense observational network over the sea and along the coast. The observational network for cyclone monitoring consists of land-based surface and upper-air stations, Doppler Weather Radars (DWRs)/Cyclone Detection Radars (CDRs), satellites, ships and buoy. The synoptic charts are prepared and analyzed every 3 h to monitor the TCs over the NIO. As the NIO is a data sparse region and a cyclone genesis takes place in mid-oceanic region, most of the cyclogenesis (location and time) and intensity are determined with the satellite observations. When the system comes closer to the coast and lies within radar range, it is monitored with radar followed by satellite. However, there are cases of genesis near the coast within the radar range which could not be detected by satellite. When the cyclone lies close to the coast, coastal observation is given more weightage followed by radar and satellite. Brief descriptions on various observational aspects are given below.

2.1 Satellite Based Observations

The geostationary satellite, Kalpana-I provides imageries in visible (VIS), infrared (IR) and water vapour (WV) channels. In addition INSAT-3A is also equipped with Charged Coupled Device (CCD) cameras capable of providing imageries in VIS, Near IR (NIR) and Short-Wave IR (SWIR) channels with greater resolution. During cyclone situation, data from Kalpana-I are processed at hourly/half hourly intervals to assess the location and intensity. In addition to above, the products like outgoing long wave radiation, quantitative precipitation estimates, sea surface temperatures (SST), cloud motion vectors, water vapour derived wind vector, and isotherm analysis on enhanced infrared images are also analyzed on operational mode for cyclone monitoring. The microwave imageries were introduced for monitoring and guidance in a subjective manner during later part of 2000s. It was used more objectively to locate the system centre since 2010 by utilizing the cyclone module available in cyclone forecasting workstation at IMD, New Delhi. Microwave imageries are more helpful in predicting the structural characteristics and intensification in short range, as the characteristics of intensification is first observed in microwave imageries unlike VIS and IR imageries (Jha et al. 2013). Apart from Indian satellites, products from other international satellites are also used for TC monitoring.

The sea surface wind as estimated by scatterometer-based satellites (ASCAT, Windsat and OSCAT) is very useful in locating the centre of the TC (Uhlhorn and Black 2003). However, it has the limitations as it provides only two observations. It also suffers from rain contamination and inability to measure the wind speed more than 50 knots (28 mps). The OSCAT-based surface winds are being used since November 2009. Gray's Parameters (Gray 1968) including SST & Ocean heat content, convective instability, wind shear, low level relative vorticity, coriolis parameter and upper level divergence are monitored for genesis and intensification and these parameters are mostly estimated by satellite technique. Past studies (Kalsi 2002; Bhatia and Sharma 2013) have built up a store-house of knowledge on satellite applications in cyclone monitoring in the NIO.

2.2 Buoy and Ship Observations

Government of India has established a National Data Buoy Programme (NDBP) at National Institute of Ocean Technology (NIOT), Chennai. Under this programme, 12 moored data buoys are deployed currently in the NIO. The data buoys are fitted with sensors to measure air pressure, air temperature, wind speed and direction and sea surface temperature among other parameters. These buoys have resulted in better monitoring and reduction of location and intensity error in association with ship and satellite observations. As per the guidelines issued to Indian voluntary observing fleet (IVOF), synoptic observations are made at the main standard times: 0000, 0600, 1200 and 1800 UTC by the ships. When additional observations are required, they are made at one or more of the intermediate standard times: 0300, 0900, 1500 and 2100 UTC. Over the NIO, 186 ships are registered under IVOF. The ship observations were quite high during pre-satellite period (before 1960s). It gradually decreased with the advent of polar orbiting satellite in 1960s and reduced further with the introduction of Indian geostationary satellites during 1980s.

2.3 Radar Observations

There are 11 numbers of S-band radar stations viz. Kolkata, Paradip, Visakhapatnam, Machilipatnam, Chennai, Sri Harikota, Karaikal, Kochi, Goa, Mumbai and Bhuj (Fig. 4a). Out of these 11 stations, five stations (Chennai, Kolkata, Sriharikota, Visakhapatnam and Machilipatnam) are using DWRs and the remaining stations have conventional S-band radars. Conventional radar provides information on reflectivity and range only, whereas a DWR provides velocity and spectral width data along with various meteorological, hydrological and aviation products which are very useful for forecasters in estimating the storm's centre, its intensity and predicting its future movement. A radar image of a matured TC consists of eye, eye wall, spiral bands, pre-cyclone squall lines and streamers.

2.4 Surface Observational Network Including Automatic Weather Stations (AWS)

IMD has a good network of surface observatories satisfying the requirement of WMO. There are 70 departmental manned surface observatories of IMD at present all along the coast and over Bay and Arabian Sea islands. There are 21 pilot balloon observatories and 15 radiosonde/radio wind (RS/RW) observatories. The meteorological data thus collected all over these stations are used on real time basis for TC monitoring.

2.5 High Wind Speed Recorders (HWSRs)

The high wind speed recorder for TC monitoring has been installed in Digha (West Bengal), Puri, Paradip and Gopalpur (Odisha), Visakhapatnam, Machilipatnam and Nellore (Andhra Pradesh), Chennai (Tamil Nadu), Karaikal (Puducherry), Mumbai (Maharashtra), Veraval and Dwarka (Gujarat). Further it is planned to install at eight more places in next 2 years. It can measure the wind speed upto 250 kmph.

3 Standard Operation Procedure for Monitoring of Cyclone

Various kinds of analytical procedure are described in Standard Operation Procedure (SOP) Manual (IMD 2003, 2013). A systematic check list is prepared for identification of location and intensity of cyclone. The procedure necessarily deals with determination of location and intensity along with other characteristics like associated MSW, estimated central pressure and pressure drop at the centre, shape and size, radius of outermost closed isobar, point and time of landfall, if any or area of dissipation etc. with the available observations in the storm region.

3.1 Monitoring of Genesis of Cyclone

Genesis parameters are evaluated in following steps to monitor the cyclogenesis:

Step I: SST, depth of 26 °C isotherm and ocean thermal energy
Step II: Conditional instability through a deep and moist atmospheric layer
Step III: Pre-existing disturbance
Step IV: Environmental conditions (vertical wind shear, low level vorticity, upper level divergence etc.,)
Step V: NWP and dynamical-statistical model forecasts for genesis

Based on synoptic, statistical, dynamical-statistical and NWP models guidance, a consensus decision is taken on genesis of depression and its likely intensification into TC.

3.2 Determination of Location of Centre and Intensity of Cyclone

The location of the centre of the TC is determined based on (a) Synoptic, (b) Satellite (INSAT/METSAT/microwave) and (c) Radar observations. When the cyclone is far away from the coast and not within the radar range, satellite estimate gets more weight, though it is modified sometimes with availability of ship and buoy observations. When the cyclone comes closer to the coast, radar estimate gets maximum preference followed by satellite. When cyclone is very close to coast or over the land surface, coastal observations get the highest preference followed by radar and satellite observations.

3.2.1 Synoptic Technique

In synoptic technique, the centre of the system is determined by considering the centroid of the wind distribution at the surface level. In the pressure field, the location of lowest mean sea level pressure is considered as the centre of the system (IMD 2003). The synoptic technique has got serious limitations over the open sea due to non-availability of sufficient observations. However, the AWS stations along coast are very useful as they provide hourly observations on real time basis (Bhatia et al. 2008). The coastal hourly observations help not only in correctly analyzing the location, but also in determining the landfall point and time and hence help in adverse weather warning. For intensity estimation, the available surface observations are taken into consideration to find out MSW and number of closed isobars at the interval of 2 hPa within a specified region around the system centre (IMD 2003).

3.2.2 Satellite Technique

In the initial stage (depression/deep depression), the centre is determined from the centre of the low cloud lines (IMD 2003). There are four types of cloud pattern (Dvorak 1984) in TC. In case of shear pattern, when the convection lies away from the centre, centre is same as the centre of low cloud lines. As the system intensifies and acquires the banding pattern, the centre is determined from the banding feature using logarithmic spiral. In the central dense overcast (CDO) pattern, the centre of CDO is the centre of the system. In the eye pattern, the centre determination is easier and accurate as it is same as the centre of the eye of the cyclone.

The intensity classification by satellite technique is based on Dvorak's technique (Dvorak 1984; Velden et al. 2006). The intensity of the tropical system is indicated by a code figure called T Number based on above pattern recognition technique. Another feature of the technique is the Current Intensity number (C.I.) which relates directly to the intensity (in term of wind speed) of the cyclone. The C.I. number may differ from the T number on some occasions to account for certain factors which are not directly related to cloud features. The empirical relationship between C.I. number and the MSW are given in Table 2. Third column of the table gives the pressure depths (peripheral pressure minus central pressure in hPa) as applicable for Indian Sea area using the relation $V_{max} = 14.2 \times SQRT$ (pressure depth) following Mishra and Gupta (1976). As there is no aircraft reconnaissance in the NIO, Dvorak's technique has not been verified and also the pressure wind relationship not verified.

Table 2 Maximum sustained wind (MSW) speed and pressure depth in relation to CI number

C.I. number	Max. wind speed (knots)	Pressure depth (in mb)
1	25	2.0
1.5	25	3.0
2	30	4.5
2.5	35	6.1
3	45	10.0
3.5	55	15.0
4	65	20.9
4.5	77	29.4
5	90	40.2
5.5	102	51.6
6	115	65.6
6.5	127	80.0
7	140	97.2
7.5	155	119.1
8	170	143.3

Comparison of satellite-based intensity and the best track estimates of IMD indicate a difference of about 0.5 T (Goyal et al. 2013). Recently the microwave imageries and brightness temperatures are also used to determine central pressure and MSW (Jha et al. 2013). However, this technique has not been validated over the NIO due to non-availability of aircraft reconnaissance.

3.2.3 Radar Techniques

The *eye* or the centre of the TC can be derived from a continuous and logical sequence of observations. The geometric centre of the echo-free area is reported as the *eye location*. If the wall cloud is not completely closed, it is still usually possible to derive an eye location with a high degree of confidence by sketching the smallest circle or oval that can be superimposed on the inner edge of the existing portion of the wall cloud. When the wall cloud is not developed fully but a centre of circulation is identifiable, this feature is observed and reported similar to the eye. When the eye or centre is indistinct or outside the range or the radar beam overshoots the inner eyewall when it does not extend very high, spiral band overlays are used to estimate the location of the centre (IMD 1976). Based on observed winds from DWR, the intensity can be determined (Raghavan 2013). As radar based wind are not available at surface level, the wind observations from these techniques are converted to 10-m wind using the suitable conversion technique like those used in case of aircraft reconnaissance technique in Atlantic.

The location estimation error has been about 55 km over the sea areas (Mohapatra et al. 2012a; Goyal et al. 2013). According to Elsberry (2003), the errors in determining the TC centre over the northwest Pacific Ocean can be upto 50 km by satellite fixes, 20–50 km by radar observations and by about 20 km by aircraft reconnaissance. The induction of DWR has reduced the error in fixing the centre of cyclones in radar range. The landfall point estimation error has been reduced to about 25 km by 2010 mainly due to installation of coastal AWS. The average error in MSW estimation has reduced over the years. It could have been T0.5 (05–20 knots or 3–10 mps) with the introduction of Dvorak's classification of intensity since 1974.

4 Cyclone Forecasting System

A variety of observational data was used in India till 1960s to forecast the track intensity and landfall of TCs. Since 1960s, satellite era, added another feature. There has been rapid development in objective techniques since 1970s and especially in recent years for forecasting tracks and intensity of TCs in the NIO. To summarise, following methods are currently used by IMD for TC track and intensity forecasting:

 (i) Statistical technique—Analogue, Persistence, Climatology, Climatology and persistence (CLIPER)
 (ii) Synoptic technique—Empirical technique
 (iii) Satellite techniques—Empirical technique
 (iv) Radar techniques—Empirical technique
 (v) Numerical weather prediction (NWP) models
 (vi) Dynamical statistical models

In the synoptic method, prevailing environmental conditions like wind shear, low to upper level wind and other characteristics are considered. All these fields in the NWP model analyses and forecasts are also considered. The development of characteristic features in satellite and radar observations is also taken into consideration. While, the synoptic, statistical and satellite/radar guidances help in short-range forecast (upto 12/24 h), the NWP guidance is mainly used for 24–120 h forecasts. Consensus forecasts that gather all or part of the numerical forecast tracks and intensity and uses synoptic and statistical guidance are utilised to issue official forecast.

4.1 NWP Models

There are three types of NWP models for cyclone forecasts, viz., individual deterministic models, multi-model ensemble (MME) and single model ensemble prediction system (EPS). Also there are dynamical statistical models for the purpose of genesis and intensity prediction.

4.1.1 Individual Deterministic Model

- **Global Forecast System (GFS)**

  The Global Forecast System (GFS), adopted from National Centre for Environmental Prediction (NCEP), at T574L64 (~25 km in horizontal) resolution (incorporating Grid point Statistical Interpolation (GSI) scheme as the global data assimilation for the forecast up to 7 days) is implemented operational at IMD, New Delhi on IBM-based High Power Computing Systems (HPCS). The model is run twice in a day (00 UTC and 12 UTC). The real-time outputs are made available to the national web site of IMD.

- **Non-hydrostatic mesoscale modelling system WRFDA-WRF-ARW**

  The mesoscale forecast system Weather Research and Forecast WRFDA (version 3.2) with 3DVAR data assimilation is being operated twice daily to generate mesoscale analysis at 27 km and 9 km horizontal resolutions using IMD GFS-T574L64 analysis/forecast as first guess. Using initial and boundary conditions from the WRFDA, the WRF (ARW) is run for the forecast up to 3 days with double nested configuration and horizontal resolution of 27 km and 9 km and 38 Eta levels in the vertical. The model mother domain covers the area between lat. 25°S to 45°N and long 40°E to 120°E and child covers whole India.

- **Quasi-Lagrangian Model (QLM)**

  The QLM, a multilevel fine-mesh primitive equation model with a horizontal resolution of 40 km and 16 sigma levels in the vertical, is being used for cyclone track prediction in IMD. The integration domain consists of 111×111 grid points centred over the initial position of the cyclone. The model includes parameterization of basic physical and dynamical processes associated with the development and movement of a cyclone by (i) merging of an idealized vortex into the initial analysis and (ii) imposition of a steering current over the vortex area with the use of a dipole. The initial fields and lateral boundary conditions are taken from the IMD GFS T574L64. The model is run twice a day based on 00 UTC and 12 UTC initial conditions to provide six hourly track forecasts valid up to 72 h.

- **Hurricane WRF Model (HWRF)**

  Recently, under Indo-US joint collaborative programme, IMD adapted Hurricane-WRF model for cyclone track and intensity forecast over NIO region. It has nested domain of 27 km and 9 km horizontal resolution and 42 vertical levels with outer domain covering the area of 800×800 and inner domain 60×60 with centre of the system adjusted to the centre of the observed cyclone. The model has special features such as vortex initialization, coupling with ocean model to take into account the changes in SST during the model integration, tracker and diagnostic software to provide the graphic and text information on track and intensity prediction for real-time operational requirement. The model is run on real time twice a day based on 00 UTC and 12 UTC initial conditions to provide six hourly track and intensity forecasts valid up to 120 h. The model uses IMD GFS-T574L64 analysis/forecast as first guess.

- **Other Models**

    IMD also makes use of NWP products prepared by some other operational NWP Centres like, European Centre for Medium Range Weather Forecasting (ECMWF), GFS-USA, GFS-NCMRWF, Japan Meteorological Agency (JMA), United Kingdom Meteorological Office (UKMO), Global Tropical model, Meteo-France etc.

4.1.2 Multi-Model Ensemble (MME)

The MME technique (Kotal and Roy Bhowmik 2011) is based on a statistical linear regression approach. The predictors selected for the ensemble technique are forecast latitude and longitude positions at 12-h interval up to 120-h of five operational NWP models. The 12-hourly predicted cyclone tracks are then determined from the respective mean sea level pressure fields using a cyclone tracking software. A collective bias correction is applied in the MME by applying multiple linear regression based minimization principle for the member models WRF(ARW), QLM, GFS(IMD), GFS(NCEP), ECMWF and JMA. There is also facility in cyclone module of forecasters' work station to develop MME using equal weightage to individual model tracks available in cyclone module.

4.1.3 Ensemble Prediction System (EPS)

As part of WMO Programme to provide a guidance on cyclone forecasts in near real-time for the ESCAP/WMO member countries, IMD implemented JMA supported software for real-time forecast over NIO during 2011. The Ensemble and deterministic forecast products from ECMWF (50+1 members), NCEP (20+1 members), UKMO (23+1 members), MSC (20+1 members) and JMA (20+1 members) are available near real-time for NIO region. These products include: Deterministic and ensemble track forecasts, strike probability maps and strike probability of cities within the range of 120 kms 4 days in advance. The super-ensemble has also been developed based on above ensembles.

In India, NCMRWF runs the global ensemble forecasting system (GEFS) configuration consisting of four cycles corresponding to 00Z, 06Z, 12Z 18Z and 10-day forecasts are made using the 00 UTC initial condition. A T190L28 control that is started with T574L64 analysis is run out to 10 days with 20 perturbed forecasts. The ensemble spread is a measure of the difference between the members and is represented by the standard deviation with respect to the ensemble mean. On an average, small (high) spread indicates a high (low) forecast accuracy. The ensemble spread is flow-dependent and varies for different parameters. It usually increases with the forecast range, but there can be cases when the spread is larger at shorter forecast ranges than at longer.

4.1.4 Statistical Dynamical Model for Cyclone Genesis and Intensity Prediction

A genesis potential parameter (GPP) for the NIO has been developed (Kotal and Bhattacharya 2013) as the product of four variables, namely vorticity at 850 hPa, middle tropospheric relative humidity, middle tropospheric instability, and the inverse of vertical wind shear. The GPP is operationally used for predicting cyclogenesis at their early development stages. The grid point analysis and forecast of the genesis parameter up to 7 days are generated on real time. Region with GPP value equal or greater than 30 is found to be high potential zone for cyclogenesis.

A statistical-dynamical model (SCIP) (Kotal et al. 2008) has been implemented for real time forecasting of 12-hourly intensity upto 72 h. The model parameters are derived based on model analysis fields of past cyclones. The parameters selected as predictors are: initial storm intensity, intensity changes during past 12 h, storm motion speed, initial storm latitude position, vertical wind shear averaged along the storm track, vorticity at 850 hPa, divergence at 200 hPa and SST. For the real-time forecasting, model parameters are derived based on the forecast fields of IMD GFS model.

4.2 Adverse Weather Forecasting

A TC causes three types of adverse weather, viz., heavy rain, gale wind and storm surge during its landfall.

4.2.1 Heavy Rainfall

The forecast/warning of heavy rainfall includes: (i) time of commencement, (ii) duration, (iii) area of occurrence and (iv) intensity of heavy rainfall. The methods for prediction of heavy rainfall include: (i) synoptic, (ii) climatological, (iii) satellite, (iv) radar and (v) NWP techniques. While NWP models provide prediction of rainfall for different lead period; satellite and radar provides quantitative precipitation estimates during past 3/12 h. The intensity and spatial distribution of rainfall estimated by satellite and radar are extrapolated to issue forecast. In synoptic and climatology method, synoptic climatology of rainfall intensity and spatial distribution are used. The final forecast is the consensus arrived from various methods as mentioned above.

4.2.2 Gale Wind

The forecast of gale wind includes: (i) time of commencement, (ii) duration, (iii) area of occurrence and (iv) magnitude of gale wind. The methods for prediction of gale wind include: (i) synoptic, (ii) climatological, (iii) satellite, (iv) radar, (v) NWP and (vi) dynamical statistical techniques. In the satellite method, region of maximum reflectivity and mesoscale vortices are assumed to be associated with higher wind. In radar technique, the direct wind observation are available through uniform wind technique, ppv2 product and radii velocity measurements. The wind estimates from satellite and radar and other observations are extrapolated to forecast the wind. MSW is also available from other sources like scatterometry wind from satellite, buoy and ships apart from estimate by Dvorak technique. Though the wind forecasts by the NWP models are underestimated the initial condition of wind from the model can be corrected based on actual observations and accordingly model forecast wind can be modified. The forecast based on dynamical statistical model also can be utilised in the similar manner.

4.2.3 Storm Surge

Storm surge is the rise of sea water above the astronomical tide due to cyclone. The storm surge depends on pressure drop at centre, radius of maximum wind, point of landfall and interaction with sea waves, astronomical tide, rainfall, river run off, bathymetry, coastal geometry etc. The forecast of storm surge includes: (i) time of commencement, (ii) duration, (iii) area of occurrence and (iv) magnitude of storm surge. The methods for prediction of gale wind include: (i) IMD Nomogram (Ghosh model), (ii) IIT Delhi Model (Dube et al. 2013) and INCOIS, Hyderabad model. INCOIS model also provides coastal inundation in terms of aerial extent and height of inundation.

4.3 SOP for Forecasting and Decision Support System (DSS)

An SOP (IMD 2003, 2013) is followed for analyzing various forecast guidance available from different sources as discussed in previous sections. There is well defined road map and check list for this purpose. The TC analysis, prediction and decision-making process is made by blending scientifically based conceptual models, dynamical and statistical models, meteorological datasets, technology and expertise. For this purpose, a decision support system (DSS) in a digital environment is used to plot and analyse different weather parameters, satellite, radar and NWP model products. In this hybrid system, synoptic method could be overlaid on NWP models supported by modern graphical and GIS applications to produce high quality analyses and forecast products. The cyclone module installed in this system has the following facilities:

- Analysis of all synoptic, satellite and NWP model products for genesis, intensity and track monitoring and prediction
- Preparation of past and forecast tracks upto 120 h
- Depiction of uncertainty in track forecast
- Structure forecasting (forecast of wind in different sectors of TC)

However, all the data are not still available in cyclone module. For better monitoring and prediction, additional help is taken of ftp and websites to collect and analyse:

- Radar data and products from IMD's radar network and neighbouring countries
- Satellite imageries and products from IMD and international centres
- Data, analysis and forecast products from various national and international centres

The automation of the process has increased the efficiency of system, visibility of IMD and utility of warning products (Mohapatra et al. 2013a).

4.4 Forecast and Warning Products

4.4.1 Track Forecast Products

Considering recent development in prediction capability, IMD introduced the objective cyclone track forecast valid for next 72 h in 2009 and upto 120 h in 2013. The TC forecast is issued six times a day at the interval of 3 h, i.e. based on 00, 03, 06, 09, 12, 15, 18 and 21 UTC observations. The forecasts are issued about 3 h after the above mentioned observation time. An example of the product during cyclone, Thane is shown in Fig. 2a.

4.4.2 Cone of Uncertainty in Track Forecast

The "cone of uncertainty (COU)"—also known colloquially as the "cone of death," "cone of probability," and "cone of error"—represents the forecast track of the centre of a cyclone and the likely error in the forecast track based on predictive skill

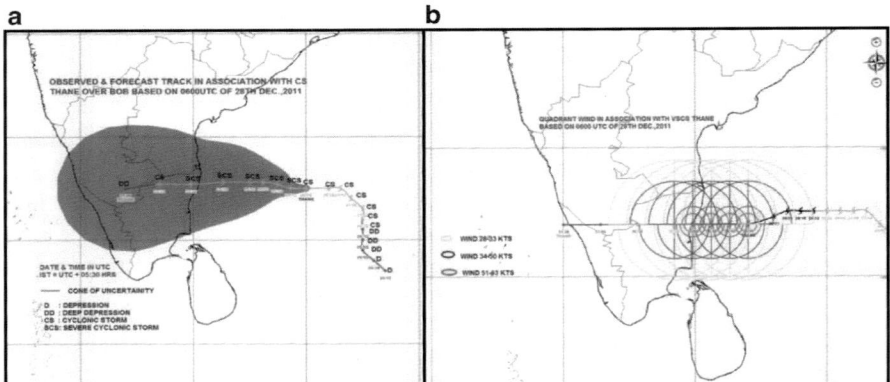

Fig. 2 (a) A typical example of observed and forecast track of depression which later on became the very severe cyclonic storm, Thane and (b) a typical graphical presentation of quadrant wind forecast during cyclone, Thane in 2011

of past years. The COU in the forecast of IMD has been introduced with effect from the TC, 'WARD' during December, 2009 (Mohapatra et al. 2012c). It is helpful to the decision makers as it indicates the standard forecast errors in the forecast. A typical example of COU forecast showing the uncertainty circles for different forecast periods are shown in Fig. 2b. The observed track lies within the forecast COU in about 60–70 % of the cases like other Ocean basins like northern Atlantic and Pacific Oceans.

4.4.3 Quadrant Wind Radii Forecasting

The cyclone wind radii representing the maximum radial extent of winds reaching 34 kts, 50 kts and 64 kts in each quadrant (NW, NE, SE, SW) of cyclone are generated as per requirement of ships. The initial estimation and forecast of the wind radii of TC is rather subjective and strongly dependent on the data availability, climatology and analysis methods. The subjectivity and reliance on climatology is amplified in NIO in the absence of aircraft observations. However, recently with the advent of easily accessible remotely sensed surface and near-surface winds (e.g. Ocean Sat., Special Sensor Microwave Imager (SSMI), low level atmospheric motion vectors and Advanced Microwave Sounder Unit (AMSU) retrieval methods, DWR, coastal wind observations and advances in real time data analysis capabilities, IMD introduced TC wind radii monitoring and prediction product in Oct., 2010. A typical example of the quadrant wind radii product is shown in Fig. 2b.

4.5 TC Forecasting Skill Accuracy over NIO

All these initiatives as mentioned in previous sections have resulted in improved cyclone warning service delivery, timeliness of the warning, and reduction in loss of lives as the outcome. The trends in forecast performance during 2003–2012 are presented here to illustrate these facts.

4.5.1 Landfall Forecast Accuracy

The landfall point forecast error has reduced significantly in recent years. The 12 and 24 h landfall point forecast errors have decreased at the rate of about 16 and 34 km per year respectively during 2003–2012 (Fig. 3). However, the rate of decrease is relatively less in case of landfall time forecast error.

4.5.2 Track Forecast Accuracy

The annual average TC track forecast errors have decreased at the rate of about 5.1 km/year and 7.2 km/year during 2003–2012 for 12 and 24 h forecasts respectively. The 36–72 h forecast errors also have decreased as shown in Fig. 4 (Mohapatra

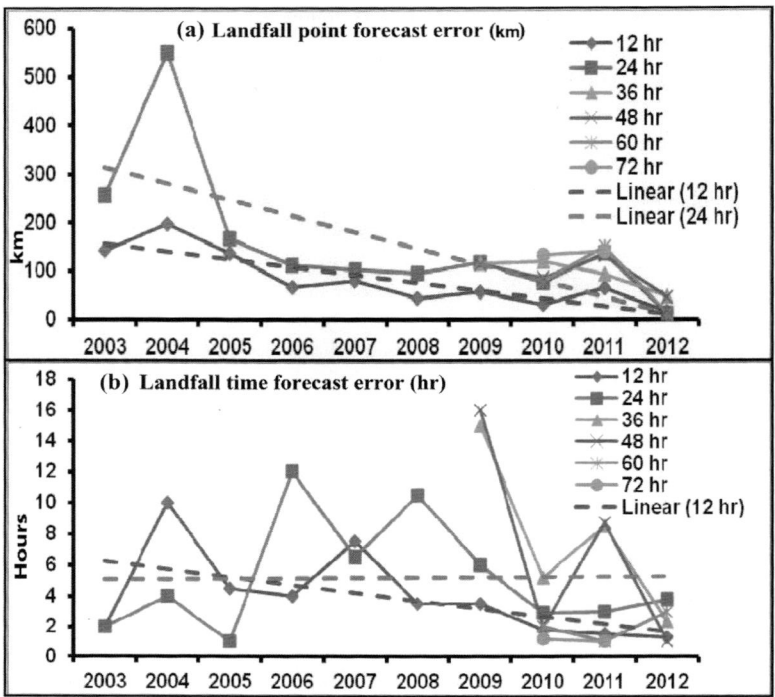

Fig. 3 Landfall point and time forecast errors of IMD during 2003–2012

et al. 2013b). The skill in tropical cyclone track forecast has increased at the rate of 8.2 % and 4.1 % for 12 and 24 h forecasts respectively during 2003–2012. There is also significant increase in skill of 36–72 h track forecasts. The average track forecast errors (skill) for 24, 48 and 72 h lead periods are 136 km (32 %), 253 km (42 %) and 376 km (50 %) respectively during 2009–2012.

4.5.3 Intensity Forecasting Accuracy

The average absolute error (AE)/root mean square error (RMSE) in intensity (wind) forecast is about 10 (13), 13 (18) and 19 (24) knots, respectively, for 24-, 48- and 72-h forecasts over the NIO as a whole during 2009–2012. The skill of intensity forecast is about 44 %/48 %, 60 %/58 % and 60 %/65 % for 24-, 48- and 72-h forecasts during 2009–2012 with respect to AE/RMSE. There is also improvement in terms of reduction in AE and RMSE of MSW forecast over the NIO (Fig. 5) like that over the northwest Pacific and northern Atlantic Oceans during 2005–2012 (Mohapatra et al. 2013c). The skill in intensity forecast compared to persistence method has significantly improved by about 6 % (10 %) and 9 % (8 %) per year, respectively, for 12- and 24-h forecasts considering the AE (RMSE).

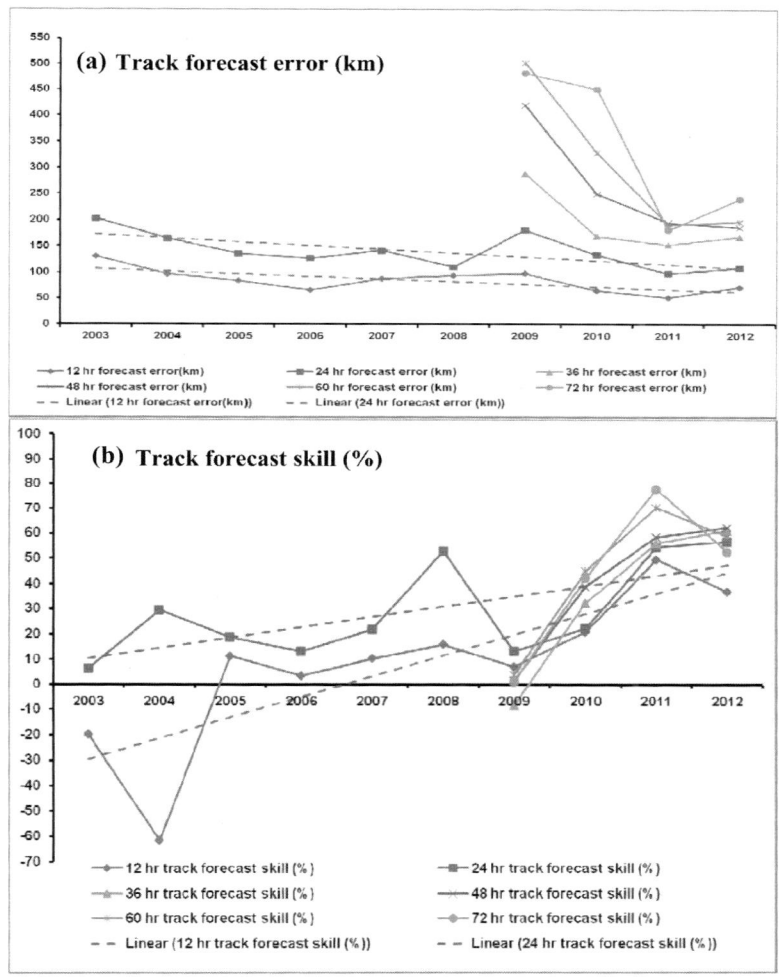

Fig. 4 (**a**) Average tropical cyclone track forecast error and (**b**) track forecast skill during 2003–2012

5 Cyclone Warning Organisation

At present, the cyclone warning organization of IMD has three-tier system to cater to the needs of the maritime states at national, regional and local levels and to carry out international responsibility.

The liaison with the Central Government organisations and other agencies as well as co-ordination and supervision of cyclone warning activities are done by

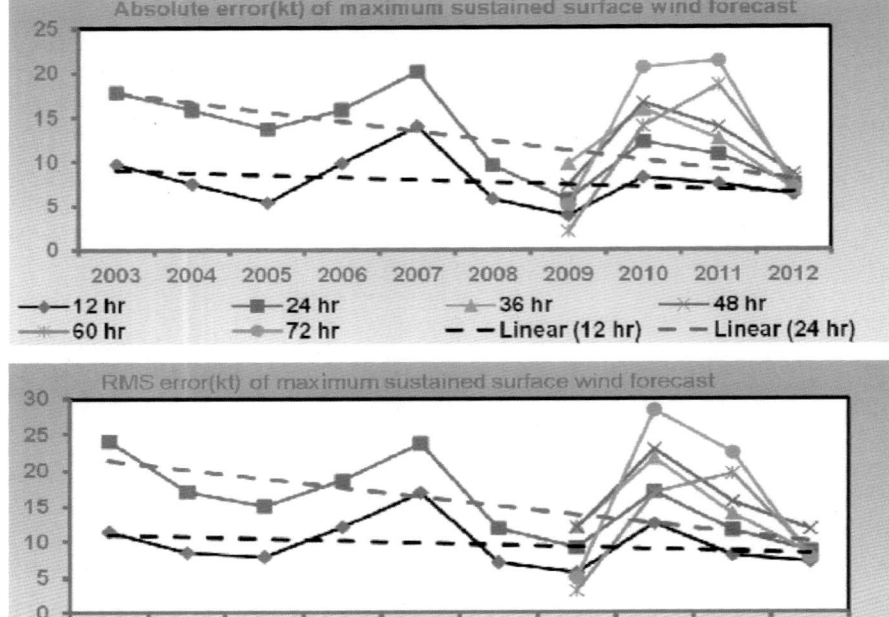

Fig. 5 Absolute and RMS errors of maximum sustained surface wind forecast issued by IMD during 2003–2012

Cyclone Warning Division (CWD) at New Delhi. CWD, New Delhi is also functioning as Regional Specialised Meteorological Centre-Tropical Cyclones (RSMC-Tropical Cyclones), New Delhi and provides the TC advisories to WMO/ESCAP Panel countries, viz., Bangladesh, Myanmar, Thailand, Sri Lanka, Maldives, Pakistan and Oman. It also acts as a TC Advisory Centre (TCAC) for international civil aviation as per the requirement of International Civil Aviation Organisation (ICAO) and issues the TC advisories to airport meteorological offices over NIO and Pacific region for issue of significant meteorological (SIGMET) information to different civil aviation authorities and airlines.

There are three Area Cyclone Warning Centres (ACWCs) at Chennai, Mumbai and Kolkata and three Cyclone Warning Centres (CWCs) at Visakhapatnam, Ahmedabad and Bhubaneswar. The ultimate responsibility for operational storm warning work for the respective area rests with the ACWCs and CWCs. Area of responsibility of various ACWCs and CWCs is shown in Table 3.

Early Warning Services for Management of Cyclones

Table 3 Area of responsibility of ACWC/CWC

Centre	Sea area	Coastal area[a]	Maritime state
ACWC Kolkata	Bay of Bengal	West Bengal, Andaman & Nicobar Islands	West Bengal & Andaman & Nicobar Islands
ACWC Chennai		Tamil Nadu, Puducherry, Kerala & Karnataka	Tamil Nadu, Puducherry, Kerala, Karnataka & Lakshadweep
ACWC Mumbai	Arabian Sea	Maharashtra, Goa	Maharashtra, Goa
CWC Bhubaneshwar	–	Odisha	Odisha
CWC Visakhapatnam	–	Andhra Pradesh	Andhra Pradesh
CWC Ahmedabad	–	Gujarat, Diu, Daman, Dadra & Nagar Haveli	Gujarat, Diu, Daman, Dadra & Nagar Haveli

[a]Coastal strip of responsibility extends upto 75 km from the coast line

5.1 Bulletins Issued for International Users

- **Tropical Weather Outlook for WMO/ESCAP panel countries** is issued once daily at 0600 UTC based on 0300 UTC observation and analysis. It contains convective activity, meteorological situation over the basin, observed lows, and their potential of intensification within the next 72 h.
- **Special Tropical Weather Outlook for WMO/ESCAP panel countries** is issued five times a day (based on 00, 03, 06, 12 and 18 UTC). It contains current location and intensity, past movement, convective activity, T number, estimated central pressure and MSW, sea condition, 120 h (00, 06, 12, 18, 24, 36, 48, 72, 96 and 120 h) forecast track and intensity (text and graph) from deep depression stage onwards till the weakening of the system, storm surge guidance and diagnostic and prognostic features.
- **Tropical Cyclone Advisory Bulletin for WMO/ESCAP panel countries** is issued every three hourly (based on 00, 03, 06, 09, 12, 15, 18 and 21 UTC). It contains current location and intensity, past movement, convective activity, T number, estimated central pressure and MSW, sea condition, 120 h (00, 06, 12, 18, 24, 36, 48, 72, 96 and 120 h) forecast track and intensity (text and graph) from deep depression stage onwards till the weakening of the system, storm surge guidance and diagnostic and prognostic features.
- **TCAC bulletin for issue of SIGMET by Met. Watch Offices** is issued as soon as any disturbance over the NIO attains or likely to attain the intensity of cyclonic storm. These bulletins are issued at six hourly intervals based on 00, 06, 12, 18 UTC synoptic charts and the time of issue is HH+03 h. These bulletins contain present location of cyclone in lat./long., max sustained surface wind (in knots), direction of past movement and estimated central pressure, forecast position in Lat./Long and forecast winds in knots valid at HH+6, HH+12, HH+18 and HH+24 h in coded form.

5.2 Bulletins Issued at National Level

- **Four-stage warning bulletin**

The cyclone warnings are issued to central and state government officials in four stages. The *First Stage* warning known as "Pre Cyclone Watch" issued at least 72 h in advance contains early warning about the development of a cyclonic disturbance in the north Indian Ocean, its likely intensification into a cyclone and the coastal belt likely to experience adverse weather. This early warning bulletin is issued by the Cyclone Warning Division and is addressed to the Cabinet Secretary and other senior officers of the Government of India including the Chief Secretaries of concerned maritime states. The *Second Stage* warning known as "Cyclone Alert" is issued at least 48 h in advance of the expected commencement of adverse weather over the coastal areas. It contains information on the location and intensity of the storm, likely direction of its movement, intensification, coastal districts likely to experience adverse weather and advice to fishermen, general public, media and disaster managers. This is issued by the concerned ACWCs/CWCs and CWD at HQ. The *Third Stage* warning known as "Cyclone Warning" issued at least 24 h in advance of the expected commencement of adverse weather over the coastal areas. Landfall point is forecast more precisely at this stage. These warnings are issued by ACWCs/CWCs/and CWD at three-hourly interval giving the latest position of cyclone and its intensity, likely point and time of landfall, associated heavy rainfall, strong wind and storm surge along with their impact and advice to general public, media, fishermen and disaster managers. The *Fourth Stage* of warning known as "Post Landfall Outlook" is issued by the concerned ACWCs/CWCs and CWD at HQ at least 12 h in advance of expected time of landfall. It gives likely direction of movement of the cyclone after its landfall and adverse weather likely to be experienced in the interior areas.

At CWD, New Delhi, the bulletins are issued from the stage of depression onwards. During the stage of depression/deep depression, it is issued based on 00, 03, 06, 12 and 18 UTC observations. When the system intensifies into a cyclonic storm, these bulletins are issued at 00, 03, 06, 09, 12, 15, 18 and 21 UTC (every three hourly interval) based on previous observations. The cyclone warnings are sent on real time basis to the Control Room in the Ministry of Home Affairs, Government of India, besides other ministries and departments of the central government, Doordarshan and All India Radio (AIR) at New Delhi and other electronic and print media and concerned state governments. Different colour codes are being used since post-monsoon season of 2006 at different stages of the cyclone warning bulletins (cyclone alert-yellow, cyclone warning-orange and post landfall outlook-red), as desired by the National Disaster Management.

- **DGM's Bulletin for high govt. officials**

DGM's bulletin for high govt. officials is issued once a day. It summarises past 24 h development in terms of track and intensity and past 24 h weather. Other contents are same as that of bulletin for India coast as discussed in previous section.

- **Tropical Cyclone (TC) vital bulletin**

The TC vital contains the vital components required to create a synthetic cyclone in NWP model. It contains the location, intensity, radius of maximum wind, radii of 28, 34, 50 and 64 knots wind threshold in four different quadrants of the cyclone. It is issued four times a day based on 00, 06, 12 and 18 UTC observation from the deep depression stage.

5.3 Bulletins Issued at Regional and Local Levels

Following user specific bulletins are issued by all ACWCs as per their area of responsibility:

- Four-stage warning for designated govt. officials up to district level
- Audio warnings through cyclone warning dissemination systems along the coast (installed in disaster managers offices)
- All India Radio bulletin
- Press bulletin
- Warnings for fishermen through All India Radio
- Sea area bulletins
- Coastal weather bulletin
- Warning for Indian navy
- Warnings for port and fisheries officials
- Warning for aviation

The CWCs issue all the above bulletins for their area of responsibility except the sea area bulletin.

6 Warning Dissemination Mechanism

Cyclone warnings are disseminated to various users through telephone, fax, e-mail, All India Radio, Television and other print and electronic media. These warnings/advisories are uploaded in the website of IMD (www.imd.gov.in). Also cyclone warning bulletins are disseminated by SMS to state and national disaster management authorities. In addition to the above network, IMD also disseminate warnings to the concerned officials and people using broadcast capacity of INSAT satellite. This is a direct broadcast service of cyclone warning in the regional languages meant for the selected areas affected or likely to be affected by the cyclone. There are 352 Cyclone Warning Dissemination System (CWDS) stations along the Indian coast; out of these 101 digital CWDS are located along Andhra coast. The ACWCs and CWCs are responsible for originating and disseminating the cyclone warnings through CWDS. The bulletins are generated and transmitted every hour in three languages viz., English, Hindi and regional language. In case of emergency, police wireless and telecommunication lines of railways and aviation authorities are also used.

The Cyclone Advisories bulletin for WMO/ESCAP panel countries and international airports are disseminated through global telecommunication system (GTS), e-mail and through ftp and TC vitals for research community and NWP modelling are disseminated through e-mail and ftp.

7 Future Plans

ESSO-IMD continuously expands and strengthens its activities in relation to observing strategies, forecasting techniques, disseminating methods and research relating to different aspects of TCs to ensure most critical meteorological support to disaster managers and decision makers not only in India but also to the WMO/ESCAP panel countries. It has the following future plans to further improve the cyclone warning system.

 (i) It is planned to further improve the observational network including buoys over the NIO, DWR and AWS along the coast through modernisation programme during 12th Five Year Plan.
 (ii) Under INSAT-3D programme, there is an advanced imager with six imagery channels (VIS, SWIR, MIR, TIR-1, TIR-2, and WV) and a 19-channel sounder (18 IR and 1 Visible) for derivation of atmospheric temperature and moisture profiles. It will provide 1 km. resolution imagery in visible band, 4 km resolution in IR band and 8 km in water vapour channel. This new satellite will provide much improved capabilities for cyclone monitoring. In preparation for the reception and processing of this data, SAC-ISRO has installed a data reception and processing system to process the data on real time mode and provide products with respect to cyclone monitoring.
 (iii) The FDP on landfalling cyclones over the Bay of Bengal has been taken up to minimise the error in prediction of TC track and intensity forecasts and hence adverse weather. During pre-pilot phase (15 Oct–30 Nov 2008–2012), several national institutions participated for joint observational, communicational and NWP activities resulting in improved forecast and delivery of services (Mohapatra et al. 2013a). With possible manned and unmanned aircraft reconnaissance during 2013–2015, it will help in improving TC track and intensity forecasting and hence the adverse weather warning, as demonstrated in Atlantic and Pacific Ocean basins.
 (iv) Currently an effort is underway in which high resolution HWRF model with the support from NCEP, USA is being used in track and intensity predictions. However, only atmospheric component is operational at present and effort will be made to operationalise the coupled model with inclusion of ocean component, which is being customized by INCOIS, Hyderabad.

(v) Attempt will be made to assimilate more observational data, especially remotely sensed satellite and DWR data as it has become necessary to provide adequate and realistic observations for frequent initialization of NWP models for short to medium range forecasting of track, intensity and associated adverse weather.

(vi) With the completion of ongoing modernisation programme and other initiatives as mentioned above, the cyclone forecast error is likely to reduce by about 20 % by 2015 and by 40 % by 2020 from the base year of 2010 according to vision document of Ministry of Earth Sciences, Govt. of India.

References

Bhatia RC, Sharma AK (2013) Recent advances in observational support from space-based systems for tropical cyclones. Mausam 64:97–104

Bhatia RC, Das S, Mohapatra M, Roy Bhowmik SK (2008) Use of satellite and AWS data for weather prediction. IMD meteorological monograph synoptic meteorology no. 6/2008. India Meteorological Department, New Delhi

BMTPC, Government of India (2006) Vulnerability atlas of India. BMTPC, Government of India, New Delhi

Dube SK, Rao AD, Poulose J, Mohapatra M, Mohapatra M, Murty TS (2013) Storm surge inundation in south Asia under climate change scenario. In: Mohanty UC, Mohapatra M, Singh OP, Bandyopadhyay BK, Rathore LS (eds) Monitoring and prediction of tropical cyclones over Indian Ocean and climate change. Capital Publishing Co, New Delhi

Dvorak VF (1984) Tropical cyclone intensity analysis using satellite data. Technical report (NOAA TR NESDIS 11), National Oceanic and Atmospheric Administration, National Environmental Satellite, Data, and Information Service

Goyal S, Mohapatra M, Sharma AK (2013) Comparison of best track parameters of RSMC, New Delhi with satellite estimates over north Indian Ocean. Mausam 64:25–34

Gray WM (1968) Global view of origin of tropical disturbance and storms. Mon Weather Rev 96:669–700

IMD (1976) Weather radar observations manual. India Meteorological Department, New Delhi

IMD (2003) Cyclone manual. India Meteorological Department, New Delhi

IMD (2008) Track of storm and depressions over the Indian Seas during 1891–2007, cyclone e-Atlas of IMD. India Meteorological Department, New Delhi

IMD (2013) Cyclone warning services: standard operation procedure. Cyclone Warning Division, India Meteorological Department, New Delhi

INCOIS, Hyderabad (2012) Coastal vulnerability atlas of India. INCOIS, Hyderabad

Jha TN, Mohapatra M, Bandyopadhyay BK (2013) Estimation of intensity of tropical cyclone over Bay of Bengal using microwave imagery. Mausam 64:105–116

Kalsi SR (2002) Use of satellite imagery for tropical cyclone intensity analysis and forecasting. Meteorological monograph, Cyclone Warning Division No. 1/2002

Kotal SD, Bhattacharya SK (2013) Tropical cyclone Genesis Potential Parameter (GPP) and its application over the north Indian Sea. Mausam 64:149–170

Kotal SD, Roy Bhowmik SK (2011) A multimodel ensemble technique for cyclone track prediction over north Indian Sea. Geofizika 28:275–291

Kotal SD, Roy Bhowmik SK, Kundu PK, Das AK (2008) A statistical cyclone intensity prediction model for Bay of Bengal. J Earth Syst Sci 117:157–168

Mishra DK, Gupta GR (1976) Estimates of maximum wind speed in tropical cyclones occurring in the Indian Seas. Indian J Meteorol Geophys 27:285–290

Mohanty UC, Osuri KK, Routray A, Mohapatra M, Pattanayak S (2010) Simulation of Bay of Bengal tropical cyclones with WRF model: impact of initial and boundary condition. Mar Geod 33:294–314

Mohapatra M, Bandyopadhyay BK, Tyagi A (2012a) Best track parameters of tropical cyclones over the North Indian Ocean: a review. Nat Hazards 63:1285–1317

Mohapatra M, Nayak DP, Bandyopadhyay BK (2012b) Evaluation of cone of uncertainty in tropical cyclone track forecast over North Indian Ocean issued by India Meteorological Department. Trop Cyclone Res Rev 2:331–339

Mohapatra M, Mandal GS, Bandyopadhyay BK, Tyagi A, Mohanty UC (2012c) Classification of cyclone hazard prone districts of India. Nat Hazards 63:1601–1620

Mohapatra M, Sikka DR, Bandyopadhyay BK, Tyagi A (2013a) Outcomes and challenges of Forecast Demonstration Project (FDP) on landfalling cyclones over the Bay of Bengal. Mausam 64:1–12

Mohapatra M, Nayak DP, Sharma RP, Bandyopadhyay BK (2013b) Evaluation of official tropical cyclone track forecast over North Indian Ocean issued by India Meteorological Department. J Earth Syst Sci 122:589–601

Mohapatra M, Bandyopadhyay BK, Nayak DP (2013c) Evaluation of operational tropical cyclone intensity forecasts over North Indian Ocean issued by India Meteorological Department. Nat Hazards. doi:10.1007/s11069-013-0624-z

Raghavan S (1997) Radar observations of tropical cyclone. Mausam 48:169–188

Raghavan S (2013) Observational aspects including weather radar for tropical cyclone monitoring. Mausam 64:89–96

Uhlhorn EW, Black PG (2003) Verification of remotely sensed sea surface winds in hurricanes. J Atmos Ocean Technol 20:99–116

Velden CS et al (2006) The Dvorak tropical cyclone intensity estimation technique: a satellite-based method that has endured for over 30 years. Bull Am Meteorol Soc 87:1195–1210

Development of NWP-Based Cyclone Prediction System for Improving Cyclone Forecast Service in the Country

S.D. Kotal, Sumit Kumar Bhattacharya, S.K. Roy Bhowmik, and P.K. Kundu

1 Introduction

India Meteorological Department (IMD) operationally runs two regional model WRF and Hurricane WRF (HWRF) model for short-range prediction and Global model T574L64 for medium range prediction (7 days). As part of WMO programme to provide a guidance of tropical cyclone (TC) forecasts in near real-time for the ESCAP/WMO member countries based on the TIGGE Cyclone XML (CXML) data, IMD also implemented JMA supported software for real-time TC forecast over North Indian Ocean (NIO).

As a part of effort to translate research to operation, and to meet the need of the operational forecaster, a NWP-based Objective Cyclone Prediction System (CPS) is developed and implemented for the operational cyclone forecasting work. The method comprises five forecast components, namely (a) Cyclone Genesis Potential Parameter (GPP), (b) Multi-Model Ensemble (MME) technique for cyclone track prediction, (c) Cyclone intensity prediction, (d) Rapid intensification and (e) Predicting decaying intensity after the landfall. Genesis potential parameter (GPP) is used for potential of cyclogenesis and forecast for potential cyclogenesis zone (Kotal and Bhattacharya 2013; Kotal et al. 2009). A multi-model ensemble (MME) forecast of NWP models is generated in real time for predicting the track of tropical cyclones over the North Indian seas using the outputs of member models IMD-GFS, IMD-WRF, GFS (NCEP), UKMO and JMA (Kotal and Roy Bhowmik 2012). SCIP (statistical cyclone intensity prediction) model is run for 12 hourly intensity predictions up to 72-h (Kotal et al. 2008). A rapid intensification index (RII) is used

---

S.D. Kotal (✉) • S.K. Bhattacharya • S.K.R. Bhowmik
India Meteorological Department, Lodhi Road, New Delhi 110003, India
e-mail: sdkotal.imd@gmail.com

P.K. Kundu
Department of Mathematics, Jadavpur University, Kolkata 700032, India

© Capital Publishing Company 2015
K. Ray et al. (eds.), *High-Impact Weather Events over the SAARC Region*,
DOI 10.1007/978-3-319-10217-7_8

for the probability forecast of rapid intensification (RI) (Kotal and Roy Bhowmik 2013). A decay model has been used for real time forecasting of decaying intensity after the landfall (Roy Bhowmik et al. 2005).

This paper describes the development strategy of the objective forecast system and performance skill of the system during 2008–2012. The performance of the system during the recent very severe cyclonic storm 'Phailin' is also discussed. Cyclone prediction system (CPS) is described briefly in Sect. 2. A brief life cycle of the cyclone 'Phailin' is described in Sect. 3. Forecast performance of CPS is presented in Sect. 4 and summary and conclusion is given in Sect. 5.

2 NWP-Based Objective Cyclone Prediction System (CPS)

During the last two decades, weather forecasting all over the world has greatly benefited from the guidance provided by the NWP models. However, limitations remain, particularly in the prediction of intensity of tropical cyclones (Elsberry et al. 2007; Houze et al. 2007). There is variation of forecasts among NWP models and requirements are also different for different forecast services. Therefore we need to generate more skillful, consensus, and requirement based products. As statistical post-processing can add skill to dynamical forecasts, various post-processed value added NWP-based special products are prepared for real time cyclone forecasting. The objective of the CPS was:

- To add skill to dynamical forecasts by statistical post-processing.
- To generate consensus forecast from different model forecasts.
- To develop a collective approach to address various components for improving cyclone forecast service.
- To demonstrate application of research for operational forecasting.

The five-step NWP-based Cyclone Prediction System (CPS) for the operational cyclone forecasting work is described below.

2.1 Step I: Genesis Potential Parameter (GPP)

The objective was to locate potential cyclogenesis zone over the sea and to understand the potential for intensification of a system at early stages of development. A cyclone genesis parameter, termed as the genesis potential parameter (GPP), for the North Indian Ocean is developed (Kotal et al. 2009). The parameter, which is defined as the product of four variables, namely vorticity at 850 hPa, middle tropospheric relative humidity, middle tropospheric instability, and the inverse of vertical wind shear, is computed based on outputs of IMD GFS T574/L64 (analysis as well as forecasts). The parameter is operationally used for distinction between non-developing and developing systems at their early development stages. The composite GPP value is found to be around three to five times greater for developing systems than for non-developing systems.

Development of NWP-Based Cyclone Prediction System

The grid point analysis and forecast of the genesis parameter up to 7 days is also generated on real time (Kotal and Bhattacharya 2013). Higher value of the GPP over a region indicates higher potential of genesis over the region. Region with GPP value equal or greater than 30 is found to be high potential zone for cyclogenesis.

2.2 Step II: Dynamical-Statistical Model for Cyclone Intensity Prediction (SCIP)

The objective of this component was intensity prediction at 12-h intervals up to 72 h. A dynamical-statistical model (SCIP) (Kotal et al. 2008) has been developed and implemented for real time forecasting of intensity at 12-hourly intervals up to 72 h. The model coefficients are derived based on model analysis of past cyclones. The parameters selected as predictors are: Initial storm intensity, Intensity changes during past 12 h, Storm motion speed, Initial storm latitude position, Vertical wind shear averaged along the storm track, Vorticity at 850 hPa, Divergence at 200 hPa and Sea Surface Temperature (SST). For the real-time forecasting, model parameters are derived based on the forecast fields of IMD GFS T574/L64.

2.3 Step III: Multi-Model Ensemble (MME) Technique for Track Prediction

As there are variations of track forecasts among different NWP models, the objective of this component was to generate a consensus track forecast of NWP models by collective bias correction. The multi model ensemble (MME) technique (Kotal and Roy Bhowmik 2012) is based on collective bias correction of NWP models by statistical linear regression approach. The predictors selected for the ensemble technique are forecasts latitude and longitude positions at 12-h intervals up to 72 h of five NWP models (IMD-GFS, IMD-WRF, NCEP GFS, UKMO and JMA). Multiple linear regression technique is used to generate weights (regression coefficients) for each model for each forecast hour (12, 24, 36, 48, 60, 72 h) based on the past data. In the case of cyclone Phailin, MME forecast tracks were generated using IMD-GFS, IMD-WRF, NCEP GFS, UKMO and JMA up to 120 h.

2.4 Step IV: Rapid Intensification (RI) Index

The Rapid Intensification (RI) is defined as an increase of intensity by 30 kt (15.4 ms^{-1}) or more during 24 h. The objective was probability forecast of Rapid Intensification. A rapid intensification index (RII) is developed for tropical cyclones over the Bay of Bengal (Kotal and Roy Bhowmik 2013). The RII uses large-scale characteristics of tropical cyclones to estimate the probability of rapid intensification (RI) over the subsequent 24 h. The RII technique is developed by combining threshold (index) values of the eight variables for which statistically significant differences are found between the RI and non-RI cases. The variables are: Storm latitude position, previous 12-h intensity change, initial storm intensity, vorticity at 850 hPa, divergence at 200 hPa, vertical wind shear, lower tropospheric relative humidity, and storm motion speed. The probability of RI is found to be increased from 0 to 100 % when the total number of indices satisfied increases from zero to eight.

2.5 Step V: Decay of Intensity After the Landfall

The objective of this final component was of decaying intensity after landfall at 6-h intervals upto 30 h. Tropical cyclones (TCs) are well known for their destructive potential and impact on human activities. The super cyclone of Orissa (1999) illustrated the need for the accurate prediction of inland effects of tropical cyclones. The super cyclone of Orissa maintained the intensity of cyclonic storm for about 30 h after landfall. Because a dense population resides at or near the Indian coasts, the decay forecast has direct relevance to daily activities over a coastal zone (such as transportation, tourism, fishing, etc.) apart from disaster management. In view of this, the decay model (Roy Bhowmik et al. 2005) has been used for real time forecasting of decaying intensity (after landfall) of TCs.

Flow diagram of the five-step objective Cyclone Prediction System (CPS) is shown in Fig. 1.

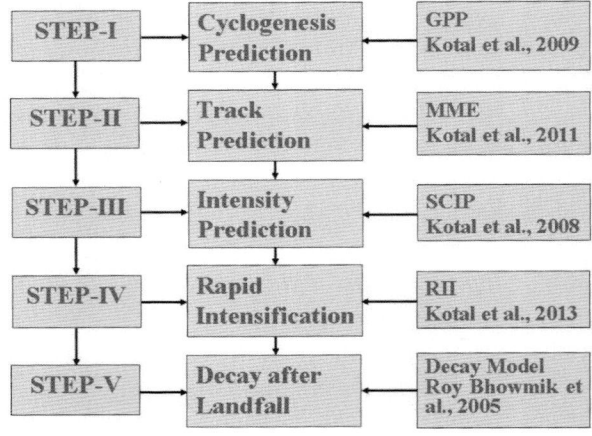

Fig. 1 Flow diagram of Cyclone Prediction System (CPS)

3 Cyclonic Storm Phailin (8–14) October 2013

The low pressure system over North Andaman Sea intensified into depression at 0300 UTC of 8 October 2013 near latitude 12.0°N and longitude 96.0°E. It moved northwestwards and intensified into a deep depression at 0000 UTC of 9 October 2013 and further intensified into a cyclonic storm (T. No. 2.5), Phailin at 1200 UTC of the same day. The cyclonic storm continued to move in northwesterly direction and intensified into severe cyclonic storm (T. No. 3.5) at 0300 UTC of 10 October 2013 and subsequently intensified into very severe cyclonic storm (T. No. 4.0) at 0600 UTC of same day. Moving northwestward direction the system further rapidly intensified to T. No. 4.5, T. No. 5.0 and T. No. 5.5 at 1200 UTC, 1500 UTC and 2100 UTC of same day (10 October 2013) respectively. At 0300 UTC of 11 October 2013 the system intensified to T. No. 6.0 and continued to move northwesterly direction with same intensity towards Odisha and crossed coast near Gopalpur at around 1700 UTC of 12 October 2013. The system maintained its intensity of very severe cyclonic storm upto 7 h after landfall and cyclonic storm intensity till 1200 UTC of 13 October 2013. The system continued to decay and weakened to deep depression at 1800 UTC of 13 October 2013 and further to depression at 0300 UTC of 14 October 2013. The observed track of the cyclone Phailin is shown in Fig. 2.

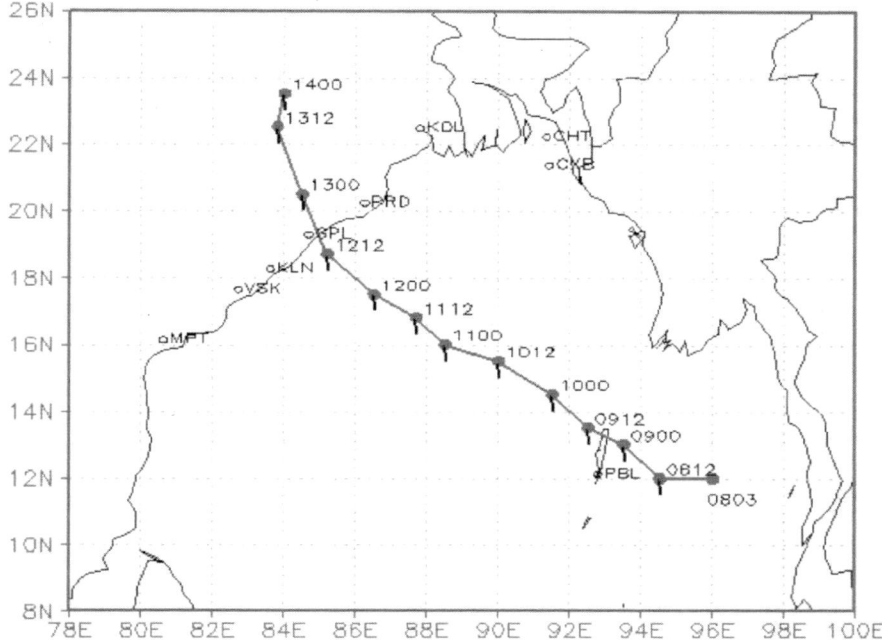

Fig. 2 Observed track of the cyclone Phailin

4 Forecast Performance of CPS

4.1 Forecast Skill of GPP for Prediction of Cyclogenesis

Three parameters, such as the probability of detection (POD), the false alarm ratio (FAR), and the Peirce skill score (PSS) are computed to evaluate the skill of the GPP for genesis forecasts during the period 2008–2012. PSS of value 1 is taken as a perfect forecast, 0 for random or constant forecasts and negative for forecasts that are worse than random forecasts.

Figure 3 illustrates the deterministic verification of GPP forecasts. The figure shows that the POD of the GPP was 77 % and the FAR was 23 % for 62 forecast events during 2008–2012. The results show that POD was much higher than FAR and the positive values of PSS indicate the skill of GPP for cyclogenesis prediction.

Grid point analysis of Genesis Potential Parameter (GPP) for cyclone Phailin [Fig. 4] shows that 168-h forecast based on 1 October 2013 (Fig. 4), 120-h forecast based on 3 October 2013 (Fig. 4), and 48-h forecast based on 6 October 2013 (Fig. 4) all valid for 00 UTC 08 October 2013, correctly indicated the location of potential cyclogenesis zone, where depression formed on that day.

Analysis and forecasts of GPP [Fig. 5] show that GPP ≥ 8.0 (threshold value for intensification into cyclone) indicated its potential to intensify into a cyclone at early stages of development (T. No. 1.0, 1.5, 2.0) of the cyclone Phailin.

4.2 Performance of MME for Track Prediction

Figure 6 shows the mean error of the MME track forecast for the period 2009–2012. Mean MME track errors during the period are about 75 km at 12 h, 95 km at 24 h, 120 km at 36 h, 160 km at 48 h, 170 km at 60 h and 205 km at 72 h. During the

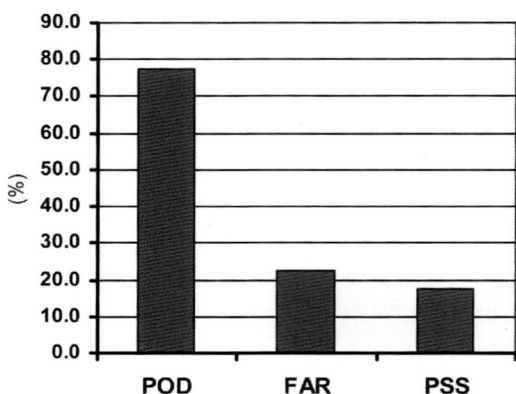

Fig. 3 Deterministic verification of GPP forecasts during 2008–2012

Development of NWP-Based Cyclone Prediction System 117

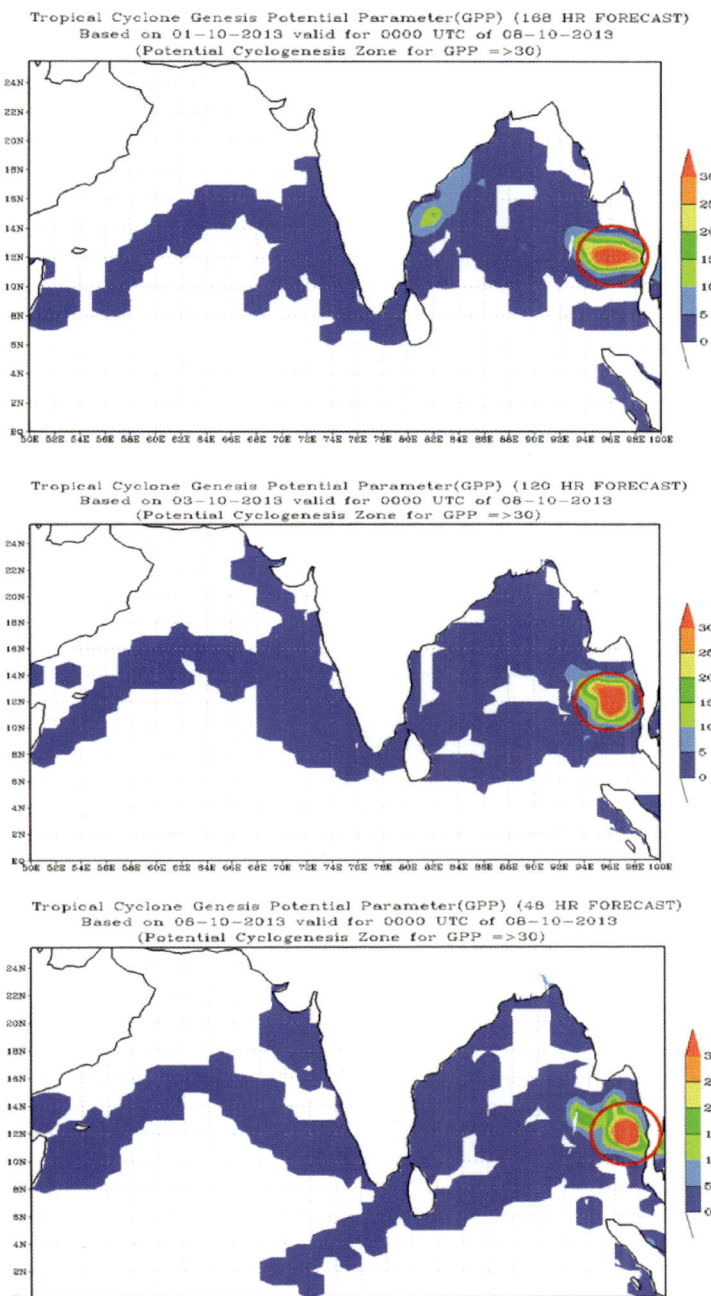

Fig. 4 Grid point analysis of Genesis Potential Parameter (GPP) for cyclone Phailin

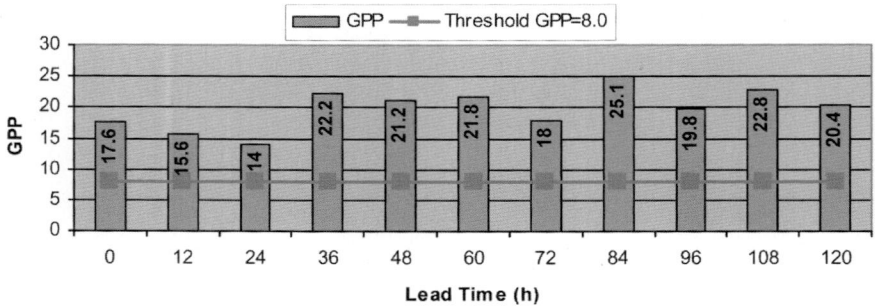
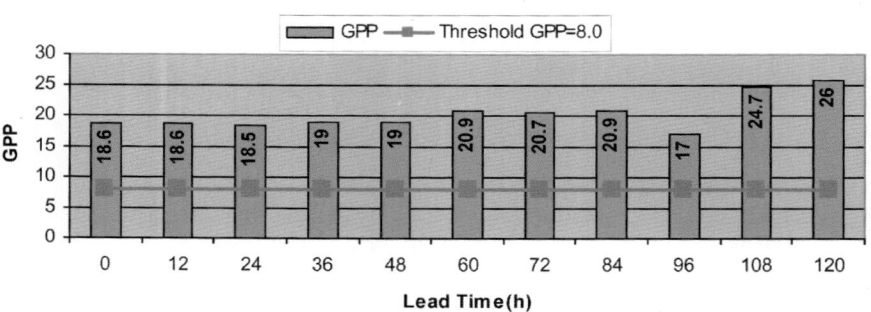
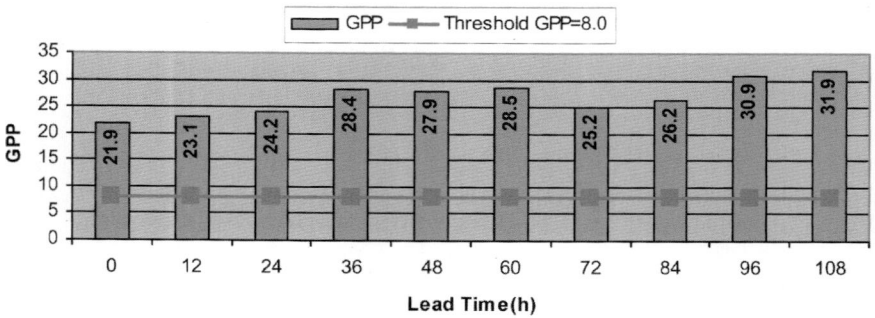

Fig. 5 Analysis and forecasts of area average Genesis Potential Parameter (GPP) of cyclone Phailin

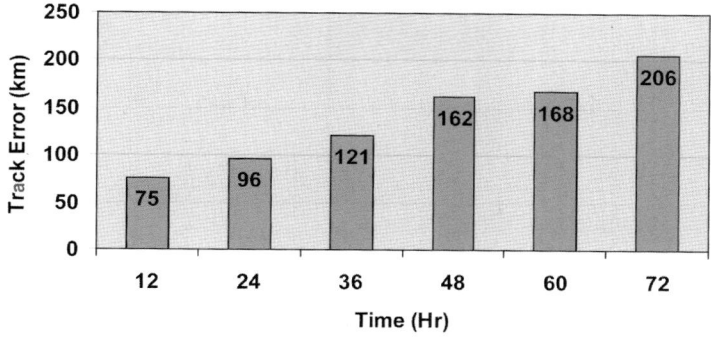

Fig. 6 Mean error of the MME track forecast for the period 2009–2012

period 2009–2012, MME track forecast errors have reduced by 38 % at 36 h, 40 % at 48 h, 58 % at 60 h and 72 h forecast and no significant improvement have occurred for the 12 h and 24 h during the period (Fig. 7).

4.2.1 MME Track Forecast Error for Phailin

The MME forecasts track based on different initial conditions along with the observed track is depicted in Fig. 8. The figure shows that from the day 1 (00 UTC 8 October to 10 UTC 10 October 2013), MME could predict correctly and consistently the landfall at Gopalpur (Odisha). The mean direct position error (DPE) of MME was about 65 km at 12 h to 150 km at 120 h (Fig. 9 along with their range).

4.2.2 Landfall Point Error (Phailin)

Landfall point forecasts errors of NWP model at different forecast lead times (Fig. 10) show that some model predicted north of actual landfall point and some predicted south of actual landfall point with a maximum limit upto about 340 km towards north and upto 215 km towards south. Under this wide extent of landfall point forecasts, MME was able to predict near-actual landfall point (Gopalpur) consistently (Table 1).

Average land fall point error (Fig. 10) shows that MME forecast error is least (20 km) compared to other models before 5 h to 113 h of landfall.

4.2.3 Landfall Time Error (Phailin)

Average land fall time error (Fig. 11) shows that MME landfall time forecast error is least (1.9 h) compared to other models.

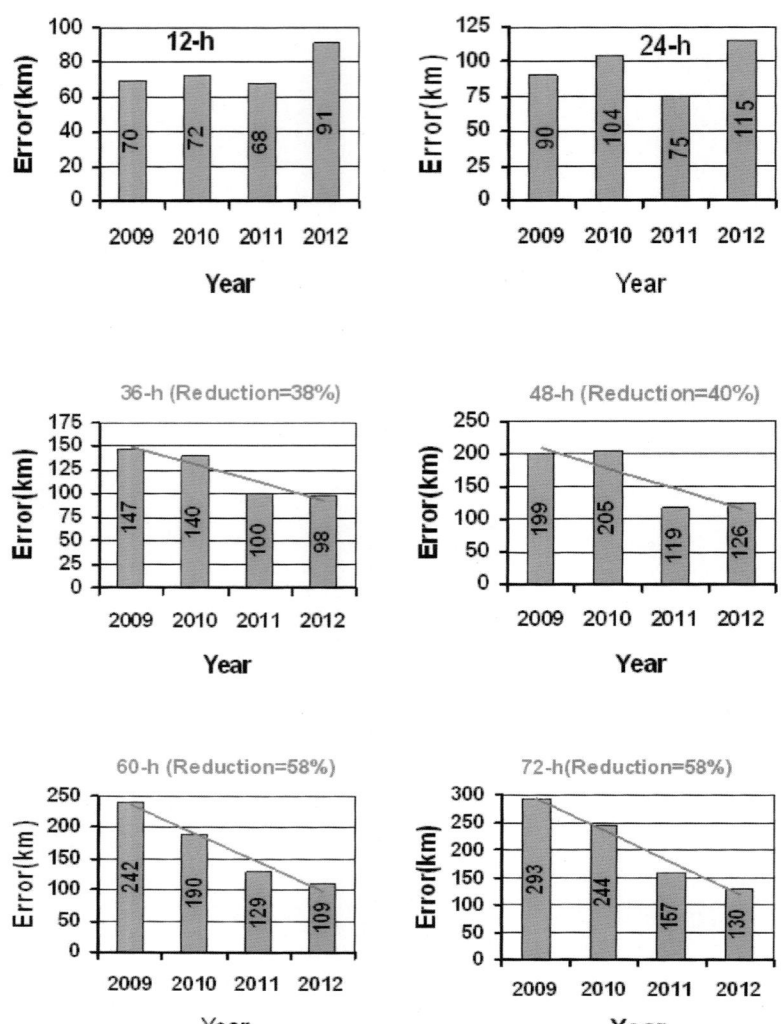

Fig. 7 Year-wise MME track forecast error reduction during 2009–2012

4.3 *Performance of SCIP for Intensity Prediction*

Figure 12 shows the mean error of the SCIP intensity forecast for the period 2008–2012. Mean forecast errors (Average absolute error) ranged from about 5 kt at 12 h to about 16 kt at 72 h. Year-wise forecast errors shows that over the past 5 years (2008–2012), 24 h to 72 h intensity forecast errors have been reduced by about 50 % with a significant improvement during last 3 years. Over the past 3 years (2010–2012), SCIP intensity forecast errors have been reduced by 50–74 % for the 36–72 h forecast periods and it was 24 % at 12 h and 7 % at 24 h (Fig. 13).

Development of NWP-Based Cyclone Prediction System

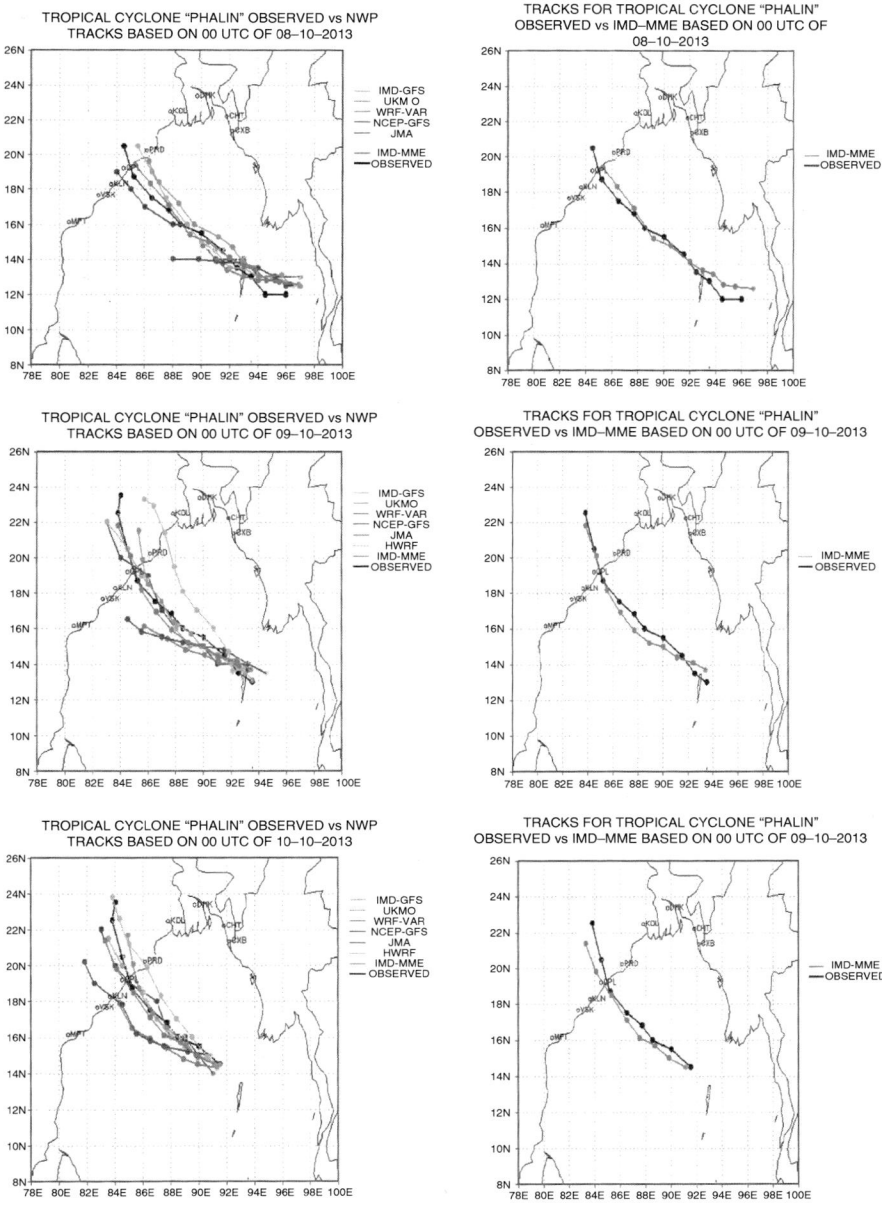

Fig. 8 MME forecasts track based on different initial conditions in case of Phailin

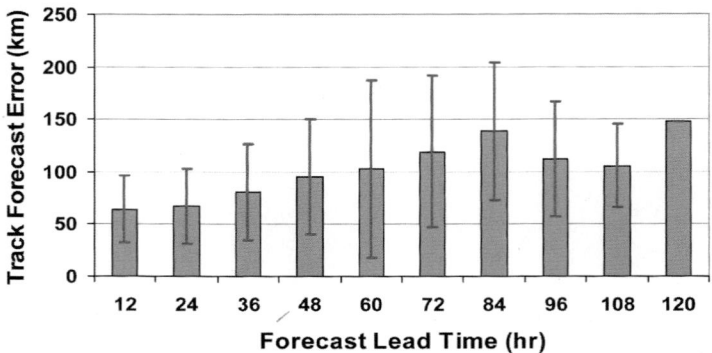

Fig. 9 Average direct position error (DPE) of MME (along with range (thick line)) in case of Phailin

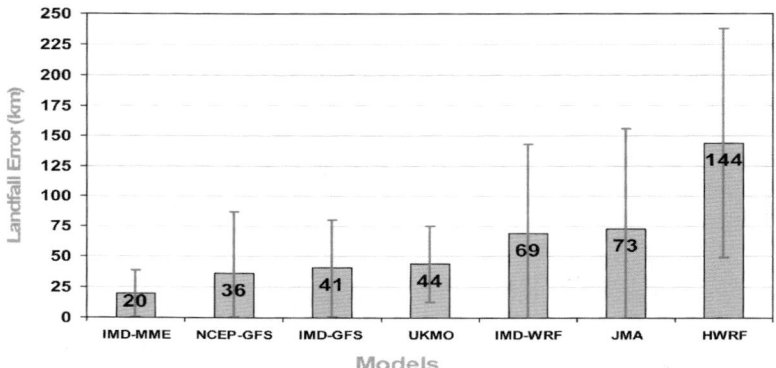

Fig. 10 Average landfall point error (km) of models (along with range (thick line)) of cyclone Phailin

Table 1 Landfall point and landfall time error of consensus NWP model (MME) forecasts

Forecast based on	Forecast lead time (hr)	Landfall point error (km)	Landfall time error (hr)
00 UTC/08.10.2013	113	39	7 h delay
00 UTC/09.10.2013	89	0	3 h delay
12 UTC/09.10.2013	77	35	3 h delay
00 UTC/10.10.2013	65	25	1 h delay
12 UTC/10.10.2013	53	0	1 h delay
00 UTC/11.10.2013	41	0	1 h delay
12 UTC/11.10.2013	29	39	0 h
00 UTC/12.10.2013	17	39	1 h delay
12 UTC/12.10.2013	5	0	0 h

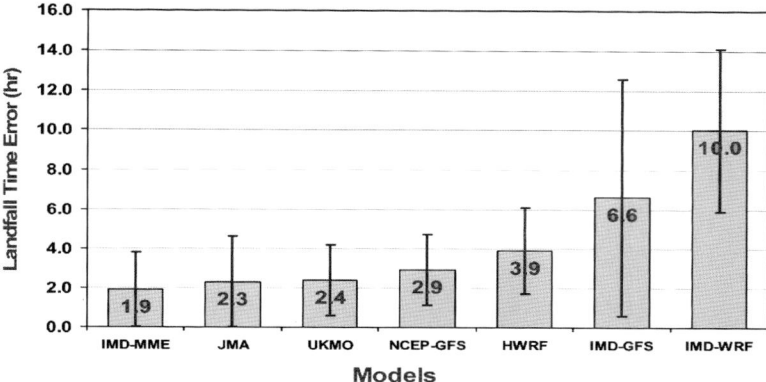

Fig. 11 Average landfall time error (hr) of models (along with range (thick line)) of cyclone Phailin

Fig. 12 Mean error of the SCIP intensity forecast for the period 2008–2012

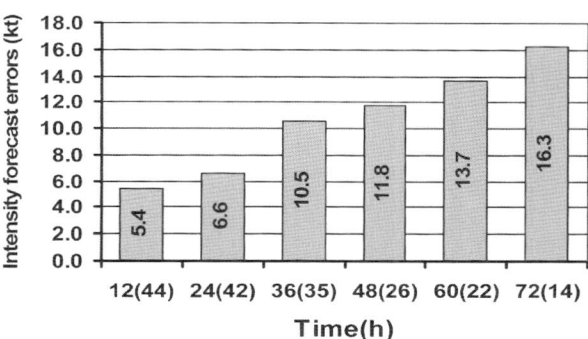

Landfall intensity predicted by SCIP model in 2–3 days before landfall (from initial cyclonic storm stage at 1200 UTC of 09 October 2013) shows that the model could predict the landfall intensity of very severe cyclonic storm with a reasonable success (Fig. 14).

4.4 Forecast Skill of RI-Index for Prediction of Rapid Intensification

The Brier score (BS) (Wilks 2006) is computed to assess the skill of the RI forecasts. The Brier score is computed using the formula

$$BS = \frac{1}{N}\sum(f-O)^2$$

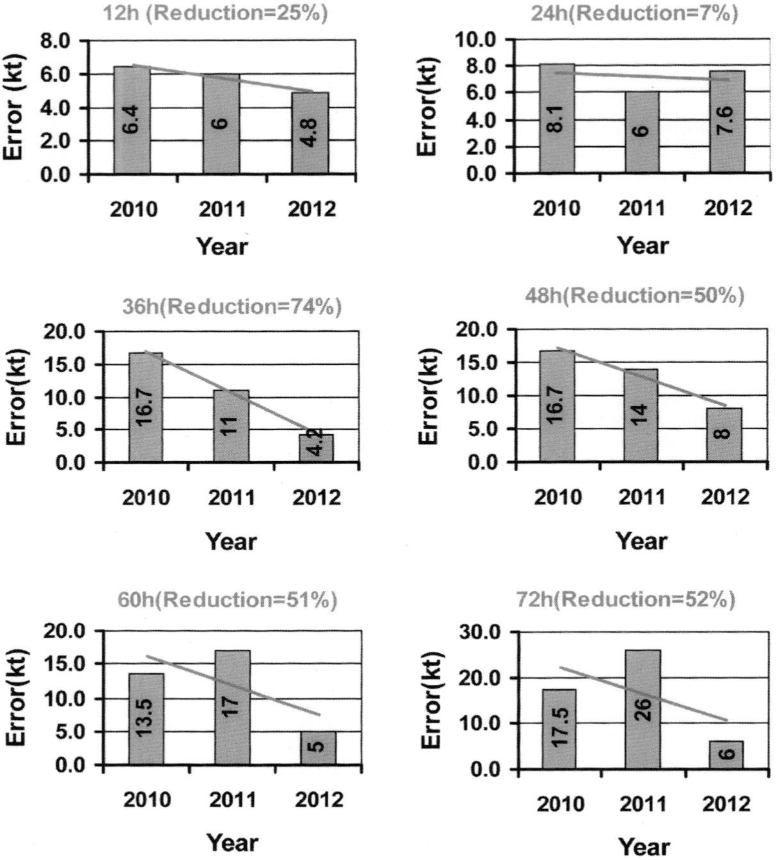

Fig. 13 Year-wise SCIP intensity forecast error reduction during 2010–2012

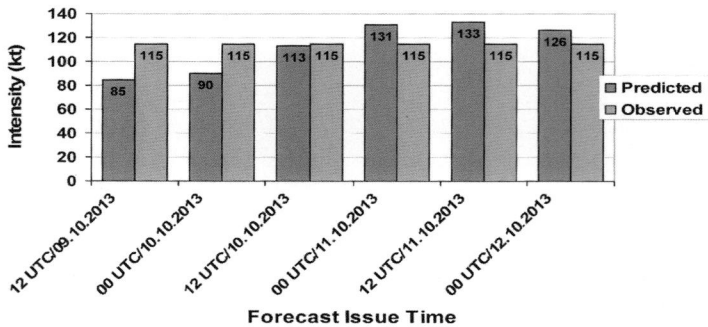

Fig. 14 Landfall intensity (kt) prediction by SCIP model of cyclone Phailin

Development of NWP-Based Cyclone Prediction System

Table 2 Probability of RI for cyclone Phailin

Forecast based on	Probability of RI predicted	Chances of occurrence predicted	Intensity changes (kt) in 24 h	Occurrence
00 UTC/08.10.2013	9.4 %	Very low	5	No
00 UTC/09.10.2013	9.4 %	Very low	15	No
12 UTC/09.10.2013	9.4 %	Very low	40	Yes
00 UTC/10.10.2013	72.7 %	High	65	Yes
12 UTC/10.10.2013	72.7 %	High	40	Yes
00 UTC/11.10.2013	72.7 %	High	5	No
12 UTC/11.10.2013	32.0 %	Moderate	0	No

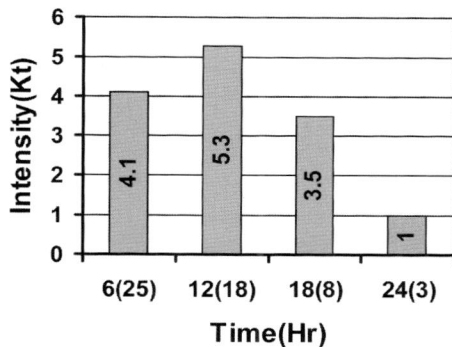

Fig. 15 Mean decay of intensity forecast errors (kt) during 2008–2012

in which f is the probability that was forecast, O the actual outcome of the event ($O = 0$ if it doesn't happen and 1 if it happens) and N is the number of forecasting instances. The BS is 0 and 1 for the best and worst score achievable respectively. During the period, 2011–2012, for 29 forecast events, the BS was found to be 0.005, which shows RII achieved a good score for RI forecasting during the period.

The probability forecasts of RI for cyclone Phailin is given in Table 2. The table shows that the RI-Index could predict occurrence as well as non-occurrence of RI of cyclone Phailin during its lifetime, except for forecast at 12 UTC of 09.10.2013 and 00 UTC of 11.10.2013.

4.5 Performance of Decay Model for Intensity Prediction After Landfall

Mean forecast errors (knots) at 6-hourly interval valid up to 24 h during the period 2008–2012 is shown in Fig. 15. The average absolute error (AAE) is ranged from 1 to 5 knots for forecasts up to 24 h. The error statistics shows that the model forecasts were reasonably good for decaying intensity after landfall.

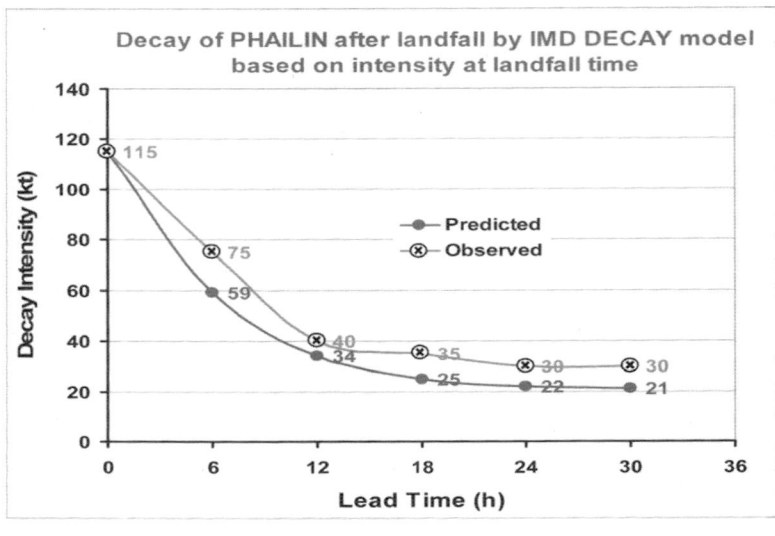

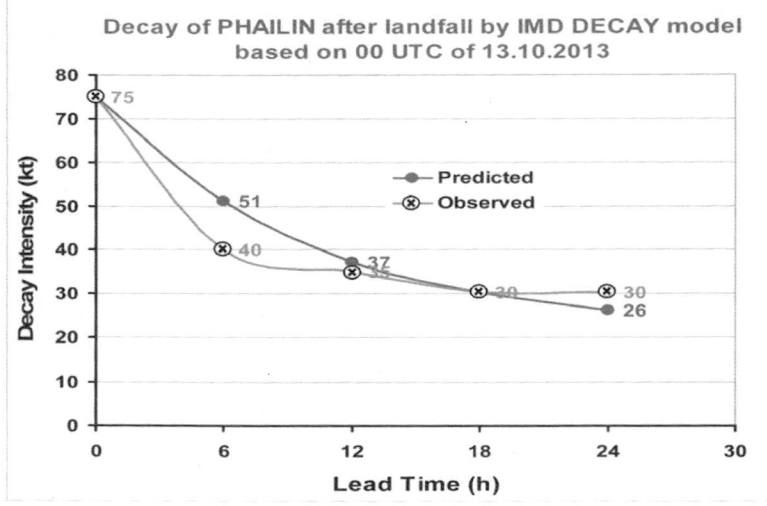

Fig. 16 Decay after landfall of cyclone Phailin

For the cyclone Phailin, decay (after landfall) prediction curve (6-hourly up to 30 h) (Fig. 16) shows slightly first decay compared to observed decay.

5 Summary and Conclusion

The performance of genesis potential parameter (GPP) shows that the POD is 77 % and FAR is 23 % during the period 2008–2012. The result shows that the genesis potential parameter (GPP) could provide the potential for intensification of a low pressure system at early development stages. For the cyclone Phailin, grid point analysis and forecasts of GPP could predict the formation and location of the system before 168 h of its formation. Analysis and forecasts of area average GPP indicated its potential to intensify into a cyclone at early stages (T. No. 1.0, 1.5, 2.0) of its development.

The mean track forecast error of MME is ranged from 75 km at 12 h to 206 km at 72 h during the period 2009–2012. For the cyclone Phailin, the average DPE for MME was about 65 km at 12 h to 150 km at 120 h. Average CTE was about 45 km at 12 h to 50 km at 120 h and average ATE was about 40 km at 12 h to 140 km at 120 h. Average land fall point error of MME was 20 km. Average land fall time error shows that MME landfall time forecast error was 1.9 h. The result shows that the MME could provide a useful consensus track forecast of NWP models.

The mean intensity forecast errors of SCIP was ranged from about 5 kt at 12 h to about 16 kt at 72 h. Over the past 5 years (2008–2012), 24–72 h intensity forecast errors have been reduced by about 50 % with a significant improvement during last 3 years. For the cyclone Phailin, average absolute error (AAE) for SCIP ranged from 10 kt at 12 h to 25 kt at 48 h, 31 kt at 60 h and 37 kt at 72 h. Landfall intensity predicted by SCIP model in 2–3 days before landfall shows that the model could predict the landfall intensity of very severe cyclonic storm with a reasonable success.

The Brier score (BS = 0.005) shows that the rapid intensification (RI) index could provide probability of rapid intensification during next 24-h. The RI-Index could predict occurrence as well as non-occurrence of rapid intensification of cyclone Phailin during its lifetime.

The decay model could predict the decaying intensity at 6-h interval upto 24-h after landfall with reasonable accuracy (decay error = 1–5 kt). Decay model could correctly predict the decaying nature of the Phailin after landfall.

Finally, the performance statistics of each component demonstrates the potential of the system for improving operational cyclone forecast service over the Indian seas.

Acknowledgements The authors are grateful to the Director General of Meteorology, India Meteorological Department, New Delhi for providing all the facilities to carry out this research work. Authors acknowledge the use of data from Cyclone Warning Division (IMD), NCEP, ECMWF, UKMO and JMA in this research work.

References

Elsberry RL, Lambert TDB, Boothe MA (2007) Accuracy of Atlantic and eastern North Pacific tropical cyclone intensity forecast guidance. Weather Forecast 22:747–762

Houze RA, Chen SS, Smull BF, Lee WC, Bell MM (2007) Hurricane intensity and eyewall replacement. Science 315:1235–1238

Kotal SD, Bhattacharya SK (2013) Tropical cyclone genesis potential parameter (GPP) and its application over the North Indian Sea. Mausam 64(1):149–170

Kotal SD, Roy Bhowmik SK (2012) A multimodel ensemble (MME) technique for cyclone track prediction over the north Indian Sea. Geofizika 28:275–291

Kotal SD, Roy Bhowmik SK (2013) Large-scale characteristics of rapidly intensifying tropical cyclones over the Bay of Bengal and a rapid intensification (RI) index. Mausam 64:13–24

Kotal SD, Roy Bhowmik SK, Kundu PK, Das AK (2008) A statistical cyclone intensity prediction (SCIP) model for Bay of Bengal. J Earth Syst Sci 117:157–168

Kotal SD, Kundu PK, Roy Bhowmik SK (2009) Analysis of cyclogenesis parameter for developing and non-developing low pressure systems over the Indian Sea. Nat Hazards 50:389–402

Roy Bhowmik SK (2003) An evaluation of cyclone genesis parameter over the Bay of Bengal using model analysis. Mausam 54:351–358

Roy Bhowmik SK, Kotal SD, Kalsi SR (2005) An empirical model for predicting decaying rate of tropical cyclone wind speed after landfall over Indian region. J Appl Meteorol 44:179–185

Interannual and Interdecadal Variations in Tropical Cyclone Activity over the Arabian Sea and the Impacts over Pakistan

Wash Dev Khatri, Zhi Xiefei, and Zhang Ling

1 Introduction

This study is to investigate the tropical cyclone (TC) activity over the Arabian Sea (AS). The idea of the topic rose from the work done by Chan (2005) on interannual and interdecadal variations of TC activity over the western North Pacific (WNP). In his study, Chan (2005) reviewed the Interannual and Interdecadal variations in tropical cyclone (TC) activity over the WNP and emphasized that the Interannual variations could largely be explained by changes in the planetary-scale flow patterns. Sea-surface temperatures (SSTs) in the WNP, however, did not contribute to such variations. Rather, SSTs in the central and eastern equatorial Pacific were significantly correlated with TC activity over the WNP. Changes in the SST in the equatorial Pacific were found to be in relation with the El Nino-Southern Oscillation (ENSO) phenomenon, and modifications of the planetary-scale flow associated with ENSO altered the conditions over the WNP and hence TC activity there.

The nature of TC activity over AS is different from TC activity over WNP, as the former has the lowest and latter has the highest yearly TC occurrence frequency. The work on the AS started in 1970s. Shukla and Misra (1977) computed the correlation coefficients among time series of SST, wind speed and rainfall. They used the data prior to 1960 and after that the TC activity over AS has more decreasing trend in contrast to previous years. Evans (1993) gave total credit to the underlying

---

W.D. Khatri
Pakistan Meteorological Department, Headquarter Office,
Sector H-8/2, Islamabad, Pakistan

SAARC Meteorological Research Center (SMRC), Dhaka 1207, Bangladesh

Z. Xiefei (✉) • Z. Ling
Nanjing University of Information Science and Technology,
5 Panxin Rd, Pukou, Nanjing, Jiangsu, China
e-mail: zhi@nuist.edu.cn

water temperature for the intensity of the TCs in different oceans, and suggested that the other factors may also have their roles in the intensity of TC activity like pressure troughs. Ramsay et al. (2008) have related the TC activity with large scale environmental factors with emphasis on SST. They have also found the correlations between the TC activity and Nino 3.4 and Nino 4. Chen et al. (2006) find out that any mechanism that can modulate the location and intensity of the monsoon trough affects the genesis location and frequency of TC. They divided the TCs into three groups, namely, low, high and very high according to their intensity. In response to tropical Pacific SST anomalies, a short wave train consisting of east-west oriented cells emanates from the tropics and progresses along the WNP rim. Population of the Group 1 TCs varies interannually in phase with the oscillation of the anomalous circulation cell northeast of Taiwan and south of Japan in this short wave train, while that of Group 3 fluctuates coherently with the tropical cell of this short wave train. Because these two anomalous circulation cells exhibit opposite polarity, the out-of-phase interannual oscillation between these two cells results in the opposite interannual variation of genesis frequency between TC of Groups 1 and 3 (Chen et al. 2006). Wu et al. (2010) found that the SST suppressed the TC activity in Indian and Pacific Oceans and similar results are obtained in case of TC activity over AS. They indicated the close relationship between the Atlantic SST and TC activity over the past 30 years, including basin-wise increases in the average lifetime, annual frequency and proportion of intense hurricanes (Wu et al. 2010).

The response of the Arabian Sea to global warming is the disruption in the natural decadal cycle in the SST after 1995, followed by a secular warming. The Arabian Sea is experiencing a regional climate-shift after 1995, which is accompanied by an increase in the occurrence of "most intense cyclones". Signatures of this climate-shift are also perceptible over the adjacent landmass of India and Pakistan (Kumar et al. 2009). Krishna (2009) found that the frequency of tropical cyclones in the north Indian Ocean has registered increasing trends during summer monsoon, which accounts for maximum number of intense cyclones. The increasing trend has been primarily due to decrease in the vertical wind shear. Thus, the future evolution of north Indian Ocean storm activity will critically depend on the warming of the sea surface waters and also the vertical wind shear. Likewise changes in ENSO statistics in the tropical pacific may become important, as they affect the SSTs in all three tropical oceans. The stronger warming of tropical NIO during recent years drove reduced vertical wind shear over the NIO and is thus responsible for the strong TC activity observed.

In the past few years Pakistan's coastal areas received a lot of economic damages due to the TC activities over AS. The coastal line of Pakistan is almost 1,200 km long along the Arabian Sea. Three main seaports of Pakistan i.e. Kaemari Karachi, Bin Qasim Karachi and Gawadar lie along this coastal belt along with large and small fish harbours and the ship breaking industry near Gadani. The coastal areas of Pakistan were frequently influenced by TC activities over AS in the past. The area in the north of the eastern coast of Pakistan is the cotton growing area and the TC activity near the area during pre-monsoon period vigorously affects the crop in case of violent rains or winds. In the same manner, during post-monsoon, the TC activity affects the wheat crop and the farmers get a big financial loss in both cases.

Although the frequency of TC landfall is much less in the coastal area of Pakistan than in other regions, the TCs near Pakistan's coastal region may have significant impact on the human life and agriculture in Pakistan. This study focusses on the analysis of tropical cyclone activity over AS, and discusses more about the impact of TC activities on Pakistan.

2 Data and Methodology

The TC data used in this study are taken from the website of Unisys weather, in which the data has been extracted from the warnings of the Joint Typhoon Warning Centre. Though the data was available from the year 1945, in this study the data used is from 1960 keeping in mind the study focussing the coastal belt of Pakistan, where the data of temperatures is generally available from this period and also various reading material suggest to use data of this period because of missing events in the past and also due to lack of proper observations and the remoteness of the region.

The data of TC are also available on the website of India Meteorological Department through RSMC, New Delhi but the problem arises for the tracks, which are only available for last 10 years.

The data for monthly rain, monthly average maximum and minimum temperatures are obtained from Pakistan Meteorological Department.

The data of Sea Surface Temperature (SST), geopotential heights, and upper atmosphere winds have been downloaded from NOAA website.

2.1 Statistical Analysis

The analysis of data is generally based on simple statistical methods, which are mostly the graphical representation of the data to find the trends in the changing frequency of the TC over the region.

The analysis has been done with two formats: (1) the actual occurrence of TCs and (2) 9 years running mean of TCs.

The first is studied under "Interannual Variations" and the later as "Inter Decadal Variations" of the TC activity over AS.

The data of TC activity is also checked in various parts of AS to find out the active region in the basin. The same has been done by dividing the area into two parts spanned in East-West directions with an imaginary line over 65°E longitude that divides the study area in almost two halves.

The monthly data of 'occurrence frequency' of TCs over the Arabian Sea is used to check the active seasons of the activity over the region and then it is examined with the SST of various months in the area.

2.2 Correlation

The data of frequency of occurrence of TCs over Arabian Sea is correlated with the precipitation, maximum temperatures and minimum temperatures of the stations in the vicinity of coastal region of Pakistan, ENSO index and SST of the key area through Correlation Matrix in accordance with Pearson's formula of correlation of product moment, which is as follows

$$r_{xy} = \frac{\sum_{i=1}^{n}(x_i - \bar{x})(y_i - \bar{y})}{(n-1)s_x s_y}$$

$$= \frac{\sum_{i=1}^{n}(x_i - \bar{x})(y_i - \bar{y})}{\sqrt{\sum_{i=1}^{n}(x_i - \bar{x})^2 \sum_{i=1}^{n}(y_i - \bar{y})^2}}$$

where x and y are the samples of X and Y, and s_x and s_y are the sample means standard deviations of X and Y.

This can also be written as:

$$r_{xy} = \frac{\sum_{i=1}^{n} x_i y_i - n \bar{x} \bar{y}}{(n-1)s_x s_y}$$

$$= \frac{n \sum_{i=1}^{n} x_i y_i - \sum x_i \sum y_i}{\sqrt{n \sum_{i=1}^{n} x_i^2 - \sum_{i=1}^{n}(x_i)^2} \sqrt{n \sum_{i=1}^{n} y_i^2 - n \sum_{i=1}^{n}(y_i)^2}}$$

The TC activity is also correlated with the geopotential heights at 500 hPa and 850 hPa level.

2.3 Wavelet Analysis

Wavelet transform is a powerful way to characterise the frequency, the intensity, the time position, and the duration of variations in a climate data series (Jiang et al. 1997; Zhi 2001; Zhang et al. 2006) which reveals the localized time and frequency information without requiring the time series to be stationary as required by the Fourier transform and other spectral methods. The advantage of wavelet transform in time-frequency analysis in comparison with conventional Fourier transform can be found in many literatures. Usage of the wavelet transform in the study of climatic changes and hydrological changes and other fields is receiving an increasing attention. Nakkan (1999) applied the continuous wavelet transforms (CWTs) to detect the temporal changing characteristics of the precipitation and the runoff processes, and their correlations and separating roles of climatic changes caused by human activities on stream flow changes. Other scholars used CWTs for analyzing stream discharge data and flood levels. In this paper, the CWT following Torrence and Compo (1998) is used. We assume that x_n is a time series with equal time spacing δt and $n=0, \ldots, N-1$. $\Psi_0(\eta)$ is a wavelet function which depends on a non-dimensional 'time' parameter η, with zero mean and is localized in both frequency and time (Farge 1992; Torrence and Compo 1998). Because Morlet wavelet provides a good balance between time and frequency localization, here we applied the Morlet wavelet defined as

$$\psi_0(\eta) = \pi^{-1/4} e^{i\omega_0 \eta} e^{-\eta^2/2}$$

where ψ_0 is the non-dimensional frequency, here taken to be 6 to satisfy the admissibility condition (Farge 1992; Torrence and Compo 1998). The continuous wavelet transform of a discrete sequence x_n is defined as the convolution of x_n with a scaled and translated version of $\psi_0(\eta)$:

$$W_n(S) = \sum_{n}^{N-1} X_{n'} \psi^* \left[\frac{(n'-n)\delta t}{s} \right]$$

where the asterisk (*) indicates the complex conjugate.

To ignore the edge effects because the wavelet is not completely localized in time, the cone of influence (COI) was introduced. Here the COI is the area where the edge effect becomes important and the wavelet power spectrum may not be regarded as real. The statistical significance of wavelet power can be assessed under the null hypotheses that the signal is generated by a stationary process by being given the background power spectrum (P_k), many geophysical series have their noise characteristics which can be modeled by a first order autoregressive (AR1) process. The Fourier power spectrum of an AR1 process with lag-1 autocorrelation α (estimated from the observed time series e.g. Allen and Smith 1996) is given by Grinsted et al. (2004) as

$$P_k = \frac{1-\alpha^2}{\left|1-\alpha e^{-2i\pi k}\right|^2}$$

where k is the Fourier frequency index.

In this study, the Morlet wavelets have been used to determine the significance in the cyclic periods in the TC activity over the period of this study.

3 Results and Discussions

3.1 Interannual Variations of the TC Activities over the Arabian Sea

Figure 1 shows the interannual variations of TC activity over AS for 50 years. It reveals a significant decreasing linear trend. Figure 2 shows that the SST has an increasing trend. The ENSO index reflects the oceanic properties in tropical central and eastern Pacific, while the IOD index reflects the oceanic properties in the east and west of Indian Ocean. Compared with the increasing trend of the SST over AS, the linear trend of the ENSO and IOD indices are not significant.

As far as the violent nature of the TC activity is concerned, the maximum sustained winds of the TCs available from 1975 till 2009 was used as an indicator. Figure 3 reveals that the duration of TC activity over AS has a decreasing trend and the violence of TCs has an increasing trend in terms of its wind speed. The increasing TC intensity may be associated with the increasing trend of the SST in the region where the TC activity occurs. Kumar et al. (2009) indicated that the Arabian Sea is experiencing a regional climate-shift after 1995, which is accompanied by a fivefold increase in the occurrence of "most intense cyclones". In general, our findings are in agreement with their result.

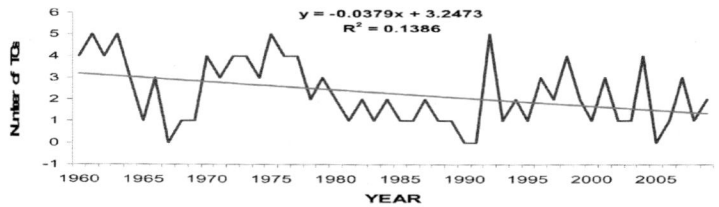

Fig. 1 Interannual variations and the trend of TC activities during the period from 1960 to 2009

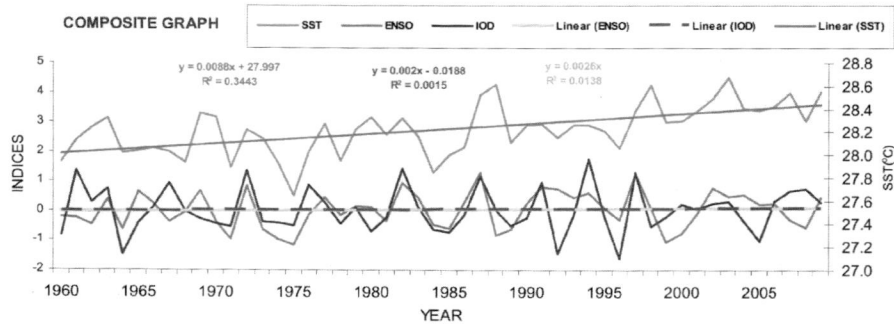

Fig. 2 SST, IOD and ENSO indices with linear trends during the period from 1960 to 2009

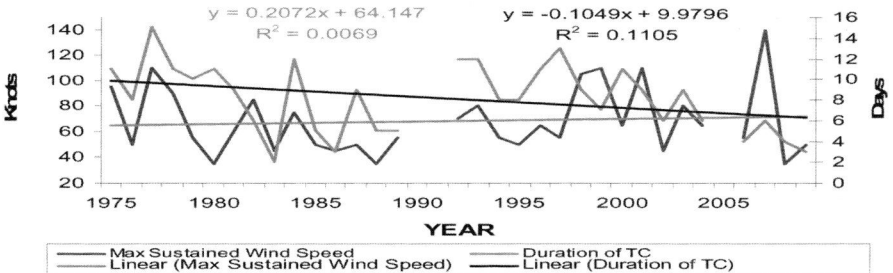

Fig. 3 Maximum sustained wind and TC duration with linear trends (*Broken lines* represent no TC activity in that year)

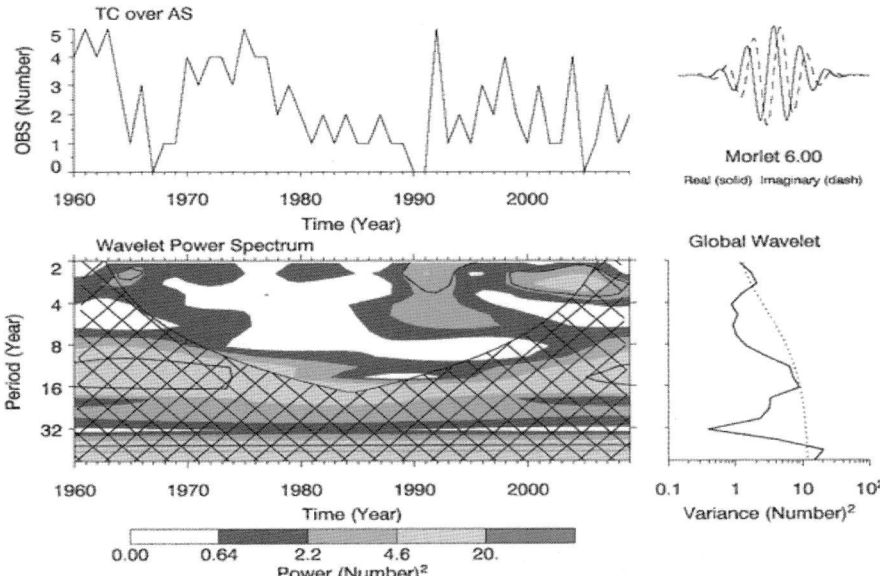

Fig. 4 The local power spectrum of the occurrence frequency of the TC activity over AS by using Morlet wavelet analysis. The *thick contour* encloses regions of greater than 90 % confidence for a red-noise process. Cross-hatched regions on either end indicate the "cone of influence," where edge effects become important

The spectral analysis of the TC occurrence frequency is done to see its power spectrum and significant period of its variations. As shown in Fig. 4, the occurrence frequency of the TC over the Arabian Sea has a significant period of 2–3 years during the periods 1962–1965, 1989–1996 and 1998–2006. The quasi-biennial oscillation (period of 2–3 years) was observed in both surface elements and upper-level fields as well as in the monsoon system in the troposphere in addition to its significant presence in the stratosphere (Walsh and Mostek 1980; Wang and Zhao 1981; Mooley and Parthasarathy 1983; Huang 1988; Zhu and Zhi 1991; Zhi 1997, 2001). Originally, the tropospheric biennial oscillation (TBO) was defined as the oscillation tendency of monsoon with the transitions occurring in the season prior to the monsoon involving coupled land–atmosphere–ocean processes over Indo-Pacific region (Meehl 1994; Meehl and Arblaster 2001). An important part of any biennial mechanism is anomalous heat storage in the ocean and the associated SST anomalies that can occur in certain regions (Meehl et al. 2003). Thus, the ocean retains the "memory" of ocean–atmosphere interaction over the course of a year to affect the atmosphere the following year (Brier 1978; Nicholls 1978; Chang and Li 2000; Li et al. 2001; Meehl 1987, 1993). Several studies have suggested that coupled ocean dynamics plays a role in the formation and maintenance of these heat content and SST anomalies associated with the TBO (Meehl 1993; Clarke et al. 1998; Webster et al. 1999, 2003; Saji et al. 1999; Meehl and Arblaster 2002a; Loschnigg and Webster 2000).

As the formation of the TCs is closely associated with the SST, it is not surprising that the occurrence frequency of the TC activity has an interannual variation with a significant period of 2–3 years.

3.2 Interdecadal Variations of the TC Activities over the Arabian Sea

The data used for discussing interdecadal variations have been processed by taking 9 years running average. Figure 5 shows the trend and the interdecadal variations of TC activity over AS for 50 years. It reveals a significant decreasing linear trend with three major peaks in 1964, 1978 and 2000. As shown in Figs. 4 and 6, the wavelet spectral analysis shows a significant period of around 16 years during the period from middle 1970s to 1989 in terms of the interdecadal variations of the TC activities over AS.

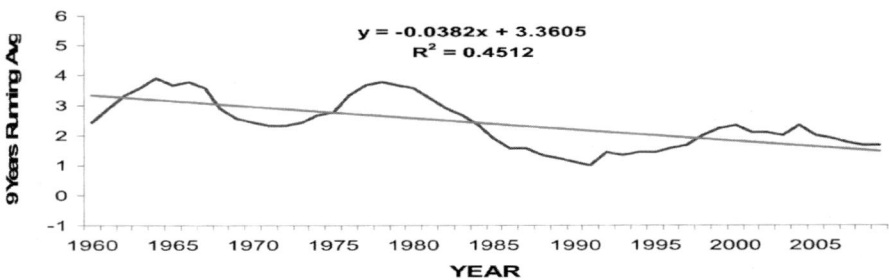

Fig. 5 Interdecadal variations and the trend of TC activities during the period from 1960 to 2009

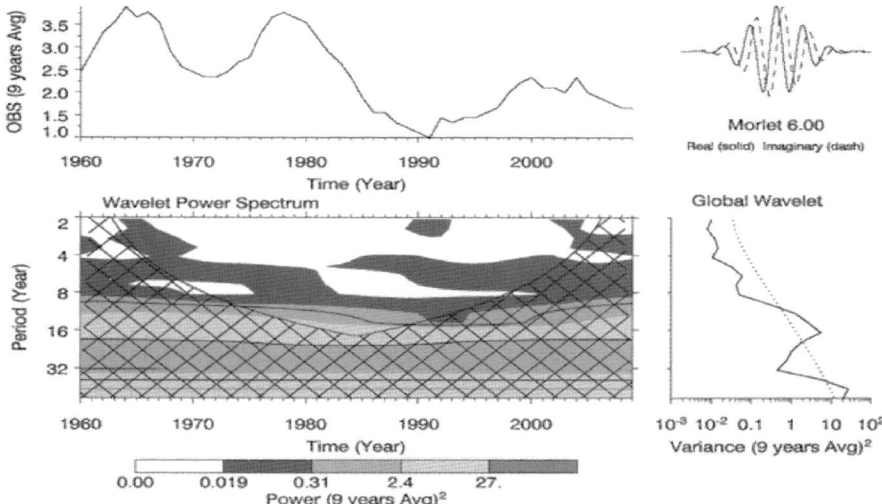

Fig. 6 The local power spectrum of the 9-year running mean occurrence frequency of the TC activity over AS by using Morlet wavelet analysis. The *thick contour* encloses regions of greater than 90 % confidence for a red-noise process. Cross-hatched regions on either end indicate the "cone of influence," where edge effects become important

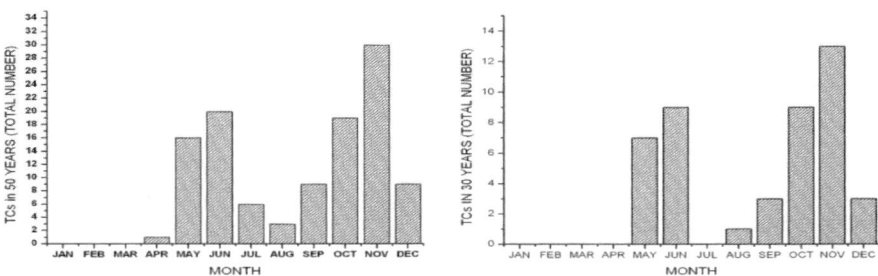

Fig. 7 Total number of the TCs in different months of a year for the period from 1960 to 2009 (*left*) and from 1980 to 2009 (*right*)

3.3 Seasonal Variation of the TC Activity over AS

Figure 7 shows that the occurrence frequency of the TCs over AS in pre-monsoon (May–June) and post-monsoon (October–November) are much higher than the rest of the months. During the past 50 years, no TC occurred in January, February and March. However, during past 30 years, there are 5 months without TC over AS.

In conclusion, there are two dominant seasons of the TC activities over AS, namely pre-monsoon season (May and June) and post-monsoon season (October and November). In other 5 months (April, July, August, September and December) there are much fewer occurrence of TCs over AS.

As shown in Figs. 7 and 8, the average monthly SST reveals that the temperatures during the two seasons are higher as compared to the months with less TC activity in the area. During TC active seasons the SST is higher than 27 °C in most areas of the Arabian Sea (Fig. 8), while the SST is lower than 27 °C in most areas of the north Arabian Sea during TC inactive season (Fig. 8). Palmen (1948) observed that tropical cyclones required ocean temperatures of at least 26.5 °C for their formation and growth. Gray (1979) also pointed out the need for this warm water to be present through a relatively deep layer (50 m) of the ocean. This 26.5 °C value is closely linked to the instability of the atmosphere in the tropical and subtropical latitudes. Above this temperature deep convection can occur, but below this value the atmosphere is too stable and little to no thunderstorm activity can be found (Graham and Barnett 1987). This indicates that the TC activity is strongly modulated by the SST underneath. If the SST in the region is lower than 27 °C, no TC or much fewer TCs will be formed there.

Figure 9 demonstrates the monthly TC activities every 10 years during the past 50 years (1960–2009). As shown in the figure, during the first two decades, there are 8 months in a year with TC activities. During last three decades, there are only 6 months with TC activities. This result coincides with the decreasing trend of the TC activity over AS during the past 50 years. In addition, it was seen that the largest occurrence frequency of TCs over the AS is in November.

3.4 Spatial Distribution of TC Activity

The TC activity over AS is checked for its variation in east-west and north-south directions to find the key region of pronounced activity. For this purpose, the study area is divided into grids of 5° × 5° and the genesis point of TCs is marked as shown in Fig. 10. Figure 10 reveals that there are much more TCs on the eastern side of 65°E than its western counterpart, and the key area for TC genesis is 5°N-25°N, 65°E-75°E. The yearly occurrence frequency of TCs over AS in the east and west of 65°E is shown graphically in Fig. 11.

Figure 11 describes the yearly distributions of TCs on the eastern and western side of 65°E. Almost 90 % of TC activity occurred in the eastern part. As can be seen in Figs. 11 and 12, the SST is generally higher on the eastern side of the Arabian Sea than the western side. This can also be seen in Fig. 12 that the mean SST during the past 50 years is higher in the eastern side of the Arabian Sea than its western side.

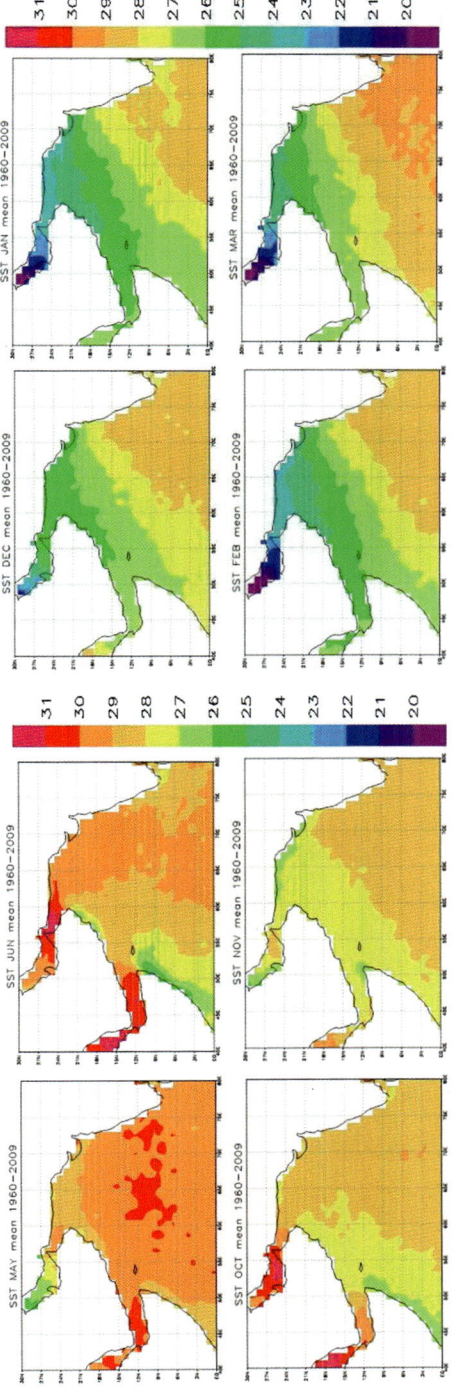

Fig. 8 Mean monthly SST averaged for the period 1960–2009 in TC active months (*left*) and inactive months (*right*)

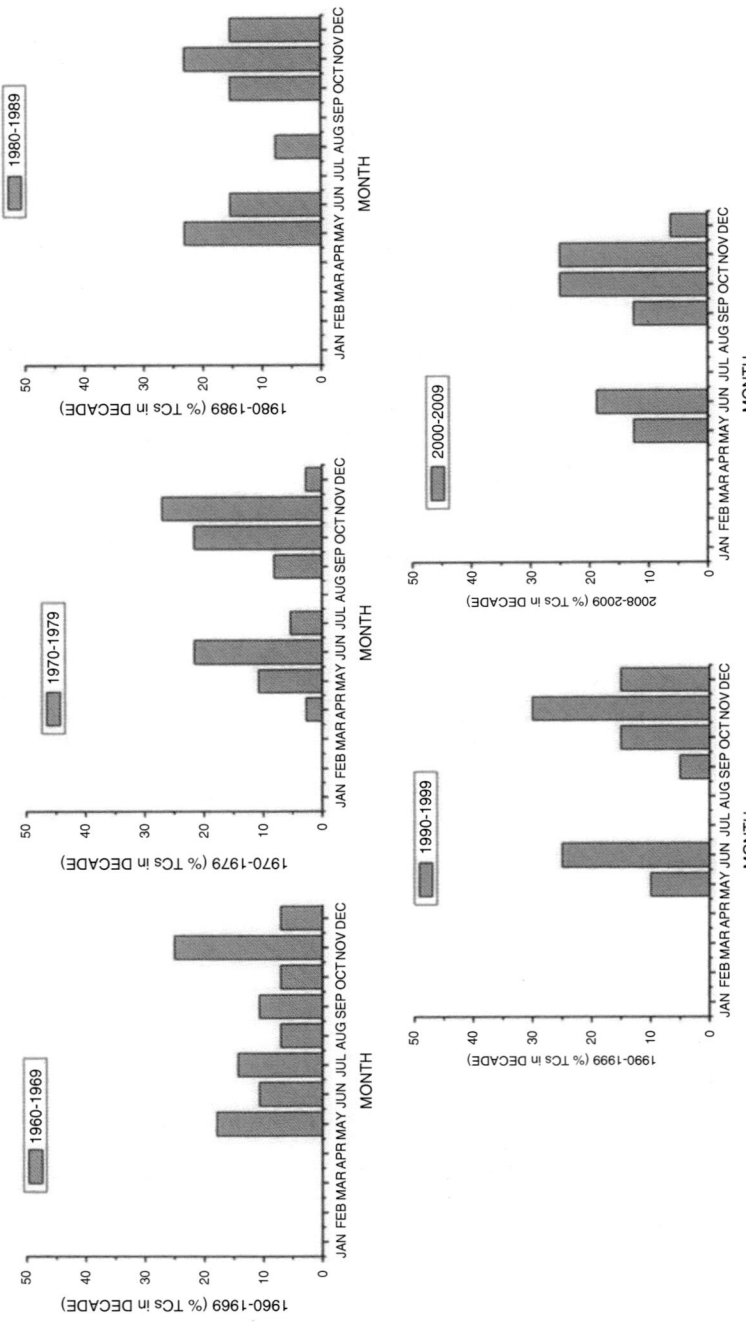

Fig. 9 Interdecadal variations of the monthly TC activities during the period from 1960 to 2009

Fig. 10 Spatial distribution of TC activity over AS (*Shaded rectangle* is the key area of TC genesis)

	45°E	50°E	55°E	60°E	65°E	70°E	75°E	80°E
25°N	0	0	0	0	0	9	4	0
20°N	0	0	0	0	5	9	18	0
15°N	0	1	2	3	8	32		4
10°N	0	0	0	0	1	7	9	4
5°N	0	0	0	0	0	1	1	2

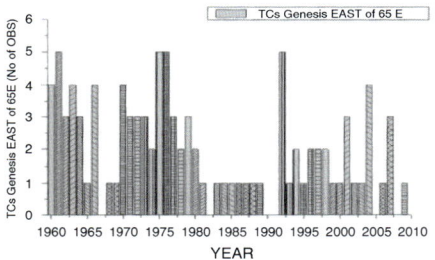 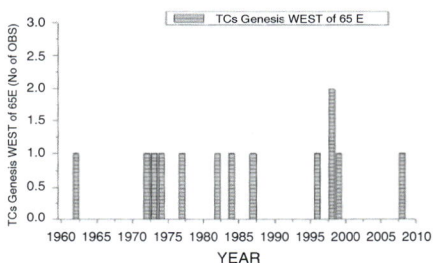

Fig. 11 Yearly distribution of TC activity over east (*left*) and west (*right*) of 65°E

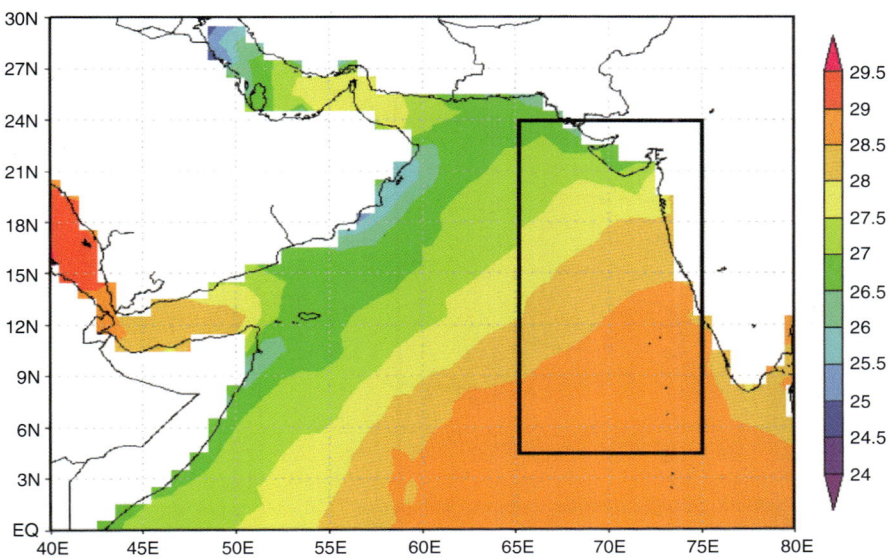

Fig. 12 The average of SST during the period from 1960 to 2009 (Note: *Rectangle* defines the key area of TC genesis)

3.5 TC Activity over AS and Rainfall Anomaly in Southern Pakistan

The TC activity over AS has a negative correlation with the rainfall anomaly in the southern region of Pakistan (Fig. 13); in other words, in case of more activity the rainfall anomaly will be negative, and vice versa. In the following, we shall give some examples as mentioned below to elaborate it.

Figure 14 shows rain anomalies of years 1996, 1998, 2001 and 2004 when there are more TC activities. It reveals that there are negative anomalies of rainfall in coastal regions of Pakistan and positive anomalies in the north of Pakistan in these years. There are 3, 4, 3 and 4 TCs, respectively in these years. The average number of TCs is 2.28 during past 50 years.

Figure 15 shows rain anomalies of years 1995, 2003, 2005 and 2006 when there are less TC activities. It reveals that there are positive anomalies of rainfall in large part of coastal regions of Pakistan. In these years the frequency of TC activity is 1, 1, 0 and 1, respectively, which is less than the average number (2.28) of TC activity during the past 50 years.

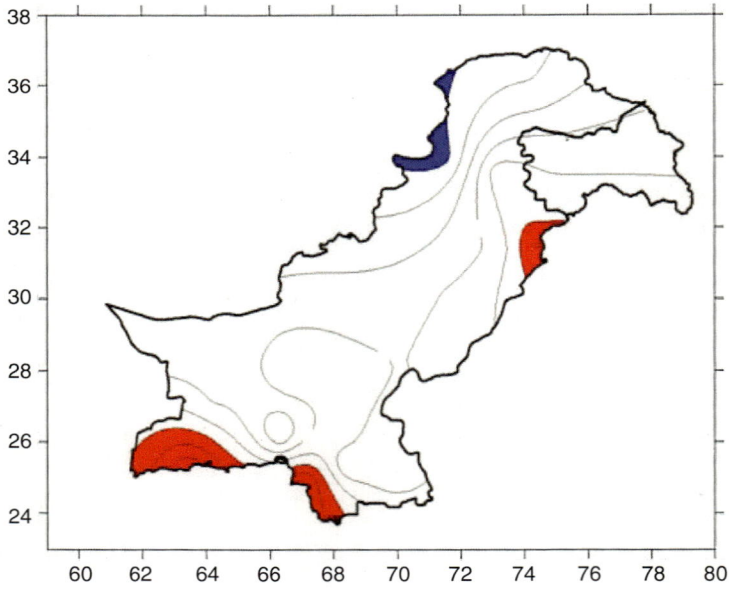

Fig. 13 Correlation coefficients between the number of TCs after removing linear trend and the rainfall anomaly in Pakistan and adjoining areas. The correlation coefficients are significant at 90 % confidence level. Significant negative (positive) correlation coefficients are shaded in *red (blue)*

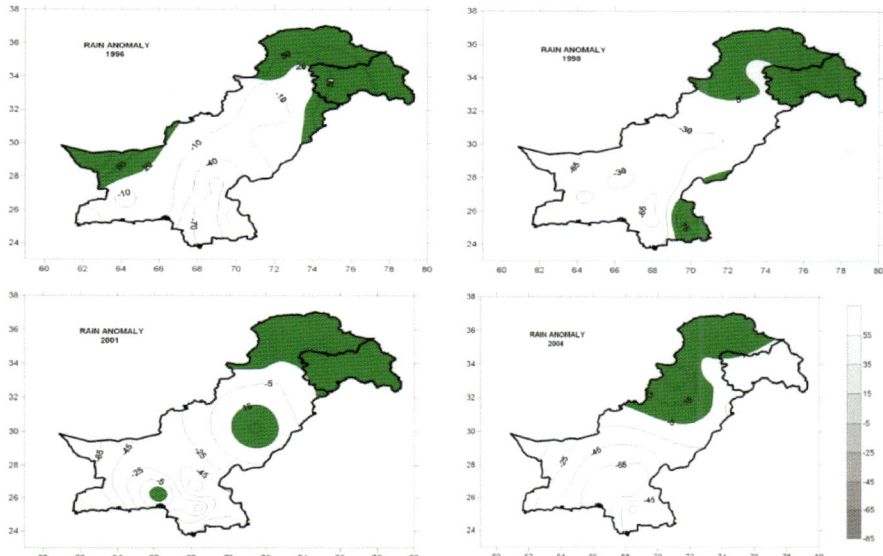

Fig. 14 Rainfall anomaly in case of more TC activities for Pakistan and adjoining areas (*shaded region* is positive anomaly)

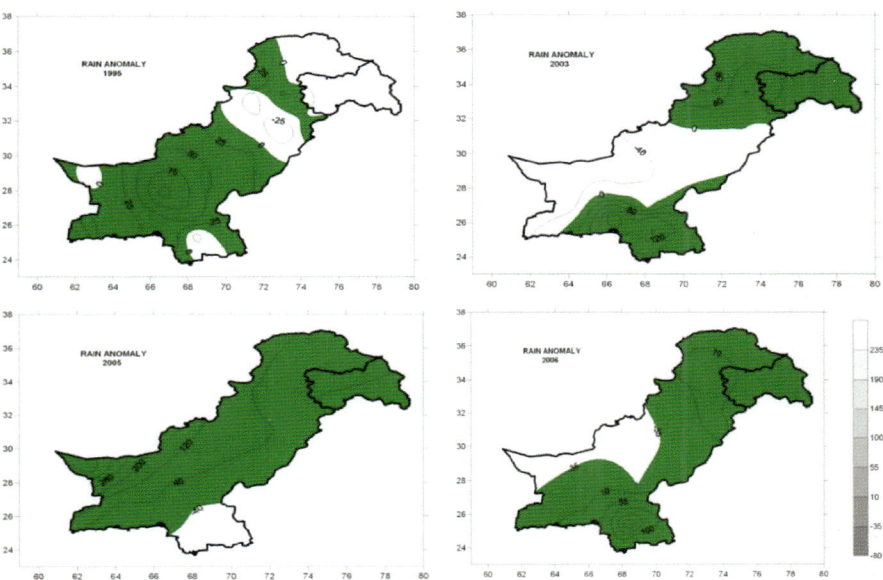

Fig. 15 Rainfall anomaly in case of less TC activity for Pakistan and adjoining areas (*shaded region* is positive anomaly)

Conclusions

The TC activity over AS during study period is found to have decreasing tendency in its frequency with a slope of −0.04 in interannual as well as interdecadal variations. It is more frequent in two spells, pre-monsoon and post-monsoon seasons, of the region of emphasis i.e. Pakistan. It is supported by the sea surface temperatures of AS, which in active seasons remain higher than the other months with less or no TC activity. The TC activity over AS is more pronounced in the region bounded by four vortices (5°N, 65°E), (25°N, 65°E), (25°N, 75°E) and (5°N, 75°E). In this region 80 % of the TCs have their point of genesis and the area is regarded as key area of AS for generation of TCs. The higher TC activity in the eastern side of 65°E longitude than the western part is in relation with the SST of open sea area. According to Morlet wavelet analysis, yearly TC activity has cyclic behaviour of 2–3 years period.

The TC activity over AS has negative impacts on the rainfall over the areas around coast or in general southern part of Pakistan. The rainfall of the area is suppressed in case of more TC activity over AS whereas in case of less TC activity the area gets more precipitation.

Acknowledgements The authors acknowledge the National Center for Environmental Prediction (NCEP) for their reanalysis data, the Department of Atmospheric and Oceanic Sciences (ATOC) at the University of Colorado at Boulder for their freely available wavelets test and Pakistan Meteorological Department for the rain fall data of Pakistan.

References

Brier GW (1978) The quasi-biennial oscillation and feedback processes in the atmosphere–ocean–earth system. Mon Weather Rev 106:938–946
Chan JCL, Je S (1996) Long-term trends and interannual variability in tropical cyclone activity over the western North Pacific. Geophys Res Lett 23:2765–2767
Chang CP, Li T (2000) A theory for the tropical tropospheric biennial oscillation. J Atmos Sci 57:2209–2224
Chen TC, Wang SY, Yen MC (2006) Interannual variation of the tropical cyclone activity over the western North Pacific. J Clim 19:5709–5720
Clarke AJ, Liu X, van Gorder S (1998) Dynamics of the biennial oscillation in the equatorial Indian and far western Pacific oceans. J Clim 11:987–1001
Evans JL (1993) Sensitivity of TCs intensity to SST. J Clim 6:1133–1140
Huang J (1988) Manifestation of QBO in rainfall over China (In Chinese). Atmo Sci Sin 12:267–272
Krishna KM (2009) Intensifying tropical cyclones over the North Indian Ocean during summer monsoon—global warming. Glob Planet Chang 65:12–16

Kumar S, Prasanna RP, Roshin J, Narvekar PK, Kumar D, Vivekanandan E (2009) Response of the Arabian Sea to global warming and associated regional climate shift. Mar Environ Res 68(5):217–222

Li T, Tham CW, Chang CP (2001) A coupled air sea-monsoon oscillator for the tropospheric biennial oscillation. J Clim 14:752–764

Loschnigg J, Webster PJ (2000) A coupled ocean-atmosphere system of SST modulation for the Indian Ocean. J Clim 13:3342–3360

Meehl GA (1987) The annual cycle and interannual variability in the tropical Indian and Pacific Ocean regions. Mon Weather Rev 115:27–50

Meehl GA (1993) A coupled air-sea biennial mechanism in the tropical Indian and Pacific regions. Regions: role of the ocean. J Clim 6:31–41

Meehl GA (1994) Coupled land-ocean-atmosphere processes and south Asian monsoon variability. Science 266:263–267

Meehl GA, Arblaster JM (2001) The tropospheric biennial oscillation and Indian monsoon rainfall. Geophys Res Lett 28:1731–1734

Mooley DA, Parthasarathy B (1983) Variability of the Indian summer monsoon and tropical circulation features. Mon Weather Rev 111:967–978

Nakkan D (1999) Wavelet analysis of rainfall-runoff variability isolating climatic from anthropogenic patterns. Environ Model Softw 14:283–295

Nicholls N (1978) Air-sea interaction and the quasibiennial oscillation. Mon Weather Rev 106:1505–1508

Ramsay HA, Leslie LM, Richman MB, Lamb PJ, Leplastrier M (2008) Interannual variability of tropical cyclones in the Australian region. J Clim 21:1083–1103

Saji NH, Goswami BN, Vinayachandran PN, Yamagata T (1999) A dipole mode in the tropical Indian Ocean. Nature 401:360–363

Shukla J, Misra BM (1977) Relationship between sea surface temperature and wind speed over the central AS with the rainfall over INDIA. Mon Weather Rev 105:998–1002

Walsh JE, Mostek A (1980) A quantitative analysis of meteorological anomaly patterns over the United States, 1900-1977. Mon Weather Rev 108:615–630

Wang S, Zhao Z (1981) Droughts and floods in China, 1470-1979, climate and history. Cambridge Press, Cambridge

Webster PJ, Moore AM, Loschnigg JP, Leben RR (1999) Coupled ocean-atmosphere dynamics in the Indian Ocean during 1997–98. Nature 401:356–360

Wu LG, Tao L, Ding QG (2010) Influence of sea surface warming on environmental factors affecting long-term. J Clim 23:5978–5989

Zhi X (1997) Quasibiennial oscillation in precipitation and its possible application to longterm prediction of floods and droughts over eastern China. Ann Meteorol 35:250–252

Zhi X (2001) Interannual variability of the Indian summer monsoon and its modeling with a zonally symmetric 2D-model. Shaker Verlag, Aachen

Zhi X, Jialu P, Zhang L (2010) An analysis of the winter extreme precipitation events on the background of climate warming in southern China. J Trop Meteorol 16(4):325–332

Zhu Q, Zhi X (1991) Quasibiennial oscillation in rainfall over China. Acta Meteorologica Sinica 5(4):426–434

Impact of Cloud Microphysics and Cumulus Parameterisation on Meso-scale Simulation of TC Sidr over the Bay of Bengal Using WRF Model

Md. Mahbub Alam

1 Introduction

Tropical cyclones are known to cause enormous damage and destruction in the coastal regions. The cyclones formed over the Bay of Bengal generally move in the north, northwest and northeast directions and cross Bangladesh, Myanmar and eastern coast of India. The ocean is very important source of energy for tropical cyclones (TCs). TCs intensify in areas that have high SST and deep oceanic mixed layer, thus having high upper-ocean heat content. The prediction of movement and intensity of TC is very much essential.

The performance of a numerical weather prediction (NWP) model greatly depends on the initial and boundary conditions as well as the physical parameterisation schemes. As the storm forms over the data sparse oceanic areas, the initial analysis obtained from the global coarse resolution model may not be adequate. Pattnaik and Krishnamurti (2007) have suggested that the track forecasts are not much influenced by microphysical modifications. However, their study has shown that the microphysical processes and microphysical parameters of hydrometeors significantly modulate the intensity forecast of the hurricane. Pattnaik et al. (2011) have suggested that the WSM6 microphysical processes have minimal impact on the storm track forecast. In different sets of explicit schemes, Fovell and Su (2007) demonstrated that hurricane track forecasts are sensitive to microphysical parameterisation schemes. Raju et al. (2011) have shown that the error in landfall time and intensity are decreasing with the delayed initial condition, suggesting that the model forecast is more dependable on the initial and boundary conditions.

Md.M. Alam (✉)
SAARC Meteorological Research Centre (SMRC), Dhaka, Bangladesh

Department of Physics, Khulna University of Engineering & Technology (KUET), Khulna, Bangladesh
e-mail: mdalam60@gmail.com

Willoughby et al. (1984) and Lord et al. (1984) studied the impact of cloud microphysics on TC structure and intensity using 2D axi-symmetric non-hydrostatic model. Their results show that the ice-phase microphysical scheme can produce a lower central sea level pressure (CSLP) than the case without the ice-phase. Wang (2001) demonstrated that the intensification rate and intensity are not sensitive to the cloud microphysical parameterisations. Davis and Bosart (2002) considered the effects of cumulus parameterisation on tropical storm track. They found that the Betts-Miller-Janjic (BMJ) and Grill schemes produced more westward track than Kain-Fritsch (KF) scheme. The KF scheme tended to intensify the storm too rapidly but produced the best track compared with observations. Zhu and Zhang (2006) informed that the hurricane track was not sensitive to cloud microphysical processes except for very weak storms.

In this research 24 experiments have been conducted by using six different MP schemes in combination of KF and BMJ schemes with two different initial conditions for the tropical cyclone Sidr that formed in the Bay of Bengal and crossed Bangladesh coast on 15 November 2007. As for Bangladesh, Alam et al. (2003) suggested that the most vulnerable month for TCs is November. The aim of this research is to investigate the suitable MP and CP schemes for the prediction of track and intensity of TC Sidr formed in the month of November over the Bay of Bengal.

2 Model Description

In the present study the Weather Research and Forecast (WRF-ARW Version 3.2.1) model consists of fully compressible non-hydrostatic equations and different prognostic variables are utilized. The model vertical coordinate is terrain following hydrostatic pressure and the horizontal grid is Arakawa C-grid staggering. The details of the model and domain configuration are given in Table 1.

3 Data and Methodology

Final Reanalysis (FNL) data ($1° \times 1°$) collected from National Centre for Environment Prediction (NCEP) is used as initial and lateral boundary conditions (LBCs) which is updated at 6 h interval i.e. the model is initialized with 0000, 0600, 1200 and 1800 UTC initial field of corresponding date. The NCEP FNL data is interpolated to the model horizontal and vertical grids and the model was integrated for 120 and 102-h period for TC Sidr. Twenty-four experiments have been conducted by using different microphysics schemes (e.g., Kessler, Lin et al., WSM3-class simple ice scheme, Ferrier, WSM 6-class graupel scheme and Thompson graupel) in combinations with Kain-Fritsch (KF) and Betts-Miller-Janjic (BMJ) schemes with the initial conditions of 0000 UTC of 11 and 12 November. The model simulated sea level pressure (SLP), maximum wind at 10 m, track, track error (km), wind at 10 m and 850 hPa level, vertical wind structure, maximum vertical

Table 1 WRF model and domain configurations

Dynamics	Non-hydrostatic
Central points of the domain	Central Lat.: 16.5°N, Central Lon.: 86.5°E
Horizontal grid distance	12 km
Number of grid points	X-direction 186 points, Y-direction 196 points
Microphysics	(1) Kessler, (2) Lin et al. (1983) scheme
	(3) WSM3-class scheme (Hong et al. 2004)
	(4) Eta (Ferrier) scheme
	(5) WSM 6-class graupel scheme and
	(6) Thompson graupel scheme
Radiation scheme	Dudhia (1989) for short wave radiation/RRTM long wave Mlawer et al. (1997)
Surface layer	Monin-Obukhov similarity theory scheme
Cumulus parameterisation schemes	(1) Kain-Fritsch scheme (KF) (Kain and Fritsch 1990, 1993)
	(2) Betts-Miller-Janjic (BMJ) (Janjic 2000)
PBL parameterisation	Yonsei University Scheme (YSU) scheme (Hong et al. 2006)

velocity and temperature departure along the centre of the TC have been analysed. Simulated track and intensity have also been compared with the IMD and JTWC observed results.

4 Synoptic Situation of Tropical Cyclone Sidr

A low pressure area formed over southeast of Andaman Islands with a weak low-level circulation near the Nicobar Islands on 9 November 2007. It moved north-northwesterly direction initially and intensified into a well-marked low over the same area. After that it transformed into depression over the southeast Bay of Bengal and adjoining Andaman Sea and lay centered at 0900 UTC of 11 November 2007 near 10.0°N and 92.0°E about 200 km south-southwest of Port Blair. The depression moved further north-northwestwards and intensified into a deep depression (DD) and lay centered 10.5°N and 91.5°E at 1800 UTC of the same day. The system further intensified into a cyclonic storm SIDR at 0300 UTC of 12 November and severe cyclonic storm (SCS) at 1200 UTC of the same day and lay centered at 11.5°N and 90°E. The system further intensified into a very severe cyclonic storm (VSCS) with the central MSLP of 986 hPa, the maximum sustained wind (MSW) of 33 m/s and the central location at about 11.5°N and 90.0°E at around 1800 UTC of 12 November. The VSCS Sidr moved in the same direction and intensified further and at 0300 UTC of 15 November its central MSLP lowered to 944.0 hPa, the MSW increased to 58.8 m/s when its central location was at about 18.0°N and 89.0°E. The VSCS Sidr moved continuously northwards and finally crossed Bangladesh coast at around 1600 UTC of 15 November 2007. After landfall the system weakened gradually by giving precipitation over Bangladesh coast. The IMD and JTWC tracks are depicted in Fig. 2.

5 Results and Discussions

Sea level pressure, maximum wind at 10 m level, track, track error (km), winds at 10 m and 850 hPa level, vertical wind structure and vertical temperature departure along the centre of the tropical cyclone simulated by the model, with six MPs coupling with two CP schemes with two initial conditions, have been discussed for the TC Sidr in the following sub-sections. Area (8–20°N and 86–96°E) averaged vorticity has been calculated from the three-hourly time profiles with the initial condition of 0000 UTC of 11 November.

5.1 Intensity of TC Sidr

Three-hourly MSLP [Fig. 1a–d] and associated MSW (m s^{-1}) at 10 m level [Fig. 1e–h] using different MP schemes coupling with CP schemes are discussed by using two different initial conditions of 0000 UTC of 11 and 12 November 2007. The simulated intensity in terms of MSLP for all MP schemes in combination with KF scheme were much higher than that of BMJ scheme with the initial condition at 0000 UTC of 11 and 12 November. IMD observed intensity (944 hPa, 58.8 m s^{-1}) is much lower than that of JTWC intensity (918 hPa, 71.5 m s^{-1}) in terms of pressure and 10 m level MSW. With the initial conditions at 0000 UTC of 11 and 12 November, the following results were obtained:

(1) The simulated pressure fall for Lin et al. scheme in combination with KF scheme were close to IMD observed results but lower than JTWC observed results. (2) The simulated MSLP for all other MP schemes in combination with KF scheme lay between IMD and JTWC observed results. (3) The simulated pressure drop and wind speed for all MP schemes in combination with BMJ scheme were much lower than that of IMD and JTWC observed pressure and maximum wind speed at 10 m level. (4) The wind speed simulated at 10 m level for all MP schemes coupling with KF scheme were lower than that of JTWC observed wind speed and matched with IMD observed wind speed all through the simulation except for maximum wind, which was less than that of IMD observed wind. (5) BMJ scheme coupling with all MP schemes had simulated wind speed at 10 m level which was much lower than that of IMD and JTWC observed wind.

5.2 Track of TC Sidr

The observed and simulated tracks of TC Sidr for different MP schemes are displayed in Fig. 2a–d. For different experiments, the track forecasts were reasonably accurate i.e., up to the landfall time (1800 UTC of 15 November 2007). For all combination of MP and CP schemes the model captured the north-northwestward movement with the initial conditions at 0000 UTC of 11 and 12 November 2007.

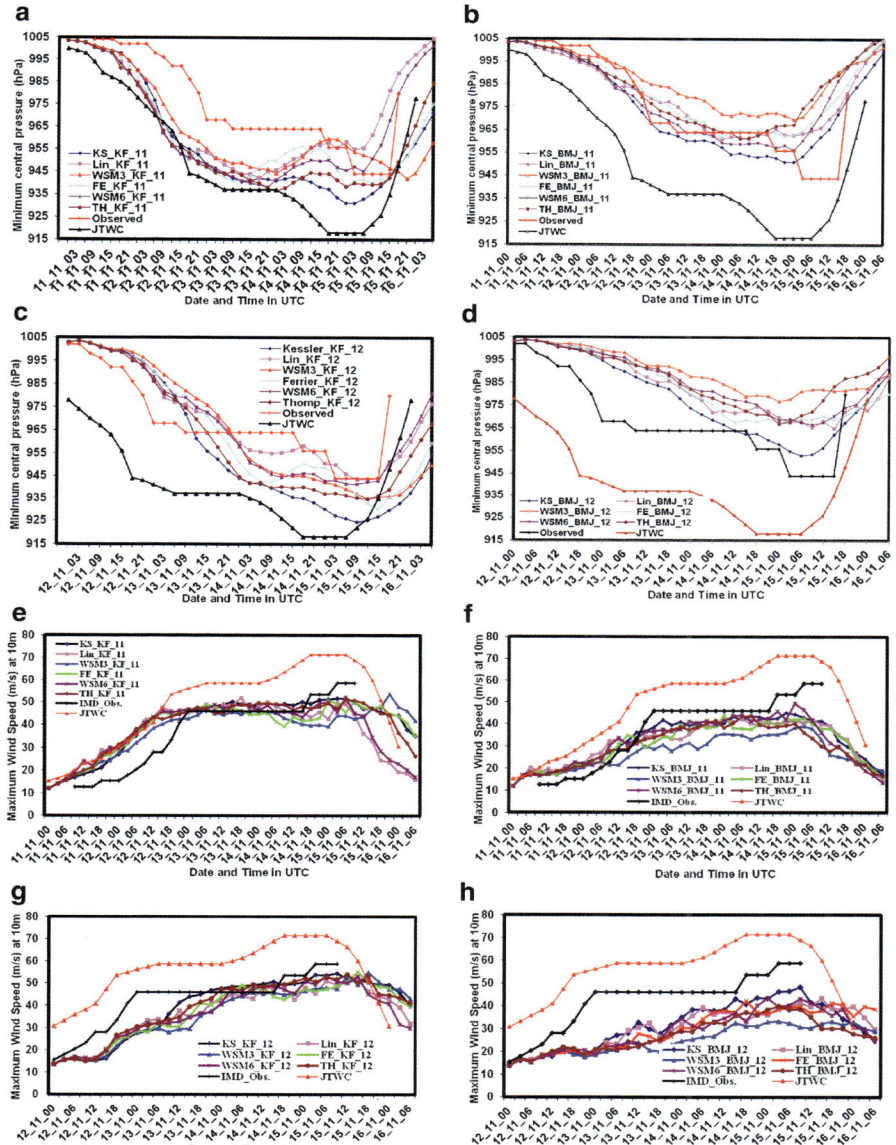

Fig. 1 Model simulated (**a–d**) MSLP (hPa) and (**e–h**) maximum sustained wind at 10 m level of TC Sidr using six different MP schemes coupling with KF and BMJ schemes with the initial conditions at 0000 UTC of 11 and 12 November 2007

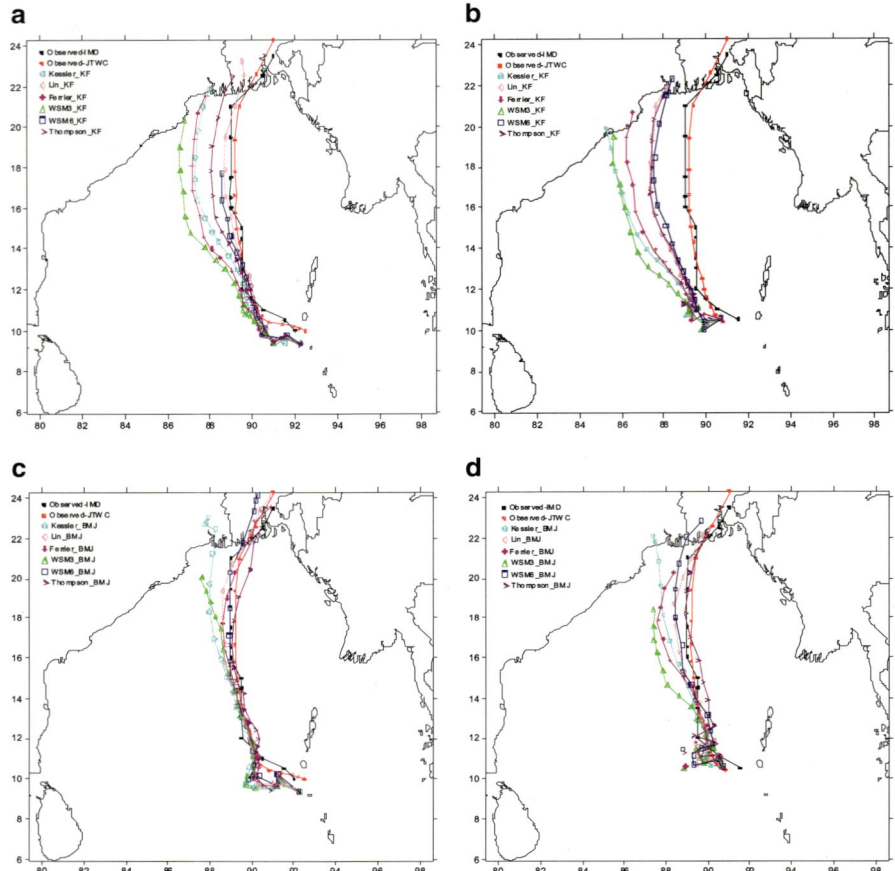

Fig. 2 Simulated, IMD and JTWC observed tracks of tropical cyclone Sidr using six different MP schemes coupling with (**a, b**) KF scheme and (**c, d**) BMJ scheme with the initial conditions of 0000 UTC of 11 and 12 November 2007

The significant deviations in tracks have been observed among different microphysical schemes in combination with CP scheme with the initial conditions of 0000 UTC of 11 and 12 November. WSM3 and Kessler schemes in combination with KF and BMJ schemes have provided most deviated track with the initial condition of 0000 UTC of 11 November and deviated towards left from the actual track of IMD and JTWC. At the time of landfall Lin et al. and WSM6 schemes in combination with KF scheme and Lin et al., Ferrier, WSM6 and Thompson schemes in combination with BMJ schemes have simulated less deviated track with the same initial conditions. The simulated tracks for all six MP schemes coupling with KF schemes have deviated significantly towards west from the original track with the initial conditions of 0000 UTC of 12 November. At the time of landfall the track deviation was minimum for Lin-BMJ and TH-BMJ combinations with initial conditions at 0000 UTC of 12 November.

There was no landfall for WSM3 scheme in combination with KF and BMJ schemes up to the simulation time of TC Sidr with the initial conditions of 0000 UTC of 11 and 12 November 2007. The landfall time has almost matched for all MP schemes in combination with BMJ scheme and Lin et al. and WSM6 schemes in combination with KF scheme for the initial condition of 0000 UTC 11 November. The other combinations are delayed by 5–15 h. The simulated landfall time by using six different MP schemes coupling with KF scheme were 0300 UTC16, 1500 UTC15, 0600 UTC16, 0600 UTC16, 1500 UTC15 and 0000 UTC16 respectively but when coupling was with BMJ scheme the landfall was at 1400 UTC15, 1500 UTC15, no landfall, 1800 UTC15, 1400 UTC15 and 1600 UTC15 respectively. The landfall time was delayed further with the initial condition of 0000 UTC 12 November. There was no landfall of TC Sidr up to 0600 UTC of 16 November for WSM3-KF, WSM3-BMJ, TH-KF and FE-BMJ combinations for initial conditions of 0000 UTC of 12 November 2007. Simulated landfall time by using Kessler, Lin et al., Ferrier and WSM6 schemes coupling with KF scheme was 0600 UTC16, 0600 UTC16, 0600 UTC16 and 0000 UTC16 respectively. But when Kessler, Lin et al., WSM6 and Thompson schemes are coupled with BMJ scheme, the landfall time was 0000 UTC16, 0600 UTC16, 0000 UTC16 and 0600 UTC16 respectively.

5.3 Track Error

The track error for all MP schemes in combination with KF and BMJ schemes increased with increase of time with the initial conditions at 0000 UTC of 11 and 12 November. WSM3 & Lin et al. and WSM3 & WSM6 schemes in combination with KF and BMJ schemes had simulated most and least deviated track with respect to IMD and JTWC track with the initial conditions at 0000 UTC of 11 and 12 November respectively. The track errors were less at the time of landfall using Lin et al. scheme in combination with KF and BMJ schemes and Ferrier scheme coupling with BMJ scheme with the initial condition at 0000 UTC of 11 November 2007.

The track errors were also less at the time of landfall using Lin et al. and Thompson schemes coupling with BMJ schemes with the initial condition at 0000 UTC of 12 November 2007. At the time of landfall the track error using Lin et al. scheme coupling with BMJ scheme were almost equal. The average track error (Fig. 3a, b) using 0000 UTC of 11 November initial condition are lower than that of 0000 UTC of 12 November initial condition. Lin-KF and Lin-BMJ combination gives the minimum average track error out of all combinations. The minimum average track error using Lin–BMJ with respect to JTWC and IMD are 54.7 km and 66.8 km. The average track error was also low using Lin-KF combination as 58.4 km and 79 km.

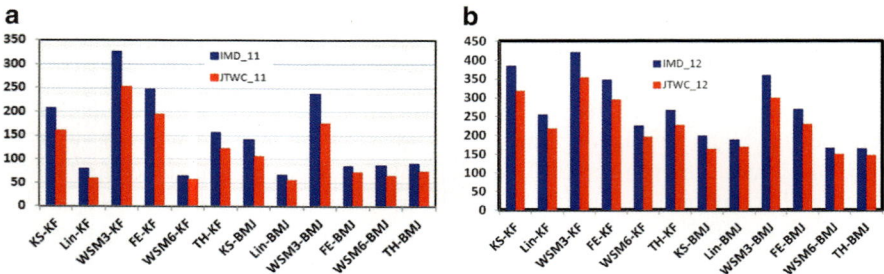

Fig. 3 Average track error (km) of TC Sidr with the initial conditions at 0000 UTC of (**a**) 11 November and (**b**) 12 November 2007

5.4 Mean Sea Level Pressure

Simulated MSLP at 0000 and 1200 UTC of 15 November using different MP schemes coupling with KF and BMJ schemes with the initial conditions of 0000 UTC of 11 and 12 November are presented in Fig. 4a–l. It is observed that for all MP schemes coupling with CP schemes with different initial conditions, the system intensified into a severe cyclonic storm with a core of hurricane intensity SCS (H). KF scheme simulated lower SLP and BMJ scheme simulated higher SLP coupling with different MP schemes with the initial conditions of 0000 UTC of 11 and 12 November. Similarly the initial condition of 0000 UTC of 11 November simulated higher pressure drops (Fig. 4a–l) than the initial condition of 0000 UTC of 12 November. It was also found that Kessler–KF combination simulated highest pressure drop with central pressure of 924.4 hPa and Ferrier-KF combination simulated maximum wind speed of 55.1 m s^{-1} at 10 m level with the initial condition of 0000 UTC of 12 November, which is less than that of Joint Typhoon Warning Centre (JTWC) pressure drop of 918.0 hPa and wind speed of 71.5 m s^{-1}.

From the simulated results it was also observed that the pressure drop (wind speed) by Kessler-KF combination is greater (less) than that of IMD observed results. The Lin-KF combination simulated pressure drop of 942.3 and 943.5 hPa for the initial conditions of 0000 UTC of 11 and 12 November, which is almost equal to the IMD observed pressure drop of 944 hPa. The other MP coupling with KF scheme simulated pressure drop (Table 2) higher than JTWC pressure drop and lower than IMD observed pressure drop but the wind speed at 10 m level was less than that of IMD and JTWC observed results. All MP schemes in combination with BMJ scheme simulated less pressure drop and wind speed (Table 2) than those obtained from IMD and JTWC. The simulated pressure drop by BMJ schemes with different initial conditions were less than that of KF scheme. Table 2 depicts that the IMD observed pressure (944 hPa) was much less than that of JTWC pressure (918 hPa).

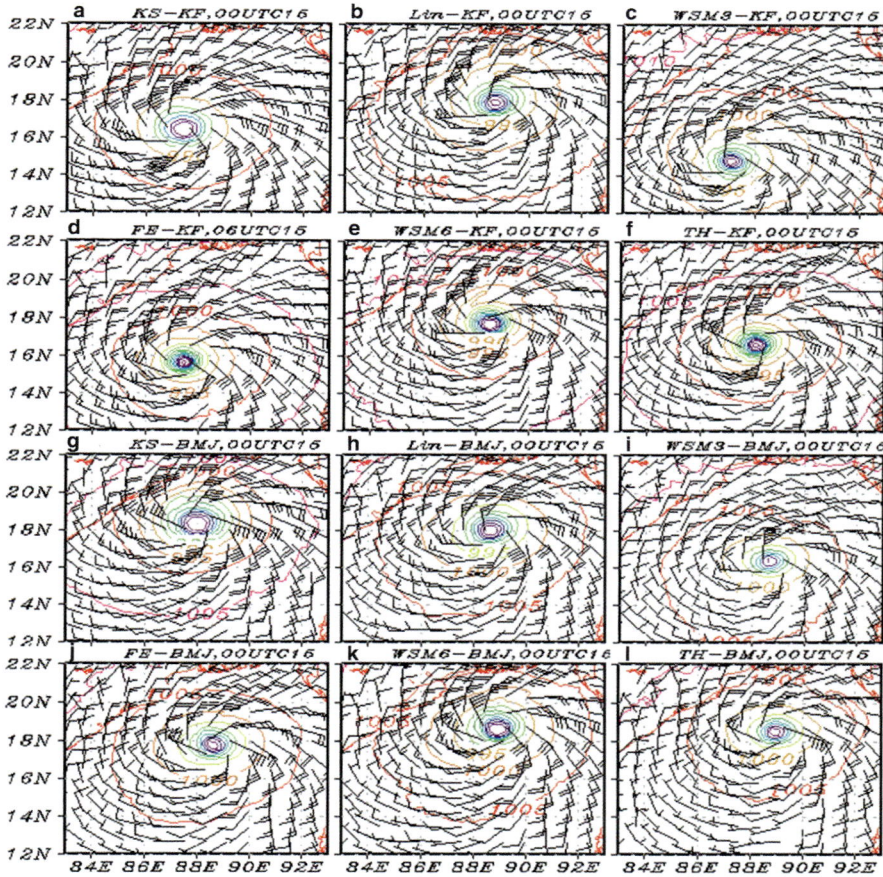

Fig. 4 Model simulated sea level pressure (*contour*) and 10 m wind (*barb*) of TC Sidr at 0000 UTC of 15 November 2007 using different MP schemes coupling with (**a–f**) KF and (**g–l**) BMJ schemes with the initial condition at 0000 UTC of 11 November

5.5 Wind

Model simulated 10 m and 850 hPa level winds at 0000 and 1200 UTC of 15 November for different MP schemes coupling with KF and BMJ schemes with the initial conditions at 0000 UTC of 11 and 12 November are presented in Fig. 4a–l. KF scheme coupling with different MP schemes simulated higher wind speed than that of BMJ scheme. Analysis for all MPs coupling with CP schemes with initial conditions of 0000 UTC of 11 and 12 November reveal that the cyclone moved in northerly direction and crossed Bangladesh coast. During this period the wind speed

Table 2 Simulated minimum sea level pressure (MSLP), time of landfall, average track error from IMD and JTWC observations and maximum wind at 10 m level of TC Sidr

MP	CP	Named as	MSLP (hPa)	Time of landfall (UTC)	IMD track error (km)	JTWC track error (km)	Max. wind at 10 m level m/s
Kessler (KS)	KF	KS_KF_11	931.2	03 UTC16	207.9	159.5	51.8
		KS_KF_12	924.4	03 UTC16	384.6	317.8	54.6
	BMJ	KS_BMJ_11	951	14 UTC15	141.7	105.1	45
		KS_BMJ_12	953	21 UTC15	198.8	163.5	48.3
Lin et al. (Lin)	KF	Lin_KF_11	942.3	15 UTC15	79	58.4	52.1
		Lin_KF_12	943.5	03 UTC16	254.4	217.9	51.9
	BMJ	Lin_BMJ_11	961	15 UTC15	66.8	54.7	42.1
		Lin_BMJ_12	965	03 UTC16	188.7	169.1	43.4
WSM3	KF	WSM3_KF_11	941.9	06 UTC16	326.8	253.3	53.8
		WSM3_KF_12	934.8	No landfall	420.3	353.1	54.8
	BMJ	WSM3_BMJ_11	969.8	No landfall	238	175.5	39
		WSM3_BMJ_12	976.8	No landfall	359.2	299.3	33.2
Ferrier (FE)	KF	FE_KF_11	938.7	06 UTC16	246.9	194.8	51
		FE_KF_12	930.1	03 UTC16	348	295	55.1
	BMJ	FE_BMJ_11	961.7	18 UTC15	85.7	72.4	43.8
		FE_BMJ_12	967.7	No landfall	270.5	230.6	42.3
WSM6	KF	WSM6_KF_11	938.2	15 UTC15	88.3	68.8	52.3
		WSM6_KF_12	941.2	00 UTC16	225.3	195.9	53.4
	BMJ	WSM6_BMJ_11	956.2	12 UTC15	86.4	65.1	49.3
		WSM6_BMJ_12	966.7	21 UTC15	166	151	43.3

(continued)

Impact of Cloud Microphysics and Cumulus Parameterisation

Table 2 (continued)

MP	CP	Named as	MSLP (hPa)	Time of landfall (UTC)	IMD track error (km)	JTWC track error (km)	Max. wind at 10 m level m/s
Thompson (TH)	KF	TH_KF_11	936.7	00 UTC16	156.3	123.1	51
		TH_KF_12	934.6	06 UTC16	267.1	226.8	54
	BMJ	TH_BMJ_11	961.2	16 UTC15	89.9	74.4	43.6
		TH_BMJ_12	967.2	03 UTC16	164.5	147.1	40
	IMD observed		944	16 UTC15			58.8
	JTWC observed		918				71.5

in the eyewall was maximum and decreased towards the centre and away from the eyewall. Cyclonic circulation persisted up to 200 hPa level for all MP schemes in combination with KF scheme and irregular pattern was also observed at this level in combination with BMJ scheme.

The vertical wind structure along the centre of TC Sidr at 0000 UTC of 15 November, 2007 using different MP schemes coupling with KF and BMJ schemes with initial condition at 0000 UTC of 11 November are presented in Fig. 5a–l. The maximum wind speed was seen from 900 to 700 hPa level for different combination of MP and CP schemes. The maximum wind extended up to 200 hPa level in the eastern side and up to 500 hPa level in the western side of the centre of TC Sidr. The simulated maximum wind speed at 850 hPa for different MP in combination with BMJ scheme was much less than that of KF combination. From the simulated vertical structure it was observed that the eye of the cyclone was almost 25–30 km in diameter and for all MP schemes in combination with KF and BMJ scheme the wind speed was maximum in the eastern side.

5.6 Temperature Departure

Simulated vertical temperature departure (°C) profile was drawn for different MPs coupling with KF and BMJ schemes along the centre of TC Sidr at different times. The temperature departure was maximum with the initial condition at 0000 UTC of 11 November as shown in Fig. 6a–l. The temperature departures were considered from the initial state of model run. The temperature departure profile was generated along the centre of the vortex, where the wind speed was minimum. Simulated temperature departure profiles using 0000 UTC of 11 and 12 November initial

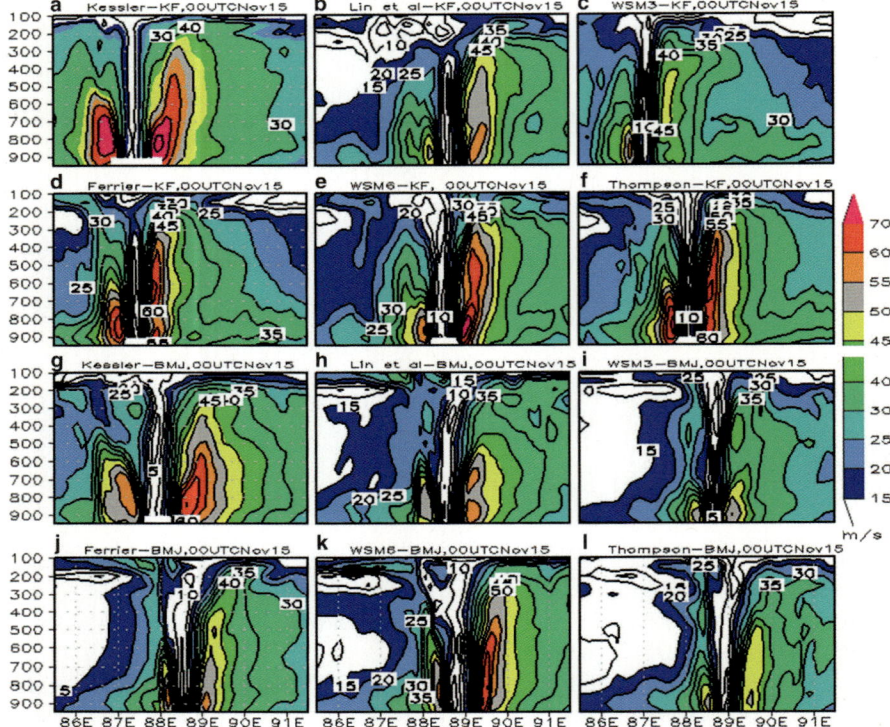

Fig. 5 Vertical wind structure along the centre of TC Sidr at 0000 UTC of 15 November 2007 using different MP schemes coupling with (**a–f**) KF and (**g–l**) BMJ schemes with the initial condition at 0000 UTC of 11 November

conditions were different. The cross section of temperature departure showed that the vortex is a warm-core system with the maximum departure at 700–300 hPa level. The maximum positive temperature departures were found along the core of TC Sidr. The maximum temperature departure were simulated in the layer 400–550 hPa and 450–800 hPa for all MP schemes in combination with KF and BMJ schemes respectively.

KF scheme simulated maximum temperature departure in the upper to middle troposphere and BMJ scheme simulated maximum temperature departure in the middle to lower troposphere. As the temperature in the upper-middle level increased, the central pressure of the vortex decreased. The KF scheme had simulated much lower pressure drop and higher wind speed, due to the higher temperature deviation in the upper-middle levels. The maximum temperature departures were almost same for KF and BMJ schemes but BMJ scheme had simulated maximum temperature departure in the middle to lower troposphere. Due to the maximum temperature increase in the lower to middle troposphere the pressure drop and wind speed was lower for BMJ scheme than that of KF scheme. Negative relationship was found between the upper level temperature departure and the central pressure of the vortex.

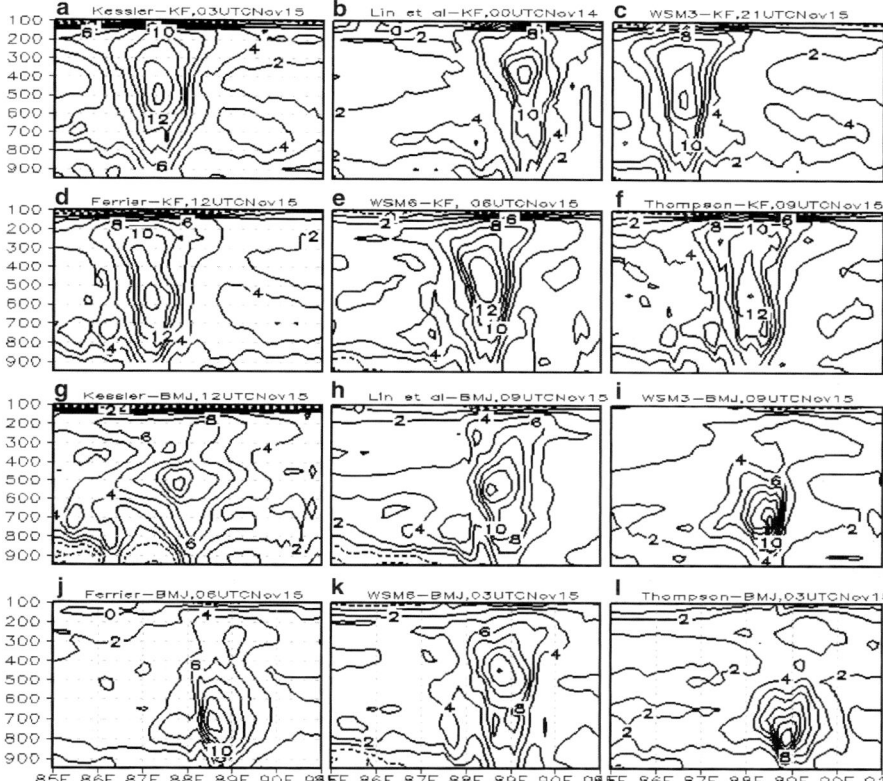

Fig. 6 Model simulated maximum temperature departure (°C) along the centre of TC Sidr at 0000 UTC of 15 November 2007 using different MP schemes coupling with (**a–f**) KF and (**g–l**) BMJ schemes with the initial condition at 0000 UTC of 11 November

5.7 Vertical Velocity

Maximum updraft (Fig. 7) has been identified during the simulated period with the initial condition of 0000 UTC of 11 November 2007 using different MPs coupling with KF and BMJ schemes. The simulated times and positions of maximum updrafts are different in different combination of MPs and CP schemes (Fig. 7). Simulated time for maximum updraft are 0000, 1800, 1800, 1500 and 0300 UTC (Fig. 7) of 14 November for Kessler, Lin et al. WSM3, Ferrier and WSM6 schemes coupling with KF scheme respectively and in case of Thompson-KF combination, the maximum updraft time is 0300 UTC of 13 November. The simulated time for maximum updraft are 1200 UTC15, 1800 UTC14, 0300 UTC15, 0900 UTC15, 0300 UTC16 and 0300 UTC15 November for Kessler, Lin et al. WSM3, Ferrier, WSM6 and Thompson schemes (coupling with BMJ scheme) respectively.

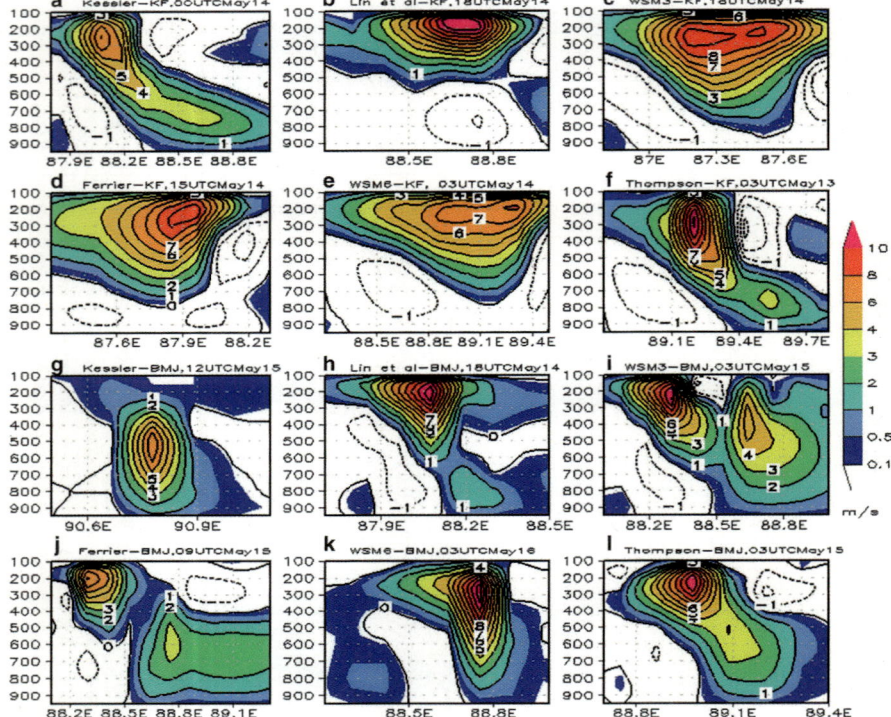

Fig. 7 Model simulated maximum vertical velocity of TC Sidr for different MP schemes coupling with (**a–f**) KF and (**g–l**) BMJ schemes with the initial condition at 0000 UTC of 11 November 2007

The simulated maximum updrafts were located at 300–200, 200–100, 350–200, 300–150, 300–150 and 400–200 hPa levels for Kessler-KF, Lin-KF, WSM3-KF, Ferrier-KF, WSM6-KF and Thompson-KF combinations respectively with the initial condition of 0000 UTC of 11 November. The significant updrafts are simulated by all MP schemes coupling with CP schemes. The downdrafts were weaker and were seen throughout the depth of the troposphere and updrafts were observed mainly in the upper troposphere. The low-level descent and high-level ascent is expected from a matured convective system and was consistent with earlier studies. The updraft in the upper troposphere could be due to the latent heat release during glaciations and vapour deposition. The increase of updraft in the upper troposphere is related to the increase in total water due to the production of cloud ice and snow in the upper troposphere and cloud graupel in the middle troposphere. From the horizontal structure of vertical velocity it was found that the maximum updraft was simulated in the southern to western periphery of the cyclone centre for all MP schemes in combination with KF scheme and was simulated in the western periphery of the cyclone centre for all MP schemes in combination with BMJ scheme.

5.8 Vorticity

Time variation of vertically integrated space averaged vorticity ($\times 10^{-5}$/s) simulated by six different MP schemes coupling with KF and BMJ schemes are presented in Fig. 8a, b. The space averaged vorticity increased continuously during 0000 UTC of 12 November to 0600 UTC of 15 November for Lin and WSM6 schemes and 1200 UTC of 15 November for Kessler and Thompson schemes and then decreased afterwards.

The space averaged vorticity increased continuously during the simulation time for WSM3 and Ferrier scheme in combination with KF scheme. All through the simulation WSM3-KF simulated minimum averaged vorticity and WSM6-KF simulated maximum averaged vorticity. For BMJ combination (Fig. 8b), the space averaged vorticity increased continuously during 0000 UTC of 12 November to 0600 UTC of 15 November and after that the vorticity decreased significantly. WSM3-BMJ combination simulated minimum averaged vorticity and Kessler-BMJ simulated maximum averaged vorticity during the period.

The vertically integrated area and time averaged (starting from 0000 UTC of 12 November to 0000 UTC of 16 November 2007) vorticity for six different MP schemes coupling with KF and BMJ schemes with the initial condition at 0000 UTC of 11 November are presented in Fig. 9a, b. The positive vorticity was simulated from surface to 300 hPa for all MPs coupling with KF and BMJ schemes with the initial condition at 0000 UTC of 11 November. For all MPs coupling with KF and BMJ schemes the vorticity simulated maximum was at 850 hPa level. Two maxima of positive vorticity were simulated at 850 and 450 hPa for KF scheme and 850 and 500 hPa level for BMJ scheme in combination with all MP schemes. The significant differences of vorticity persisted among all the MPs coupling with KF and BMJ schemes. The maximum (minimum) vorticity was simulated by Kessler (WSM3) coupling with KF and BMJ schemes with the initial condition at 0000 UTC of 11 November. It was observed that Lin and WSM6 schemes in combination with CP schemes had simulated almost equal amount of vorticity during the period. KF scheme had simulated higher (lower) vorticity in the lower (upper) troposphere than that of BMJ scheme.

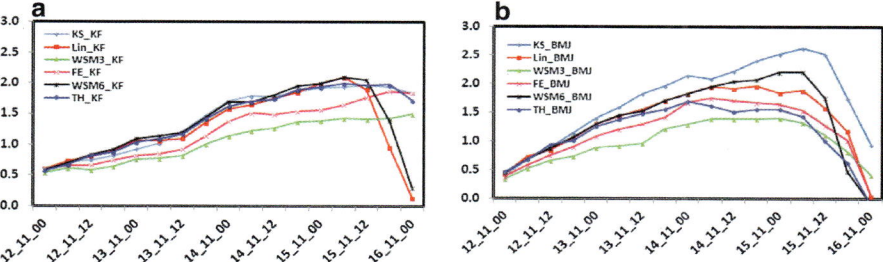

Fig. 8 Vertical profile of space averaged vorticity of TC Sidr simulated by six different MP schemes coupling with (**a**) KF and (**b**) BMJ schemes with the initial condition of 0000 UTC of 11 November 2007

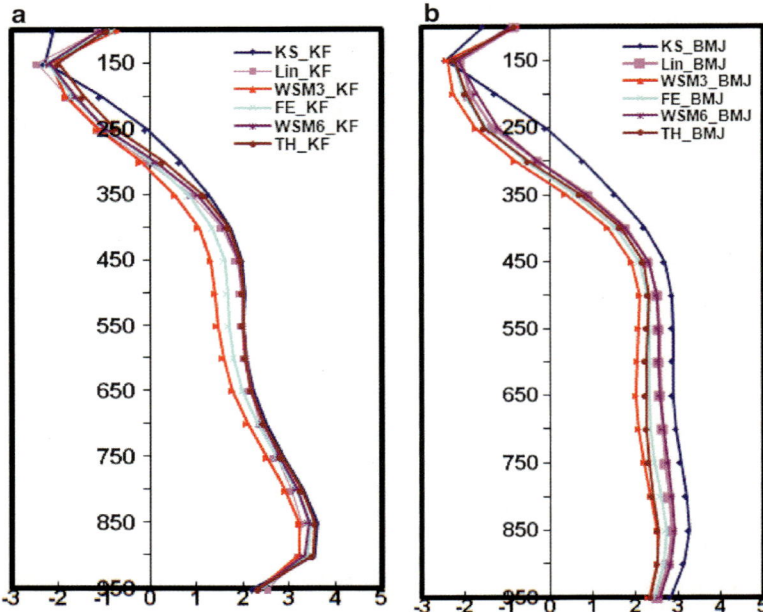

Fig. 9 Vertical profile of time and space averaged vorticity of TC Sidr simulated by six different MP schemes coupling with (**a**) KF and (**b**) BMJ schemes with the initial condition of 0000 UTC of 11 November 2007

Conclusions

The intensity of simulated tropical cyclone Sidr was comparable with observed intensity in all of the WRF runs with the initial conditions at 0000 UTC of 11 and 12 November 2007 regardless of the MP schemes in combination with KF scheme. The simulated intensity was less than that of observed for all MP schemes in combination with BMJ scheme for both initial conditions. The simulated central pressure fall agreed well with observations (944 hPa) by using Lin et al. in combination with KF scheme with the initial conditions of 0000 UTC of 11 and 12 November. The simulated tropical cyclones were strongest prior to landfall and began to weaken after landfall, which was in good agreement with observations. The forecast track error was minimum with respect to landfall position for Lin, Ferrier, WSM6 and Thompson schemes in combination with BMJ scheme with the initial condition of 0000 UTC of 11 November and Lin scheme in combination with BMJ scheme with 0000 UTC of 12 November 2007 initial condition. The forecast landfall time for KS-BMJ (14 UTC15), Lin-KF (15 UTC15), Lin-BMJ (15 UTC15),

(continued)

FE-BMJ (18 UTC15) and WSM6-KF (15 UTC15) for the initial condition of 0000 UTC of 11 November almost matched with the observed landfall time (1600 UTC15). The analogous wind speed and pressure fall for Lin et al. and WSM6 schemes in combination with KF scheme were simulated well with the observed wind speed and pressure fall. Lin-KF and Lin-BMJ combinations provided the most precise intensity and track of TC Sidr respectively.

Acknowledgements Author is grateful to the Director of SMRC, Md. Shah Alam for providing opportunity to take up this programme at SMRC. The author acknowledges Mesoscale and Microscale Meteorology Division of NCAR for providing WRF-ARW modelling system for the present study. The Grid Analysis and Display System software (GrADS) was used for analytical purposes and displaying figures.

References

Alam MM, Hossain MA, Shafee S (2003) Frequency of Bay of Bengal cyclonic storms and depressions crossing different coastal zones. Int J Climatol 23:1119–1125
Davis C, Bosart LF (2002) Numerical simulations of the genesis of Hurricane Diana (1984). Part II: Sensitivity of track and intensity prediction. Mon Weather Rev 130:1100–1124
Dudhia J (1989) Numerical study of convection observed during the winter monsoon experiment using a mesoscale two-dimensional models. J Atmos Sci 46:3077–3107
Fovell RG, Su H (2007) Impact of cloud microphysics on hurricane track forecasts. Geophys Res Lett 34:L24810
Hong SY, Dudhia J, Chen SH (2004) A revised approach to ice microphysical processes for the bulk parameterization of clouds and precipitation. Mon Weather Rev 132:103–120
Hong SY, Noh Y, Dudhia J (2006) A new vertical diffusion package with an explicit treatment of entrainment processes. Mon Weather Rev 134:2318–2341
Janjic ZI (2000) Comments on "Developments and evaluation of a convection scheme for use in climate models". J Atmos Sci 57:3686
Kain JS, Fritsch JM (1990) A one-dimensional entraining/detraining plume model and its application in convective parameterization. J Atmos Sci 47:2684–2702
Kain JS, Fritsch JM (1993) Convective parameterization for mesoscale models: the Kain-Fritsch scheme. The representation of cumulus convection in numerical models. Meteorological monographs, no. 46. American Meteorological Society
Lin YL, Farley RD, Orville HD (1983) Bulk parameterization of the snow field in a cloud model. J Clim Appl Meteorol 22:1065–1092
Lord SJ, Willoughby HE, Piotrowicz JM (1984) Role of a parameterized ice-phase microphysics in an axisymmetric, non-hydrostatic tropical cyclone model. J Atmos Sci 41:2736–2748
Mlawer EJ, Taubman SJ, Brown PD, Lacono MJ, Clough SA (1997) Radiative transfer for inhomogeneous atmosphere: RRTM, a validated correlated-k model for the longwave. J Geophys Res 102(D14):16663–16682
Pattnaik S, Krishnamurti TN (2007) Impact of cloud microphysical processes on hurricane intensity. Part 2: Sensitivity experiments. Meteorol Atmos Phys 97:126–147
Pattnaik S, Inglish C, Krishnamurti TN (2011) Influence of rain-rate initialization, cloud microphysics, and cloud torques on hurricane intensity. Mon Weather Rev 139:627–649

Raju PVS, Potty J, Mohanty UC (2011) Sensitivity of physical parameterizations on prediction of tropical cyclone Nargis over the Bay of Bengal using WRF model. Meteorol Atmos Phys 113:125–137

Wang Y (2001) An explicit simulation of tropical cyclones with a triply nested movable mesh primitive equation model: TCM3. Part 2: Model refinements and sensitivity to cloud microphysics parameterization. Mon Weather Rev 130:3022–3036

Willoughby HE, Jin H-L, Lord SJ, Piotrowicz JM (1984) Hurricane structure and evolution as simulated by an axisymmetric non-hydrostatic model. J Atmos Sci 41:1169–1186

Zhu T, Zhang DL (2006) Numerical simulation of Hurricane Bonnie (1998). Part II: Sensitivity to varying cloud microphysical processes. J Atmos Sci 63:109–126

Eddy Angular Momentum Fluxes in Relation with Intensity Changes of Tropical Cyclones Jal (2010) and Thane (2011) in North Indian Ocean

S. Balachandran and B. Geetha

1 Introduction

Tropical Cyclone (TC) intensity change involves complex interaction of processes involving multiple spatial-temporal scales. Some of the well known processes identified to be associated with TC intensity change are vertical wind shear (VWS), Sea Surface Temperature (SST), upper level divergence, land interactions, eddy angular momentum fluxes etc. and a good review on this is given by Rhome and Raman (2006). Several studies have been undertaken to understand the large/synoptic scale eddy forcings on TC intensification (Molinari and Vollaro 1989; DeMaria et al. 1993; Yu and Kwon 2005).

Riehl (1950) noted that some force is needed to ensure that the upper level outflow does not sink in the immediate storm environment and hence destroy the radial temperature gradient required to intensify the storm. He proposed that the strength and orientation of outflow channels depend upon the presence of, and interactions with, pre-existing upper anticyclones or extratropical waves. Pfeffer (1958) proposed that the calculation of lateral transport of angular momentum by azimuthal eddies provide a quantitative basis for evaluating the role of environmental asymmetries in Atlantic hurricanes. Holland and Merrill (1984) showed that eddy forces in the lower troposphere cannot penetrate to the vortex centre as the inertial stability arising out of the rapid rotation of the balanced vortex would produce strong resistance to horizontal motion. McBride and Zehr (1981), based on a study of composited TC parameters, found that developing storms contain large inward eddy fluxes of cyclonic angular momentum at outer radii, which confine almost entirely to the outflow layer of the composited storms. On the other hand non-developing tropical disturbances contain much weaker and more diffused cyclonic eddy momentum fluxes. Pfeffer and Challa (1981) conducted numerical simulation experiments and showed that developing storms forced by associated eddy momentum fluxes rapidly intensify to hurricane strength while non-developing storms with its weaker eddy fluxes do not

S. Balachandran (✉) • B. Geetha
Regional Meteorological Centre, Chennai, India
e-mail: balaimd@gmail.com

intensity. They discussed the mechanism for tropical cyclone intensification through radial-vertical circulation in response to eddy momentum forcing. Upper level outer Eddy Momentum Flux (EMF)/Eddy Flux Convergence (EFC) at the inner radii can serve as catalyst to organise diabatic sources through secondary radial circulations which excite internal instabilities in tropical cyclones. EFC >10 ms^{-1} day^{-1} has been found to be associated with eddy interaction (DeMaria et al. 1993).

Different types of environmental interactions have been brought out by various case studies. Eddy forcing in the form of westerly trough interaction has been well documented for Atlantic hurricanes (Molinari and Vollaro 1989; Hanley et al. 2001) and Pacific typhoons (Yu and Kwon 2005). Ramage (1974) and Rucker (1992) noted that Pacific typhoon Joan (October 1970) and Flo (September 1990) intensified dramatically under the influence of weakening of the upper anti-cyclone.

The present study is focussed on understanding the role of eddy forcing in the intensity changes of two TCs that formed over the Bay of Bengal during the cyclone season of October-December viz., Jal (04–08 November 2010) and Thane (25–31 Dec 2011). The analysis reveals contrasting evolution characteristics of eddy interactions associated with intensification of these cyclones.

2 Data and Methodology

Figure 1 presents the tracks of the two cyclones considered for the study (source: Cyclone eAtlas – IMD). These two TCs are chosen as sample for systems traversing over the same area (southwest Bay of Bengal) and during the same season

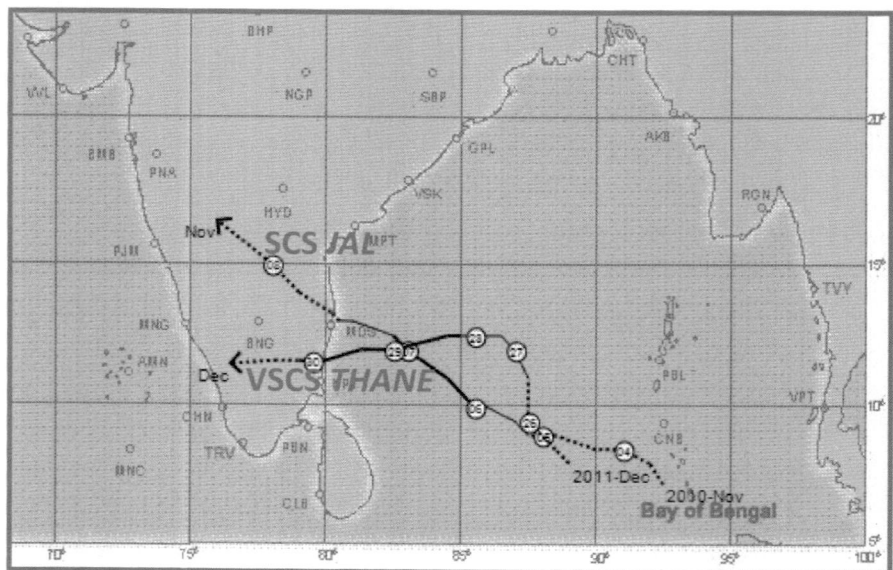

Fig. 1 Tracks of the VSCS THANE and the SCS JAL (Source: Cyclone eAtlas –IMD: www.rmc-chennaieatlas.tn.nic.in)

(post-monsoon) but undergo different types of intensity changes. Whereas Jal attained the intensity of severe cyclonic storm [SCS; maximum sustained wind speed (MSW) of 48–63 knots as per the classification of low pressure systems by the India Meteorological Department (www.imd.gov.in)] and then started weakening over the southwest Bay of Bengal before crossing coast as a deep depression (MSW: 28–33 knots) over Thiruvallur district of north coastal Tamil Nadu, Thane intensified into a very severe cyclonic storm (VSCS, MWS: 64–119 knots) over the same area before land falling over Cuddalore district of north coastal Tamil Nadu as VSCS.

It is rather interesting to note the climatological paradox of SCS Jal, which formed during the main cyclone month of November (1st week), weakening into a deep depression before making landfall, but, VSCS Thane, which formed during the fag end of the season, (last week of December) intensifying into a very severe cyclonic storm.

In the present study, the environmental influences on the intensification of VSCS Thane and the weakening of the SCS Jal are analyzed by following the methodology similar to that of the earlier workers on Atlantic hurricanes and Pacific typhoons (Molinari and Vollaro 1989). All computations are based on azimuthal averaging as this procedure removes the symmetric component arising due to the vortex and retains only the asymmetric component due to the environmental flow.

The EMF at the 200 hPa level is computed for each 1.0° radial distance from the TC centre upto 10° (about 1,100 km from the centre) at 6-h intervals for the life period of the TC when it was over the ocean according to the equation (Molinari and Vollaro 1989).

$$\frac{\partial M}{\partial t} = \frac{2\pi r^2}{g} \int_{p_2}^{p_1} \overline{u'v'} \, dp \tag{1}$$

The EFC at the inner radii is calculated as follows:

$$\left[\frac{\partial \overline{v}}{\partial t}\right]_{eddy\,flux} = \frac{1}{r^2}\frac{\partial}{\partial r} r^2 \overline{u'v'} \tag{2}$$

where u and v refer to storm relative radial and tangential velocities and prime refers to the deviations from the azimuthal mean. r is the radial distance from the TC centre and dp is the thickness of the 200 hPa level (taken to be of 200 hPa thick).

The basic data used for analysis is the NCEP's $1° \times 1°$ FNL data at 6-hourly intervals. From the conventional zonal and meridional wind data, radial (u) and tangential (v) velocities are computed for the entire life period of both the storms at 6-hourly intervals in a storm relative Lagrangian frame of reference. For this purpose, the storm movement, determined from the IMD's best track data at 6-hourly intervals, is subtracted from the actual radial and tangential velocities at each grid point.

For subsequent analysis, the vertical wind shear (VWS) over the TC region is also computed for various instances by averaging the 200–850 hPa winds over a 5° radial distance from the TC centre. The SST is analysed using INCOIS data (www.incois.gov.in).

3 Results and Discussions

3.1 VSCS Thane (25–31 Dec 2011)

Synoptic history of VSCS Thane is as follows: A trough of low at mean sea level organised into a low pressure area over the southeast Bay of Bengal and neighbourhood on 24 Dec 2011 and concentrated into a depression on 25/1200 UTC centered near 8.5°N/88.5°E. It moved northwestwards and intensified into a deep depression centered near 9.5°N/87.5°E on 26/0000 UTC. Moving northwards, it intensified into cyclonic storm (Thane) centered near 11.0°N/87.5°E on 26/1800 UTC. Moving westward, it intensified into a severe cyclonic storm on 28/0900 UTC near 12.5°N/85.0°E and further intensified into a very severe cyclonic storm on 28/1200 UTC centered near 12.5°N/84.5°E. It further moved west southwestwards and crossed north Tamil Nadu coast close to and to the south of Cuddalore between 0100 and 0200 UTC of 30 December. It continued to move westwards and weakened into a severe cyclonic storm on 30/0300 UTC and then into a depression on 30/1200 UTC over north Tamil Nadu close to Salem. Moving westwards, it further weakened into a well marked low pressure area over north Kerala and neighbourhood on 31 December at 0000 UTC.

Figure 2a–c presents the azimuthally averaged radial velocity (in m s^{-1}), EMF (in units of $\times 10^{17}$ kg-m^{2}s^{-2}) and EFC (in units of ms^{-1}day^{-1}). The x-axis of the plots depict the radial distance from the TC centre in degrees (1° ≈ 110 km) and the y-axis corresponds to the date and time (UTC). From Fig. 2a, it may be noted that the positive radial velocity (outflow) maximum at the outer radii (600–800 km) from the TC centre gradually shifts inwards with time from 28/0600 UTC onwards. From Fig. 2b, it is observed that the EMF at 600–800 km from the TC centre changes from negative to positive on 28/0000 UTC and shifts inwards with time. From Fig. 2c it may be noted that the EFC at the inner radii (300–500 km) also turns positive and crosses the threshold of 10 ms^{-1}day^{-1} for eddy interaction (DeMaria et al. 1993) at about 28/0600 UTC. Within 300–400 km from the TC centre, the EFC varied from −40 ms^{-1}day^{-1} to +40 ms^{-1}day^{-1} during intensification phase from depression stage to severe storm stage. The outflow maximum near the centre coincided with the EMF maximum at the same time. Thus, the inward shift of radial outflow maximum is concomitant with the inward shift of EMF at 600–800 km and the EFC at the inner radii (300–500 km).

To further substantiate the results obtained, EFC (averaged over 300–600 km radial distance), maximum sustained wind speed around the TC and the central pressure fall during the last 6 h are plotted at 6-hourly intervals and presented in Fig. 3a–c. The dotted lines on the plots correspond to the time when EFC reached the threshold value of 10 ms^{-1}day^{-1} required for eddy interaction. It may be noted that the EFC at the inner radii (300–600 km) turned positive at 28/0000 UTC and crossed the threshold of 10 ms^{-1}day^{-1} at 28/0600 UTC. Around the same time, between 28/0000 and 28/0600 UTC the central pressure fall sharply increased from 0 to 4 hPa and between 28/0600 and 28/1200 UTC, there was a 6 hPa pressure fall

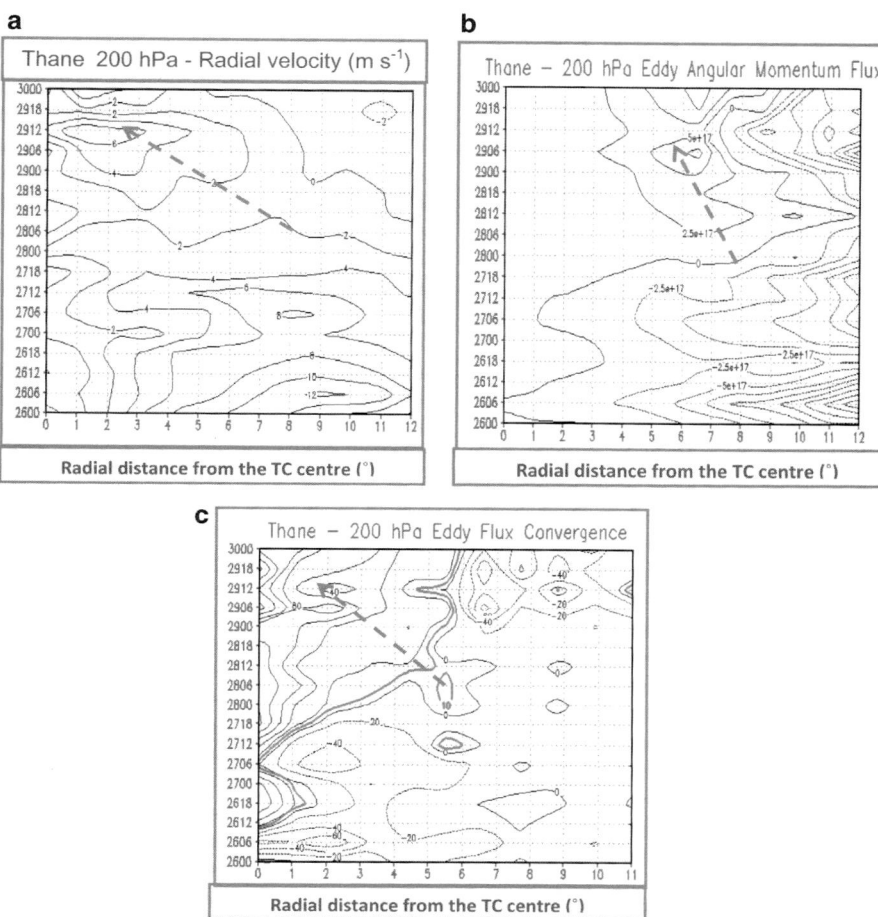

Fig. 2 (**a–c**): Azimuthally averaged (**a**) radial velocity (in m s^{-1}), (**b**) Eddy momentum flux (in units of ×1,017 kg-m^{2}s^{-2}) and (**c**) Eddy flux convergence (in units of ms^{-1}day^{-1}) during the life period of VSCS THANE. X-axis depicts the radial distance from the TC centre in degrees (1° ≈ 110 km) and the Y-axis corresponds to the date and time (UTC) at 6-hourly intervals

at the centre. The MSW wind speeds drastically increased from 45 to 65 kts during the period 28/0600 to 28/1200 UTC when the system intensified from the stage of a cyclonic storm (CS; MSW: 34–47 kts) to a VSCS. Thus, it is evident that when the EFC at 300–600 km radial distance from the TC centre increased to 10 ms^{-1} day^{-1}, within about 6 h, there was an enhanced intensification of the TC.

Next, to understand the large/synoptic scale environmental interaction feature responsible for the eddy forcing, the streamline flow patterns during 26/0000 UTC and 28/1200 UTC are presented in Fig. 4a, b. It may be noted that during the initial stages of the growth of the TC (26/0000 UTC), the TC is located to the south of the sub-tropical ridge (STR) at 200 hPa level and the outflow over the TC

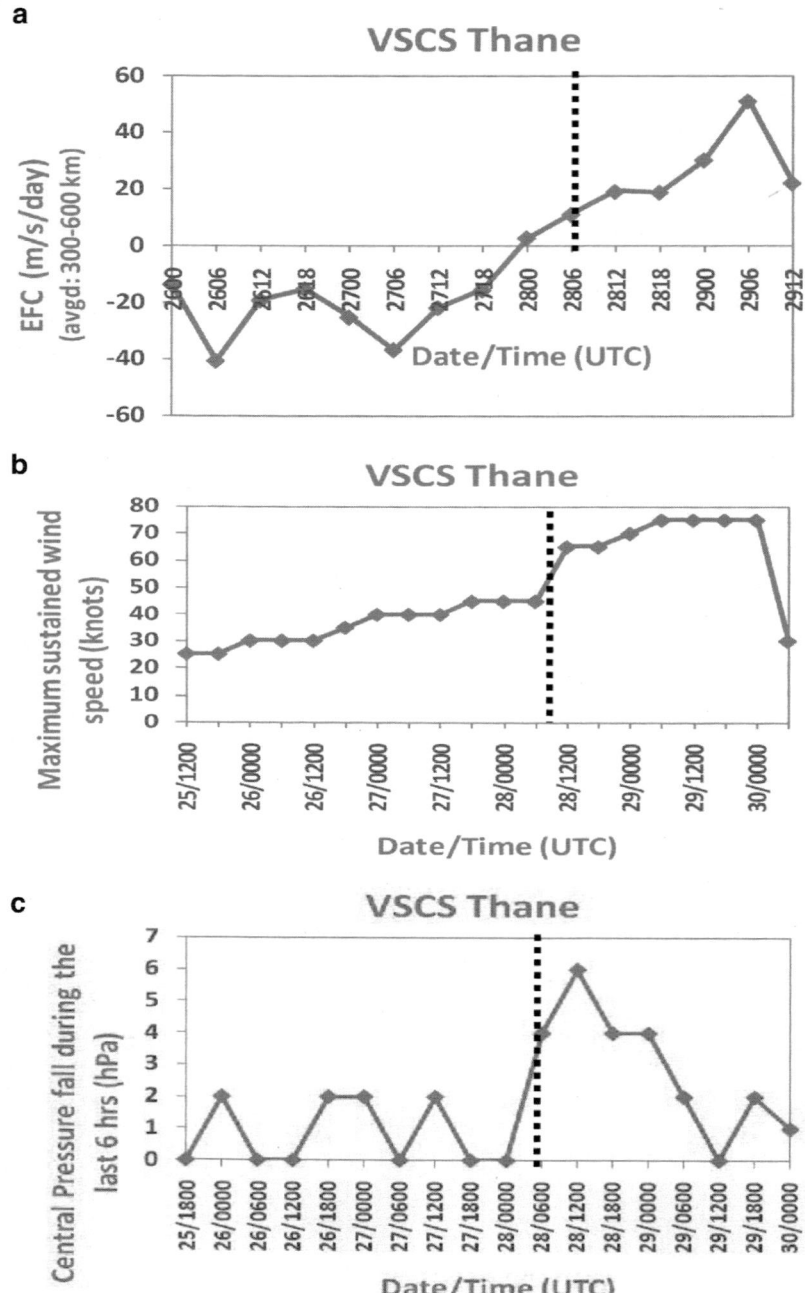

Fig. 3 (**a–c**): Plots of (**a**) Eddy flux convergence (averaged over 300–600 km radial distance), (**b**) Maximum sustained wind speed around the TC and (**c**) the central pressure fall during the last 6 h at 6-hourly intervals. The *dotted lines* correspond to the time when EFC reached the value of 10 ms^{-1}day^{-1}

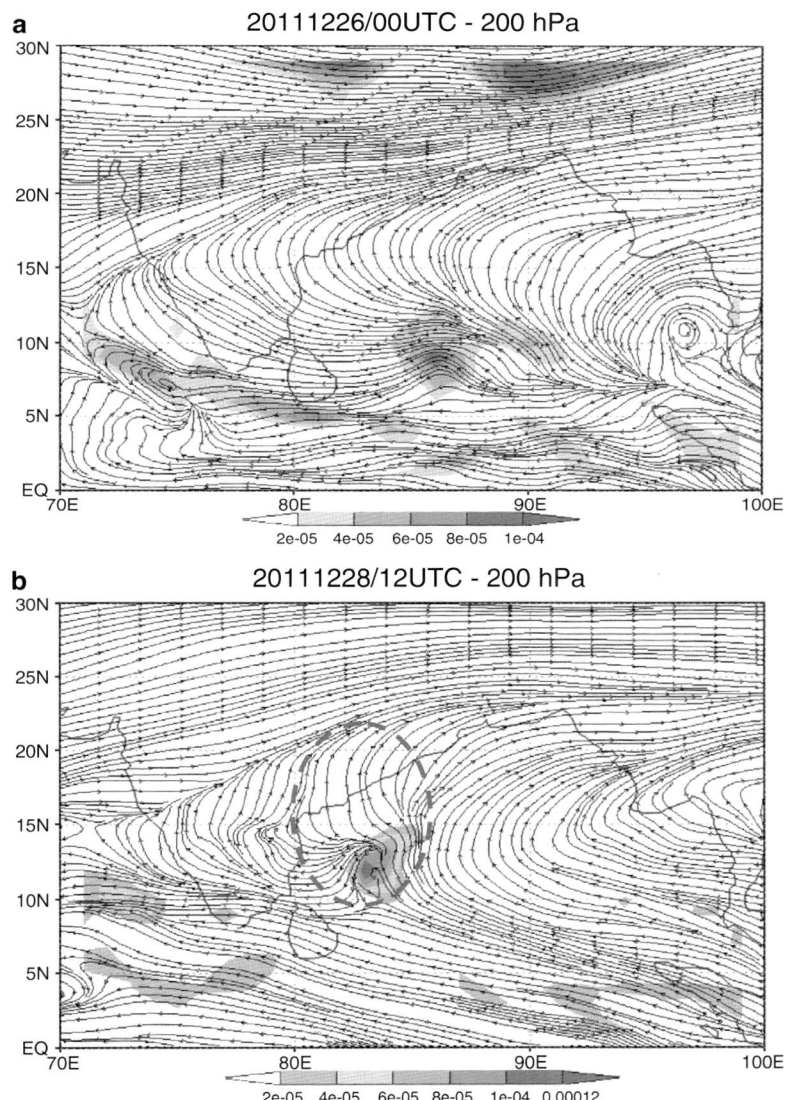

Fig. 4 (a, b): Streamline flow patterns at 200 hPa level during (**a**) 26th/0000 UTC and (**b**) 28th/1200 UTC of Dec 2012. Positive values of vorticity (cyclonic vorticity) (in units of s^{-1}) *shaded*

centre is westwards. However, at 28/1200 UTC, when the TC undergoes an enhanced intensification, the TC is co-located with the STR and to the west of the anticyclone. The anticyclone over the TC centre has weakened and cyclonic circulation exists over the TC centre. This synoptic situation provides an enhanced poleward as well as equatorward outflow channel above the TC centre in the upper troposphere. The enhanced outflow over the TC centre causes deepening of the central pressure and hence intensification of the TC (Rucker 1992).

Whereas earlier works for Atlantic hurricanes (e.g. Molinari and Vollaro 1989) depicted a time lag of the order of 27–33 h between EFC crossing the threshold value and the intensification of the TC, in the present case such large time lag between the commencement of eddy interaction and TC intensification is not noticed. This aspect needs to be analyzed with more cases.

3.2 SCS Jal (04–08 Nov 2010)

The synoptic history of the SCS Jal is as follows: A low pressure area formed over the Andaman Sea and neighbourhood on 2nd November 2010 after emerging from the east. It concentrated into a depression over the southeast Bay of Bengal near 8.0°N/92.0°E on 04th/0000 UTC and further into a deep depression on 05/0000 UTC near 9.0°N/88.5°E. It intensified into a cyclonic storm (Jal) on 05/0600 UTC near 9.0°N/87.5°E and further into SCS on 05/2100 UTC near 10.0°N/86.0°E. Moving northwestwards over the southwest Bay of Bengal, it weakened into a CS on 07th/0600 UTC and further into a deep depression on 07/1200 UTC before crossing north Tamil Nadu-south Andhra Pradesh coast, close to north of Chennai on 17th/1600 UTC. Moving west-north-westwards, it weakened into a depression on 08/0000 UTC over Rayalaseema.

Figure 5a–c depict the azimuthally averaged radial velocity, EMF and EFC at 6-hourly intervals for the SCS Jal. It may be noted, that in contrast to VSCS Thane, wherein, the outflow maximum shifted inwards, in the case of SCS Jal, the outflow maximum shifts outwards from the TC centre. The EMF and the EFC are negative throughout up to 06/1800 UTC. Within 300–400 km from the TC centre, the EFC values are mainly negative and varied from −40 to −10 ms^{-1}day^{-1} during intensification phase from depression to storm stage. Thus, the response to the eddy forcing in the case of SCS Jal is detrimental to intensification of the TC. However, it should be noted that the EMF and the EFC at 500–700 km from the TC centre turned positive on 07/0000 UTC, but could not contribute to sustain the intensity of the TC which started weakening by 07/0600 UTC.

The synoptic situations associated with negative eddy interaction are presented in Fig. 6a–c. It may be noted that, unlike the case of VSCS Thane, wherein, the upper level anticyclone weakened over the TC centre and paved way for enhanced poleward outflow, in the case of SCS Jal, the upper level anticyclone continued to remain dominant and the TC was located to the south of the STR throughout with no enhanced poleward outflow channel.

To understand the roles of other environmental features associated with the initial intensification and final weakening of SCS Jal, time series of vertical wind shear (200–850 hPa wind averaged over a 5° radial distance from the TC centre), EFC averaged over 500–700 km radial distance from the TC centre and spatial plots of daily SST during 04–07 Nov 2010 are presented in Fig. 7. It may be noted that SCS Jal was in an environment of strong vertical wind shear throughout (>20 ms^{-1}) which also was not favourable for intensification. However, on 5 Nov, when the intensification of the TC from deep depression to CS (at 0600 UTC) and

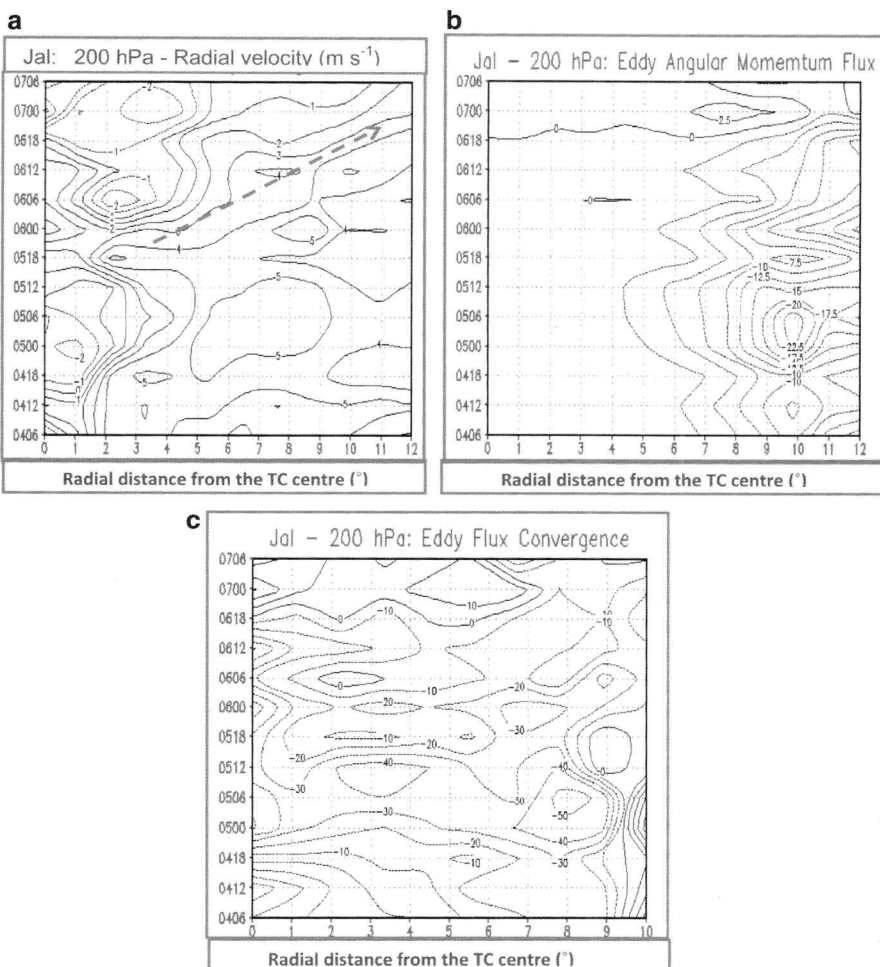

Fig. 5 (a–c): Same as Fig. 2(a–c), but for SCS JAL

then to SCS (at 2100 UTC) occurred, EFC was strongly negative (−40 ms^{-1}day^{-1}), but, the VWS slightly decreased from 25 to 20 ms^{-1}, and, the TC was moving towards warmer waters over the sea. Hence, the intensification of the TC up to the stage of SCS would have been due to internal dynamics as well as partly due to warmer SST. Subsequently, on 6 Nov, colder waters surfaced over the region of the TC and VWS increased to 30 ms^{-1} at 6/1800 UTC. Thus, on 6 Nov, colder SST, along with strong vertical wind shear and negative EFC was associated with subsequent weakening of the TC on 07 Nov at 0000 UTC. On 7 Nov when the TC was close to land and positive EFC upto 10 ms^{-1}day^{-1} could not help to sustain the intensity of the TC, Jal weakened into a deep depression before making landfall over north of Chennai.

Fig. 6 (**a–c**): Streamline flow patterns at 200 hPa level during (**a**) 05th/0000 UTC, (**b**) 06th/0000 UTC and (**c**) 06th/1200 UTC of Nov 2010. Cyclonic vorticity (in units of s^{-1}) *shaded*

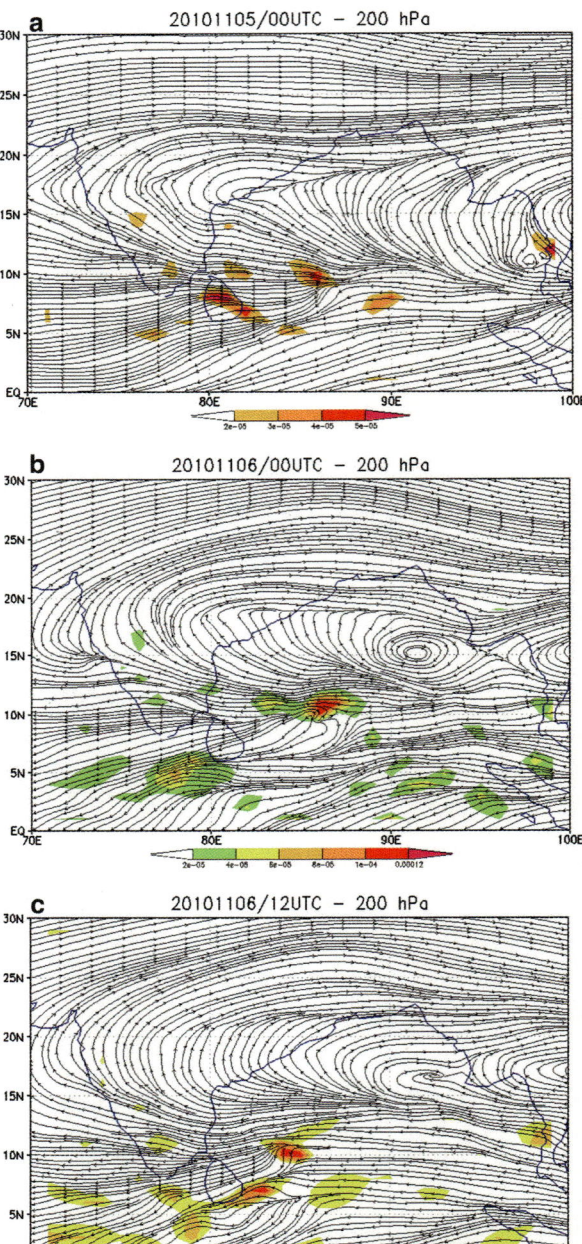

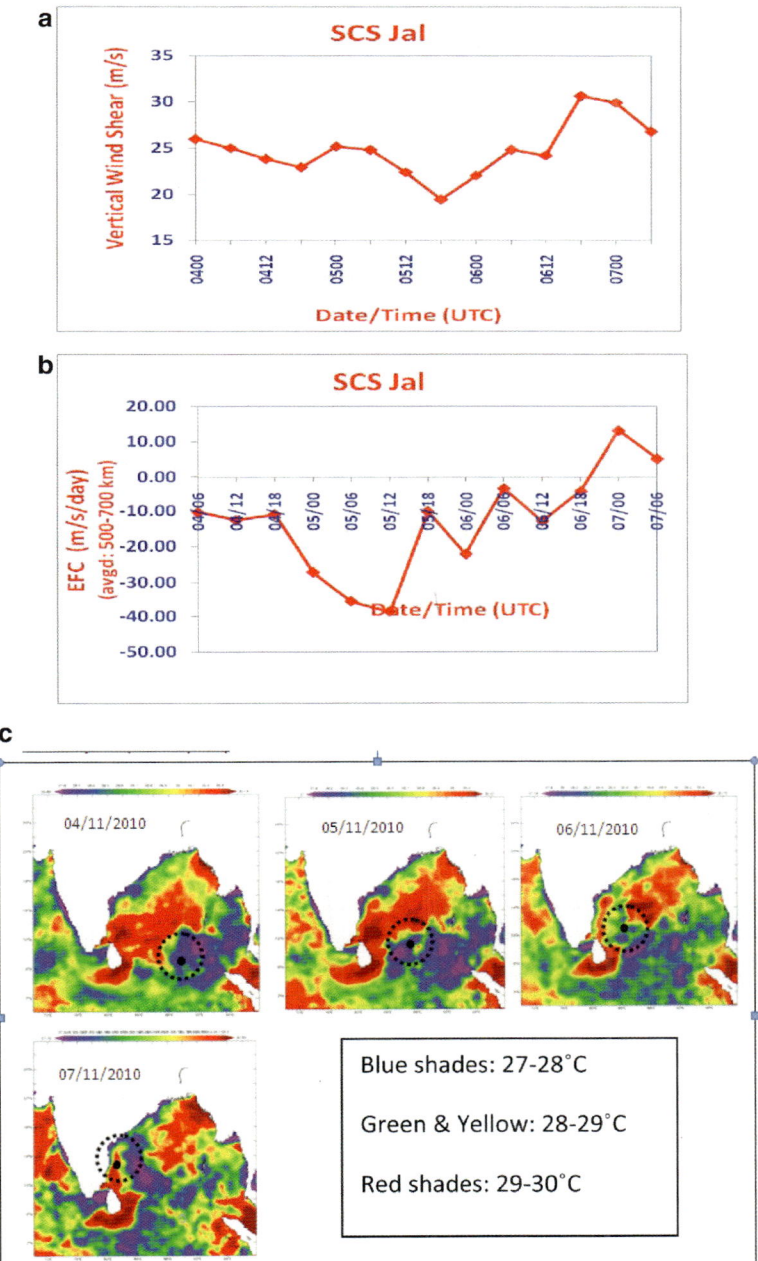

Fig. 7 (**a–c**): Time series of (**a**) vertical wind shear and (**b**) EFC averaged over 500–700 km from the TC centre during 4–7 Nov 2010 (**c**) Plots of daily SST during 4–7 Nov 2010. Location of the TC centre is marked as a *black dot* inside a *black dotted circle* (Source: www.incois.gov.in)

The above discussions have brought out the contrasting roles of eddy forcing on the intensity changes of two TCs of NIO. Whereas in the first case, EFC was associated with enhanced intensification of VSCS Thane, in the second case, EFC has not played any role in the intensity change of SCS Jal.

4 Summary

Based on the study of eddy angular momentum flux and eddy angular momentum flux convergence during the life cycles of SCS Jal and VSCS Thane the following were noted:

(i) Environmental interaction through transport of angular momentum flux has been observed to either help or not in the intensity changes of the two TCs over NIO as noted in earlier studies for TCs of Atlantic and Pacific. Whereas in the case of VSCS Thane, eddy forcing was associated with enhanced intensification, in the case of SCS Jal, eddy forcing has not played any role in the intensity change.
(ii) In the case of VSCS Thane, the eddy interaction is manifested in the intensity change with inward shift of radial outflow maxima and concomitant shift of positive eddy momentum flux towards the inner radii as well as eddy flux convergence at the inner radii.
(iii) In the case of SCS Jal, eddy momentum forcing were mainly negative and radial outflow maxima shifted towards outer radii. Moreover, the system was under strong vertical shear environment and cooler waters were surfacing over the region of the TC.
(iv) Within 300–400 km from the TC centre, in case of VSCS Thane, the EFC varied from -40 ms^{-1}day^{-1} to $+40$ ms^{-1} day^{-1} while in the case of SCS Jal the EFC values are mainly negative and varied from -40 to -10 ms^{-1}day^{-1}.

While the present analysis is a case study for two cyclones, similar computations for more TC cases would help in understanding the general characteristics and influence of eddy fluxes in the intensification processes of TCs of NIO. Also eddy heat flux computations along with other factors like inner core dynamics would add more insight into the dynamical and thermo dynamical processes associated with time evolution of the system.

Acknowledgements The authors thank the Deputy Director General of Meteorology, Regional Meteorological Centre, Chennai for providing facilities to undertake this study.

References

DeMaria M, Baik JJ, Kaplan J (1993) Upper-level eddy angular momentum fluxes and tropical cyclone intensity change. J Atmos Sci 50(8):1133–1147
Hanley D, Molinari J, Keyser D (2001) A composite study of the interactions between tropical cyclones and upper-tropospheric troughs. Mon Weather Rev 129:2570–2584

Holland GJ, Merrill RT (1984) On the dynamics of tropical cyclone structure changes. Q J R Meteorol Soc 110:723–745

McBride JL, Zehr R (1981) Observational analysis of tropical cyclone formation. Part II: Comparison of non-developing versus developing systems. J Atmos Sci 38:1132–1151

Molinari J, Vollaro D (1989) External influences on hurricane intensity, Part I: Outflow layer eddy angular momentum fluxes. J Atmos Sci 46(8):1093–1105

Pfeffer RL (1958) Concerning the mechanism of hurricanes. J Meteorol 15:113–120

Pfeffer RL, Challa M (1981) A numerical study of the role of eddy fluxes of momentum in the development of Atlantic hurricanes. J Atmos Sci 38:2393–2398

Ramage CS (1974) The typhoons of October 1970 in the South China sea: intensification, decay and ocean interaction. J Appl Meteorol 13(7):739–751

Rhome JR, Raman S (2006) Environmental influences on tropical cyclone structure and intensity: a review of past and present literature. Indian J Mar Sci 35(2):61–74

Riehl H (1950) A model of hurricane formation. J Appl Meteorol 9:917–925

Rucker JH (1992) Upper-tropospheric forcing on the intensification rates of tropical cyclones Flo and Ed based on TCM-90 observations. Masters' thesis. Naval Postgraduate School, Monterey, CA

Yu H, Kwon HJ (2005) Effect of TC-trough interaction on the intensity change of two typhoons. Weather Forecast 20:199–211

Impact of Initial and Boundary Conditions on Mesoscale Simulation of Bay of Bengal Cyclones Using WRF-ARW Model

K.S. Singh and M. Mandal

1 Introduction

Land falling tropical cyclones (TCs) that form over the BOB cause disaster along the east coast of India, Bangladesh and Myanmar. The destruction is mainly due to the strong wind, heavy rainfall and associated storm surges. It causes huge damage to property. So, it is very important to predict the TCs as accurately as possible to reduce the loss of life and damage to the property. The track and intensity predictions of the tropical cyclone remain a challenging task for atmospheric scientists and operational forecasters. Several studies suggested that incomplete representation of physical processes, insufficient model spatial resolution (Demaria and Kaplan 1999; Zhang et al. 2011) and inaccurate IBCs are possible reasons for poor forecast of the tropical cyclones (Cacciamani et al. 2000).

Several numerical studies were conducted over the BOB using mesoscale models to evaluate the performance of the model with respect to physical parameterization, resolution, initial conditions and impact of data assimilation, etc. (e.g. Mandal et al. 2004; Bhaskar Rao and Prasad 2006; Srinivas et al. 2007, 2012; Deshpande et al. 2010; Osuri et al. 2011, 2012; Pattanayak and Mohanty 2008; Pattanayak et al. 2012; Mohanty et al. 2010; Singh and Mandal 2014). Pielke et al. (2006) suggested that even a small error in initial condition may contribute large error in subsequent forecast. The accuracy of boundary conditions also play important role because the atmospheric waves and disturbances generated at the boundary can rapidly propagate throughout the domain. Majewski (1997) documented that the forecast skill of the model is influenced by the large-scale flow advected from the boundary. So, it is important to study the impact of initial and lateral boundary condition.

K.S. Singh • M. Mandal (✉)
Centre for Oceans, Rivers, Atmosphere and Land Sciences,
Indian Institute of Technology Kharagpur, Kharagpur 721 302, India
e-mail: mmandal@coral.iitkgp.ernet.in

In this study, ARW-WRF model is used to examine the impact of IBCs derived from the NCEP FNL and high resolution GFS datasets towards mesoscale simulation of BOB cyclones. The model description and configuration of WRF model used in this study is described in Sect. 2. The numerical experiments and data used are discussed in Sect. 3. The results obtained from the model simulations and related discussions is provided in Sect. 4 followed by conclusion in Sect. 5.

2 Model Description and Configuration

The advanced research core of WRF model is developed at the National Center for Atmospheric Research (NCAR) in collaboration with a number of agencies viz., the National Oceanic and Atmospheric Administration (NOAA), the National Center for Environmental Prediction (NCEP) and various universities. It is based on an Eulerian solver for the fully compressible nonhydrostatic equations with complete Coriolis and curvature terms. The solver uses the 2nd and 3rd order Runge-Kutta time integration scheme, 2nd to 6th order advection in both horizontal and vertical directions and time-splitting technique for using smaller time steps for acoustic and gravity-wave modes. The terrain-following hybrid sigma-pressure is used as vertical coordinate. The grid staggering is the Arakawa C-grid. A detailed description of the model is available in Skamarock et al. (2008).

The double nested two-way interactive ARW-WRF model with 27 km and 9 km horizontal resolutions is used in the study. There are 35 vertical sigma levels with higher resolution in the boundary layer with the model top at 10 hPa. In this study the physical parameterization schemes used in the model are New Kain-Fritsch (Kain 2004) convection, Yonsei University (YSU) PBL (Hong et al. 2006); Lin microphysics (Lin et al. 1983); the Rapid Radiative Transfer Model (RRTM) for longwave radiation (Mlawer et al. 1997) and Dudhia's scheme (Dudhia 1989) for shortwave radiation.

3 Numerical Experiments and Data Used

In this study, seven land-falling BOB cyclones are simulated to study the impact of IBCs. These include three severe cyclonic storms (if MSW is between 48 and 63 kt) and four very severe cyclonic storms (if MSW is between 64 and 119 kt) that crossed Bangladesh, Myanmar and different parts of east coast of India. The details of synoptic situations and best-fit track data of these cyclones are obtained from IMD Regional Specialized Metrological Centre (RSMC) reports from 2006, 2007, 2008, 2009, 2010 and 2011. The best-fit track of these storms with intensity as obtained from IMD is provided in Fig. 1. The time of initialization and length of model integration in each of these seven cyclone cases is provided in Table 1. Two numerical simulations

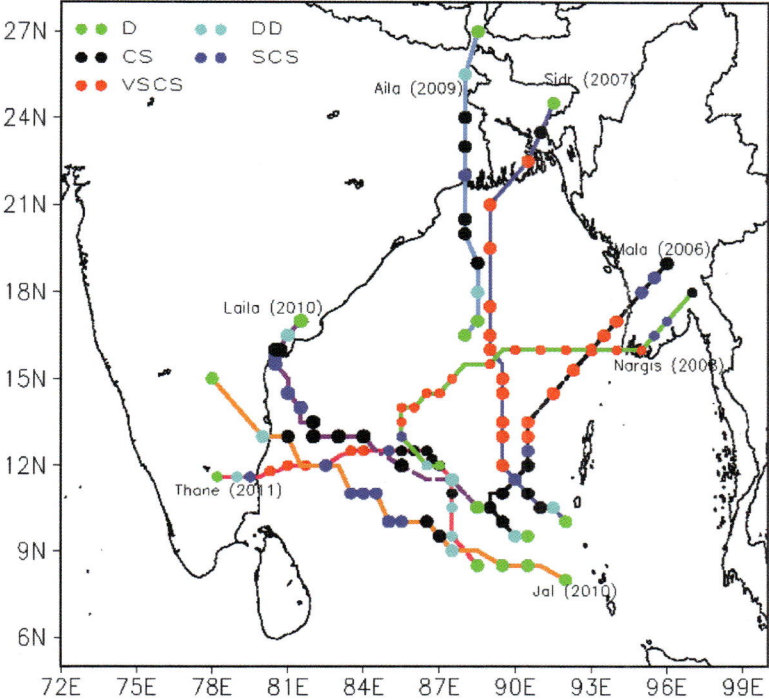

Fig. 1 IMD best-fit tracks of BOB cyclones used in the present study, *solid line colour* represents name of the cyclone and colours of the legend represent intensity of cyclones

Table 1 BOB cyclones considered in the study with time of model initialization and length of simulation

S. No	Cyclone names	Initialization time	Simulation length (h)
1	Thane (2011)	26 Dec. 1200 UTC	96
2	Laila (2010)	18 May 0000 UTC	72
3	Jal (2010)	05 Nov. 0000 UTC	72
4	Aila (2009)	24 May 0000 UTC	51
5	Nargis (2008)	29 April 0000 UTC	96
6	Sidr (2007)	12 Nov. 0000 UTC	96
7	Mala (2006)	25 April 1200 UTC	96

are conducted for each of the cyclones. In the first simulation (FNL simulation), the initial and boundary conditions for the model is provided every 6 h from FNL analysis. In the second simulation (GFS simulation), the initial condition is taken from GFS analysis and the boundary conditions from GFS forecast every 3 h. The model configuration is kept unchanged in all aspects in both the simulations.

The topography for the model domains are derived from United States Geographical Survey (USGS) global topography datasets at 10-min and 5-min resolutions respectively. The best-fit track data obtained from IMD is used for comparison of the model simulated track, intensity and landfall time & position of the cyclone.

4 Results and Discussions

The impact of IBCs on mesoscale simulation of BOB cyclones are mainly focused on track including landfall time & location and intensity of the storm. The model simulated tracks are compared with IMD best-fit track datasets. The vector displacement errors (VDEs) in model simulated track, error in landfall time and location are computed and discussed. The mean VDE of seven cyclone cases are also presented and discussed. The model simulated intensity in terms of CSLP and MSW and its time evolution is compared with IMD best-fit track. Mean absolute error in intensity prediction in terms of CSLP and MSW are also presented and discussed.

4.1 Track

Figure 2 shows the tracks of seven cyclone cases obtained from model simulations (GFS and FNL simulations) along with IMD best fit track. Figure 3 represents the VDEs in individual cyclones and mean of VDEs of seven cyclone cases. In general, the simulated tracks of all the cyclones are in good agreement with observed track throughout the simulation period in both simulations. Figure 3a shows that the track of the cyclone Mala is slightly better simulated with FNL IBCs. In both simulations the track of the storm is to the left of the best-fit track and movement is faster. This led to 19 h and 21 h advance landfall of the storm in GFS and FNL simulations respectively. The landfall occurred 21 h before the actual landfall at a location 78 km away to the left of the actual landfall point in FNL simulation.

In case of cyclone Sidr, the track of the storm is well simulated by the model in both simulations throughout the simulation period but movement of the storm was slower than observation. The location and time of landfall of the storm is better simulated by model with GFS IBCs. The landfall was delayed by 6 h and 7 h with location error of 12 km and 35 km respectively in GFS and FNL simulations.

In case of cyclone Nargis, the model simulated tracks of the storm are evenly distributed to the left and right of the IMD best fit track. The track of the storm is better simulated by the model in GFS simulation. The landfall time and location of the storm is also better simulated with GFS IBCs. The storm crossed coastline 14 and 16 h before the actual landfall time at a location 189 and 234 km away to the left of the actual landfall point in GFS and FNL simulations respectively. In case of cyclone Thane the movement of the storm is well simulated by the model with GFS IBCs. In first 36 h, the track of the storm is better simulated with FNL IBCs but then

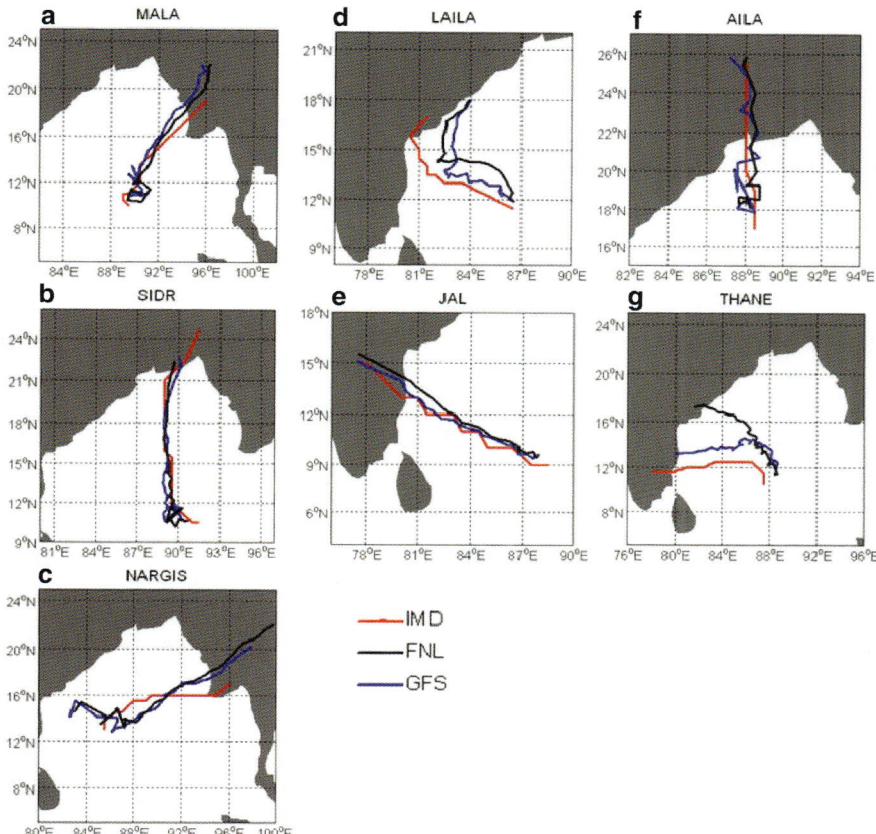

Fig. 2 Model simulated tracks and IMD best-fit track (**a**) Mala, (**b**) Sidr, (**c**) Nargis, (**d**) Laila, (**e**) Jal, (**f**) Aila and (**g**) Thane

the curving of the storm to the left is not captured in FNL simulation. The curving of the storm and its movement is better simulated by the model with GFS IBCs. This may be due to updates of boundary condition more frequently. The storm hit the coastline 6 and 4 after at 162 and 691 km away to the right of the actual landfall point in GFS and FNL simulations respectively.

In case of cyclone Aila, the model simulated tracks closely followed the observed track and are always to its right. The track of the storm is better simulated by the model throughout the simulation period with FNL IBCs. It may be due to less initial positional error. It is observed that the landfall time and location is better simulated by the model in GFS simulation and the storm crossed coastline 1 h before at a location 40 km away to the right of the actual landfall point. In case of cyclone Laila, the model could able to capture the movement of the cyclone in both simulations but did not make landfall. In this case, the track of the storm is better predicted in

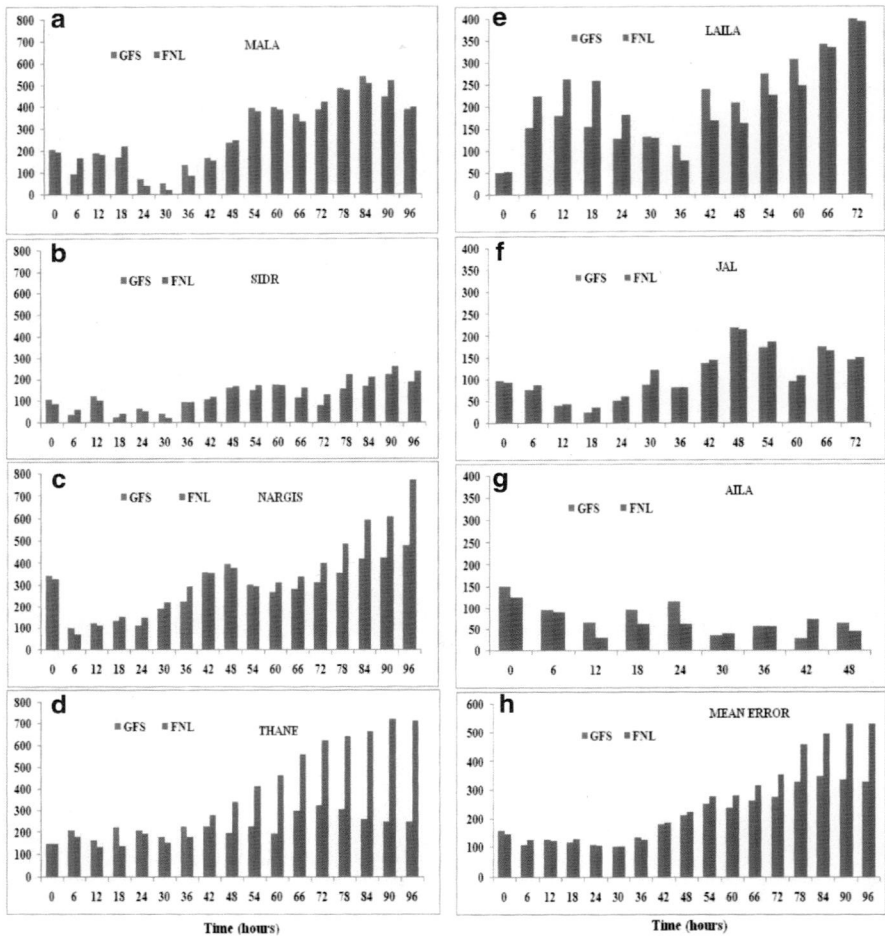

Fig. 3 Variation of VDEs (in km) with time for (**a**) Mala, (**b**) Sidr, (**c**) Nargis, (**d**) Thane, (**e**) Laila, (**f**) Jal, (**g**) Aila and (**h**) mean of VDEs

GFS simulation in first 36 h and after that it is better simulated with FNL IBCs. The track of cyclone Jal is better simulated by the model in GFS simulation throughout the simulation period. The storm crossed the coastline 4 h and 1 h before at a location 11 km and 94 km away to the right of the actual landfall point in GFS and FNL simulations respectively.

The mean VDEs of track of seven cyclone cases at every 6-h intervals are presented in Fig. 3h. This clearly indicates that the average error in track forecast is less in GFS simulation. It is also found that there is not much difference in mean VDEs in first 48 h simulation but after that track is significantly better predicted by

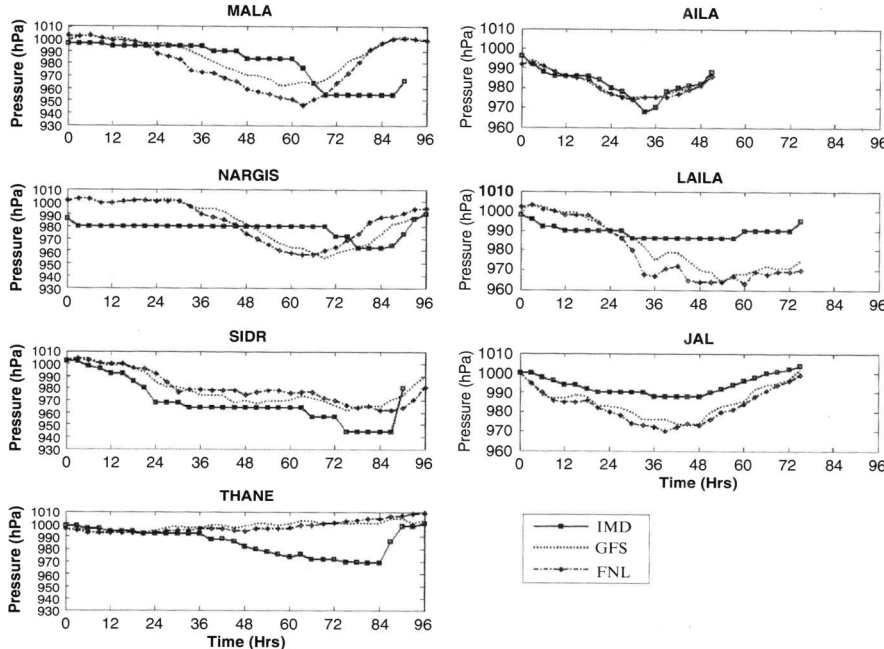

Fig. 4 Variation of central SLP (hPa) with time from model simulations and IMD best-fit track

the model with GFS IBCs. The mean VDE at 24, 48, 72 and 96 h model simulations are 106, 212, 276 and 326 km respectively with GFS IBCs. Whereas in FNL simulation these errors are 106, 223, 354 and 531 km respectively and the mean error in landfall locations are around 90 km and 198 km respectively in GFS and FNL simulations.

4.2 Intensity

The time evolution of intensity (simulated and observed) in terms of CSLP and MSW for the seven simulated BOB cyclones are shown in Figs. 4 and 5 respectively. In case of cyclone Mala the intensity in terms of CSLP and MSW is better simulated by the model in GFS simulation up to 66 h simulation. In both simulations, the model could not capture the dissipation of the storm. The peak intensity is under-predicted and intensification is advanced compared to the observation.

In case of cyclone Sidr, the intensity of the storm is reasonably better simulated by model in GFS simulation. But the rate of intensification was slow, particularly, in first 2 days. The model could not simulate sharp dissipation of the storm in both simulations. The intensity of storm is under-predicted by the model throughout

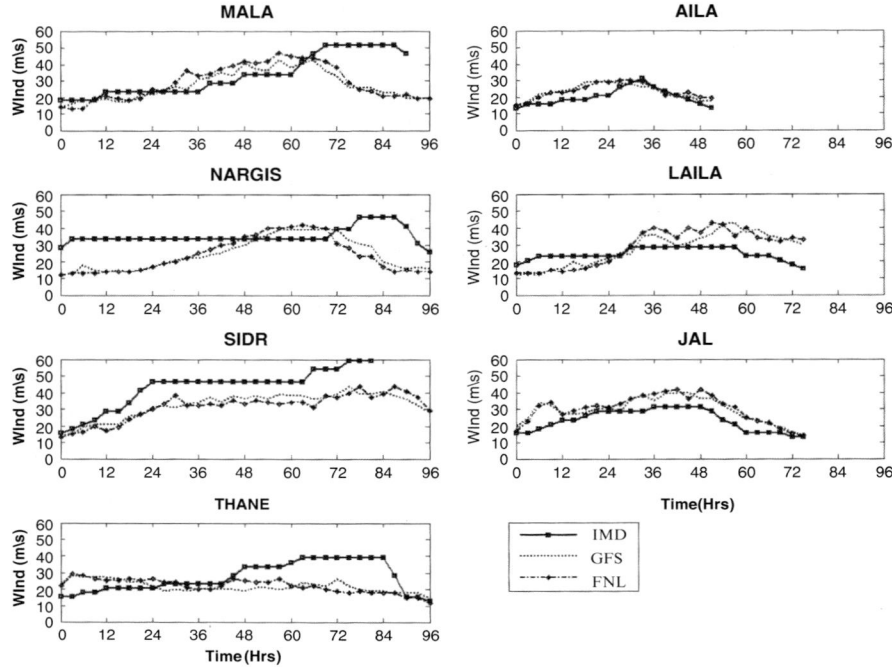

Fig. 5 Variation of maximum surface wind (m/s) with time from model simulations and IMD best-fit track

the simulation period. Overall intensity of the storm is better predicted in GFS simulation. In case of cyclone Nargis, the peak intensity of the storm in terms of CSLP is reasonably well simulated by the model in GFS simulation but the peak intensity is attended 9 h earlier than that in the observation. In model simulation, the storm reached peak intense stage at 2100 UTC on 01 May with CSLP 958 hPa whereas in the observation the peak intensity is recorded at 0600 UTC on 02 May 2008 with CSLP 962 hPa. In GFS simulation, the whole process of intensification and dissipation of the storm appeared to be advanced. The model simulated MSW shows similar features. The intensification and dissipation in both simulation show similar pattern.

In case of Thane cyclone, the trend of intensification and dissipation is not well captured by the model in both simulations. The intensity in term of CSLP and MSW are under-predicted. In case of cyclone Aila, the rate and trend of intensification and dissipation is better simulated by the model in both simulations though the peak intensity is under-predicted by about 7 hPa. The time evolution of MSW shows that the peak intensity of storm is better simulated by the model though MSW is over-predicted before the peak is reached. It is found that the trend of intensification is similar. In case of cyclone Laila, the intensity of the storm is reasonably better simulated by the model with GFS IBCs than with FNL IBCs during entire simulation

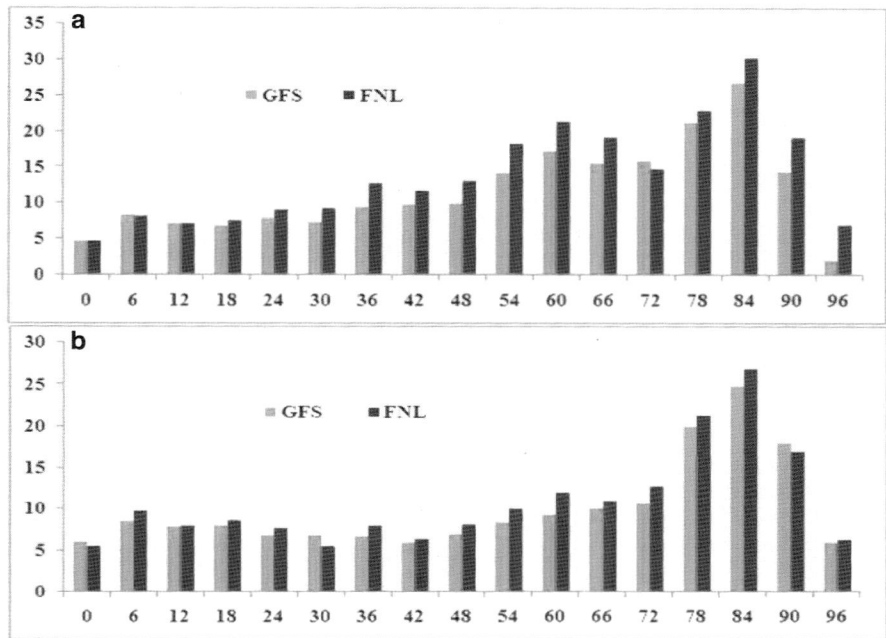

Fig. 6 Variation of mean absolute error of intensity with time in terms of (**a**) CSLP (in hPa) and (**b**) MSW (in m/s)

period but over-predict the intensity after 30 h. The time evolution of MSW shows that maximum wind is better simulated in GFS simulation in first 54 h. In last 15 h it is better simulated with FNL IBCs. In case of Jal cyclone, the intensity of the storm both in terms of CSLP and MSW is over-predicted by model with GFS and FNL datasets. The intensification and dissipation are well captured by the model in both simulations. Overall, in this case also the intensity is better predicted by the model in GFS simulation. The results discussed above indicate that the model simulated intensity in terms of CSLP and MSW is better simulated by the model and close to the observation in GFS simulation. The trend of intensification and dissipation of the storm are also better simulated by the model in GFS simulation with few exceptions.

Figures 6a, b show mean absolute error in intensity for the seven cyclone cases at every 6 h interval upto 96 h in terms of CSLP and MSW respectively. This clearly indicates the intensity of the storms both in terms of CSLP and MSW significantly better simulated by the model with GFS IBCs. The mean absolute error in simulation of intensity in terms of CSLP in 24, 48, 72 and 96 h are 8, 10, 16 and 2 hPa respectively and in terms of MSW are 7, 7, 11 and 6 m/s respectively in GFS simulation, while the initial error in intensity is about 6 hPa and 5 m/s in terms of CSLP and MSW respectively.

Conclusions

In this study, a double nested WRF-ARW model is used to investigate the impact of IBCs derived from GFS analysis & forecast and FNL analysis datasets towards mesoscale simulation of seven BOB cyclones. The results presented and discussed in the previous section can be summarized as follows.

The model simulated track and intensity of the storm is better simulated by the model with GFS IBCs. The track of the storms is better simulated by the model with GFS IBCs in most of the cases. The exceptions are Aila and Mala case, where track error is more in GFS simulation. This is due to more initial positional error in the GFS analysis. The mean errors in landfall locations are around 90 km and 198 km respectively in GFS and FNL simulations. Overall, the track of the Bay of Bengal cyclones is better simulated by the model with GFS IBCs. The intensity of the storm both in terms of CSLP and MSW is better simulated by the model with GFS IBCs in most of the cases. The mean absolute error in intensity clearly indicates that the forecast of intensity is better predicted with GFS IBCs upto 96 h.

Acknowledgements The authors sincerely acknowledge Council of Scientific & Industrial Research (CSIR) and MoES for funding the research activity. The authors are also sincerely acknowledging the IMD for providing the best-fit track data, NCEP for their FNL analysis and GFS forecasted datasets and NCAR for the WRF model.

References

Bhaskar Rao DV, Prasad DH (2006) Numerical prediction of the Orissa super cyclone (1999): sensitivity to the parameterization of convection, boundary layer and explicit moisture processes. Mausam 57(1):61–78

Cacciamani C, Cesari D, Grazzini F, Paccagnella T, Pantone M (2000) Numerical simulation of intense precipitation events south of the Alps: sensitivity to initial conditions and horizontal resolution. Meteorol Atmos Phys 72:147–159

Demaria M, Kaplan J (1999) An updated Statistical Hurricane Intensity Prediction Scheme (SHIPS) for the Atlantic and eastern North Pacific basins. Weather Forecast 14:326–337

Deshpande M, Pattnaik S, Salvekar PS (2010) Impact of physical parameterization schemes on numerical simulation of super cyclone Gonu. Nat Hazards 55:211–231. doi:10.1007/s11069-010-9521-x

Dudhia J (1989) Numerical study of convection observed during the winter monsoon experiment using a mesoscale two-dimensional model. J Atmos Sci 46:3077–3107

Hong SY, Noh Y, Dudhia J (2006) A new vertical diffusion package with an explicit treatment of entrainment processes. Mon Weather Rev 134:2318–2341

Kain JS (2004) The Kain-Fritsch convective parameterization: an update. J Appl Meteorol 4:170–181

Lin YL, Farley RD, Orville HD (1983) Bulk parameterization of the snow field in a cloud model. J Clim Appl Meteorol 22:1065–1092

Majewski D (1997) Operational regional prediction. Meteorol Atmos Phys 63:89–104

Mandal M, Mohanty UC, Raman S (2004) A study on the parameterization of physical processes on prediction of tropical cyclones over the Bay of Bengal with NCAR/PSU mesoscale model. Nat Hazards 31(2):391–414

Mlawer EJ, Taubman SJ, Brown PD, Iacono MJ, Clough SA (1997) Radiative transfer for inhomogeneous atmosphere: RRTM, a validated correlated-k model for the longwave. J Geophys Res 102(D14):16663–16682

Mohanty UC, Osuri KK, Routray A, Mohapatra M, Pattanayak S (2010) Simulation of Bay of Bengal tropical cyclones with WRF model: Impact of initial and boundary conditions. Mar Geod 33:294–314. doi:10.1080/01490419.2010.518061

Osuri KK, Mohanty UC, Routray A, Makarand AK, Mohapatra M (2011) Customization of WRF-ARW model with physical parameterization schemes for the simulation of tropical cyclones over North Indian Ocean. Nat Hazards. doi:10.1007/s11069-011-9862-0

Osuri KK, Mohanty UC, Routray A, Mahapatra M (2012) The impact of satellite-derived wind data assimilation on track, intensity and structure of tropical cyclones over the North Indian Ocean. Int J Remote Sens 33(5):1627–1652

Pattanayak S, Mohanty UC (2008) A comparative study on performance of MM5 and WRF models in simulation of tropical cyclones over Indian seas. Curr Sci 95:923–936

Pattanayak S, Mohanty UC, Osuri KK (2012) Impact of parameterization of physical processes on simulation of track and intensity of tropical cyclone Nargis (2008) with WRF-NMM model. Sci World J 2012, Article ID 671437:18. doi:10.1100/2012/671437

Pielke RA, Matsui T, Leoncini G, Nobis T, Nair U, Lu E, Eastman J, Kumar S, Peters CL, Tian Y, Walko R (2006) A new paradigm for parameterizations in numerical weather prediction and other atmospheric models. Natl Weather Dig 30:93–99

Singh KS, Mandal M (2014) Sensitivity of mesoscale simulation of Aila cyclone to the parameterization of physical processes using WRF model. doi:10.1007/978-94-007-7720-0_26

Skamarock WC, Klemp JB, Dudhia J, Gill DO, Barker DM, Wang W, Powers JG (2008) A description of the advanced research WRF version 3, NCAR technical note

Srinivas CV, Venkateshan R, Bhaskar Rao DV, Prasad DH (2007) Numerical simulation of Andhra severe cyclone (2003): model sensitivity to the boundary layer and convection parameterization. Pure Appl Geophys 164:1465–1487

Srinivas CV, Bhaskar Rao DV, Yesubabu Y, Baskaran R, Venkatraman B (2012) Tropical cyclone predictions over the Bay of Bengal using the high-resolution Advanced Research Weather Research and Forecasting (ARW) model. Q J R Meteorol Soc. doi:10.1002/qj.2064

Zhang F, Weng Y, Gamache JF, Marks FD (2011) Performance of convection-permitting hurricane initialization and prediction during 2008-2010 with ensemble data assimilation of inner core airborne Doppler radar observations. Geophys Res Lett 38:L15810. doi:10.1029/2011GL048469

Performance of Global Forecast System for the Prediction of Intensity and Track of Very Severe Cyclonic Storm 'Phailin' over North Indian Ocean

V.R. Durai, S.D. Kotal, S.K. Roy Bhowmik, and Rashmi Bharadwaj

1 Introduction

Tropical cyclone (TC) formation involves interaction of a variety of processes, both on the synoptic scale as well as the mesoscale. Gray (1979) identified several large-scale conditions as necessary for tropical cyclogenesis, including preexisting low-level relative vorticity and high mid-tropospheric humidity. TCs are one of the most dangerous natural calamities throughout the globe. The Bay of Bengal TC disaster is the costliest and deadliest natural hazard in the Indian sub-continent. It has a significant socio-economic impact on the countries bordering the Bay of Bengal, especially India, Bangladesh and Myanmar. Every year, they cause considerable loss of life and do immense damage to property. India and Bangladesh have a coastline of more than 8,000 km, which is prone to very severe cyclone formations in the Arabian Sea and Bay of Bengal. Therefore, reasonably accurate prediction of these storms has great importance to avoid the loss of valuable lives.

Prediction of intensity and track of TC continues to be a forecasting challenge. Current operational models have difficulty in accurately forecasting TC formation. There are a number of comparative studies on the performance of the mesoscale models for severe weather events triggered by convection. Rama Rao et al. (2010) made a comparative study on the performance of WRF and QLM models for the track forecast. Also sensitivity experiments were conducted with the WRF model to test the impact of various microphysical and cumulus parameterization schemes in capturing the track and intensity of two severe cyclonic storms namely super cyclone 'Gonu' over Arabian Sea and very severe cyclonic storm 'Sidr' over Bay of Bengal.

The Cyclone Warning Division (CWD) at New Delhi of India Meteorological Department (IMD) functions as a Regional Specialized Meteorological Centre

---

V.R. Durai (✉) • S.D. Kotal • S.K.R. Bhowmik
India Meteorological Department, Lodhi Road, New Delhi 110003, India
e-mail: durai.imd@gmail.com

R. Bharadwaj
Guru Gobind Singh Indraprastha University, Dwarka, New Delhi 110078, India

(RSMC) for TC forecasting, as recognized by the World Meteorological Organization (WMO). According to WMO's Tropical Cyclone Programme (TCP), one of the major responsibilities of RSMC, New Delhi is to provide TC advisories to the member countries in the north Indian seas, apart from its national responsibilities. Cyclone advice for the member countries, which begins from the cyclone stage, includes information related to present and forecast track and intensity based on the use of the sophisticated Numerical Weather Prediction (NWP) models i.e. global and regional mesoscale models. Forecasted fields of mean sea-level pressure (MSLP), 850 hPa wind, vorticity and divergence are examined for the track and intensity prediction of very severe cyclonic storm (VSCS) 'Phailin' formed over Bay of Bengal (BoB) during 8–12 Oct 2013. In particular, day 1–day 7 forecasts of these fields are generated from the GFS. The initial analysis of a low-pressure system by the CWD is taken as the time of cyclogenesis. In most cases, this initial analysis occurred several hours prior to classification as a tropical depression. The model forecasts are verified against the surface analyses produced by GDAS. Forecasts of mean sea level pressure (mslp), 10 m wind, 850 hPa wind and vorticity fields are verified against the corresponding GDAS and satellite analyses from the Kalpana-1 meteorological satellite images.

The main objective of this study is to investigate the performance skill of Global Forecast System (GFS) for the prediction of the track and intensity of very severe cyclonic storm (VSCS) 'Phailin' formed over Bay of Bengal (BOB) in the medium range (day 1–day 7) time scale. The performances of the models have been evaluated and compared with observations and verifying analyses. A brief description of the mesoscale models along with the numerical experiments and data used for the present study are given in Sect. 2. The synoptic situation for the above mentioned cyclone used in the present study is described in Sect. 3. The results are presented in Sect. 4 in order to evaluate the performance of the models and the conclusions are in Sect. 5.

2 Model Description

The Global Forecasting System (GFS) is a primitive equation spectral global model with state of art dynamics and physics (Saha et al. 2010). Inter-comparisons of physics and dynamics options of GFS T574 is shown in Table 1. Details about the GFS model are available at http://www.emc.ncep.noaa.gov/GFS/doc.php. The GFS T574L64 (~25 km in horizontal over the tropics), adopted from National Centre for Environmental Prediction (NCEP), was implemented at IMD, New Delhi on IBM based High Power Computing Systems (HPCS) (Durai and Roy Bhowmik 2013). The assimilation system (for GFS T574) is a global 3-dimensional variational technique, based on NCEP Grid Point Statistical Interpolation (GSI 3.0.0) (Kleist et al. 2009) scheme, which is the next generation of Spectral Statistical Interpolation (SSI) (Parrish and Derber 1992). The major changes incorporated in T574 GDAS compared to T382 GDAS are: use of variational quality control, flow dependent

Table 1 Physics and dynamics options of GFS T382 and T574

Physics and dynamics	T574L64
Surface fluxes	Monin-Obukhov similarity
Turbulent diffusion	Non-local closure scheme (Lock et al. 2000)
SW radiation	Rapid Radiative Transfer Model (RRTM2) (Mlawer et al. 1997)—aerosols included—invoked hourly
LW radiation	Rapid Radiative Transfer Model (RRTM1) (Mlawer and Clough 1997)—aerosols included—invoked hourly
Deep convection	SAS convection (Han and Pan 2006)
Shallow convection	Mass flux scheme (Han and Pan 2010)
Large scale condensation	Large scale precipitation (Zhao and Carr 1997; Sundqvist et al. 1989)
Cloud generation	Based on Xu and Randall (1996)
Rainfall evaporation	Kessler (1969)
Land surface processes	NOAH LSM with four soil levels for temperature and moisture (Ek et al. 2003)
Air-Sea interaction	Roughness length by Charnock (1955), Observed SST, Thermal roughness over the ocean is based on Zeng et al. (1998). 3-layer Thermodynamic Sea-ice model (Winton 2000)
Gravity wave drag and mountain blocking	Lott and Miller (1997), Kim and Arakawa (1995), and Alpert et al. (1996)
Vertical advection	Flux-Limited Positive-Definite Scheme (Yang et al. 2009)

re-weighting of background error statistics, use of new version of Community Radiative Transfer Model (CRTM 2.0.2), improved TC relocation algorithm, changes in the land, snow and ice skin temperature and use of some new observations in the assimilation cycle. In the operational mode, the Global Data Assimilation (GDAS) cycle runs four times a day (00 UTC, 06 UTC, 12 UTC and 18 UTC). The analysis and forecast for 7 days are performed using the HPCS installed in IMD Delhi. One GDAS cycle and 7 days forecast (0–168 h) at T382L64 (~35 km in horizontal over the tropics) takes about 30 min on IBM Power 6 (P6) machine using 20 nodes with seven tasks (seven processors) per node, while the same for GFS T574 (~25 km in horizontal over the tropics) is approximately 1 h 40 min. Details of data presently being processed for GFS at IMD are available at http://www.imd.gov.in/section/nhac/ dynamic/ data_coverage.pdf.

3 System Description

A low pressure system that formed over North Andaman Sea on 7 October 2013 intensified into depression at 0300 UTC of 8 October 2013 near latitude 12.0°N and longitude 96.0°E. It moved northwestwards and intensified into a deep depression at 0000 UTC of 9 October 2013 and further intensified into a cyclonic storm (T. No. 2.5), Phailin at 1200 UTC of the same day. The cyclonic storm continued to move in northwesterly direction and intensified into severe cyclonic storm (T. No. 3.5) at 0300 UTC of 10 October 2013 and subsequently intensified into very severe cyclonic storm (T. No. 4.0) at 0600 UTC of same day. Moving northwestward direction the system further rapidly intensified to T. No. 4.5, T. No. 5.0, and T. No. 5.5 at 1200 UTC, 1500 UTC and 2100 UTC of same day (10 October 2013) respectively. At 0300 UTC of 11 October 2013 the system intensified to T. No. 6.0 and continued to move northwesterly direction with same intensity towards Odisha and crossed coast near Gopalpur at around 1700 UTC of 12 October 2013. The system maintained its intensity of very severe cyclonic storm upto 7 h after landfall and cyclonic storm intensity till 1200 UTC of 13 October 2013. The system continued to decay and weakened to deep depression at 1800 UTC of 13 October 2013 and further to depression at 0300 UTC of 14 October 2013.

4 Result and Discussions

The ridge line at 200 hPa was around 20°N, which was quite north of the system centre. This feature was helpful in moving the system as its normal NW direction. The GDAS analysis of wind at middle level (500 hPa) also supports the circulation on 8–12 October 2013. The GDAS analysis of wind and vorticity at 850 hPa (Fig. 1) captured the position and intensity of the VSCS Phailin with reasonable accuracy during 7–12 October 2013. The observed cyclone track (red dot line) of 'Phailin' is superposed in Fig. 1. On 7th October a trough at 850 hPa over north Andaman Sea was seen in the analysis. It became a cyclonic circulation with wind speed of about 25–30 knots on 8th October. The 850 hPa vorticity maximum of the order of 15×10^{-5}/s on 9 October was situated around 13.5°N/93.5°E. In all days from 7 to 12 October the GDAS analyzed cyclonic centre with circulation pattern is matching very much with the observed track. The magnitude of 850 hPa vorticity value gradually increases from 7 to 12 October along the track in the northwest direction. The GDAS analysis (Fig. 1) based on 00 UTC of 07 October 2013 shows that the Depression near Andaman Sea would intensify into a Deep Depression at 00 UTC of 09 October 2013 and became a cyclonic storm (CS) at 12 UTC of 09 October 2013. GDAS analysis of 06 UTC of 10 October 2013 shows the severe CS (SCS) and at 06 UTC of 11 October 2013 it became Very Severe Cyclonic storm (VSCS).

4.1 Track Forecast

The real time track forecast and the corresponding track errors of the VSCS 'Phailin' using GFS model at 25 km resolution starting from the initial conditions of 9 Oct 2013 at 00 UTC is shown in Fig. 2. The model 12 hourly direct positional errors are calculated as the geographical distance between the observed and forecast point. The 850 hPa wind and vorticity forecast valid at 24–120 h takes the system along

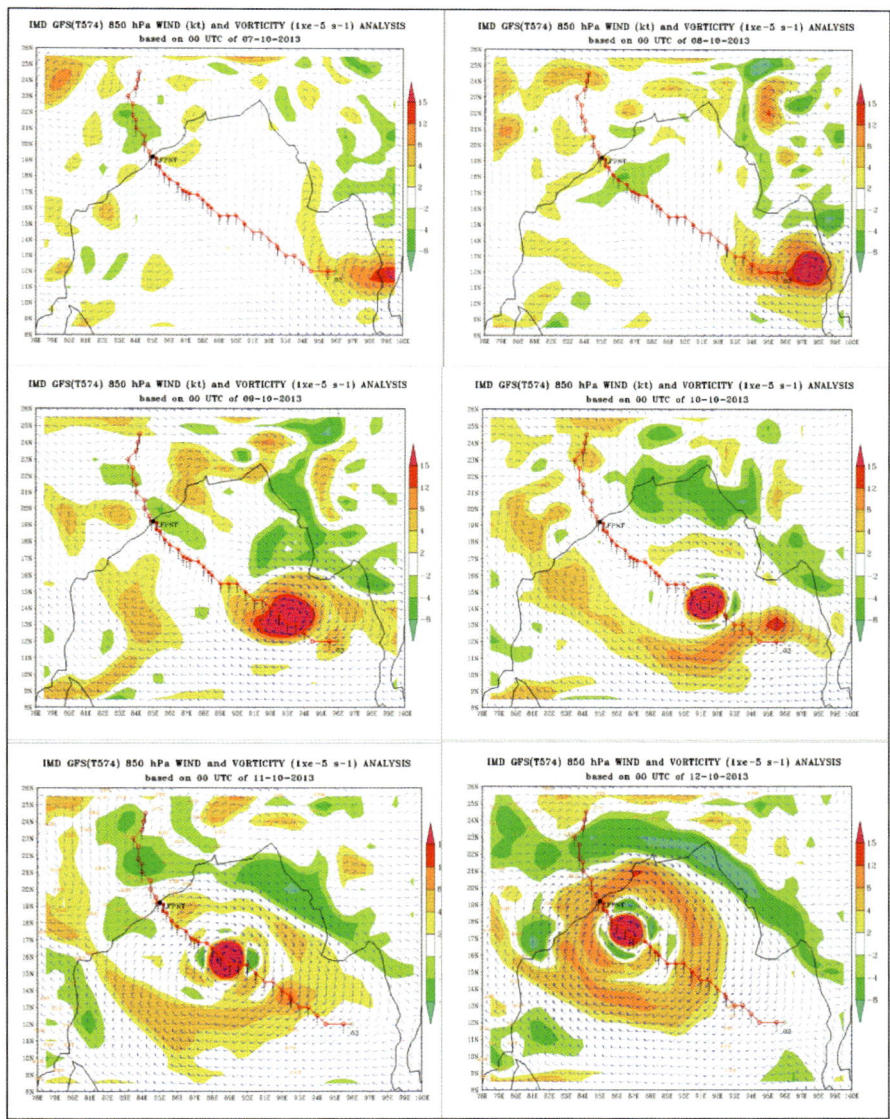

Fig. 1 The GDAS analysis of 850 hPa wind and vorticity during 7–12 October 2013

with the observed track. The 850 hPa vorticity maximum of the order of 15×10^{-5}/sec on 10 Oct was situated around 13.5°N/93.5°E. Like the observed track, the GFS day-1 to day-5 forecast based on 9 Oct 2013 initial condition (Fig. 2) also clearly indicated that the system was going to hit the Odisha coast around 17 UTC of 12 Oct 2013. The 12 hourly forecast track error of the cyclone as given in Fig. 2b based

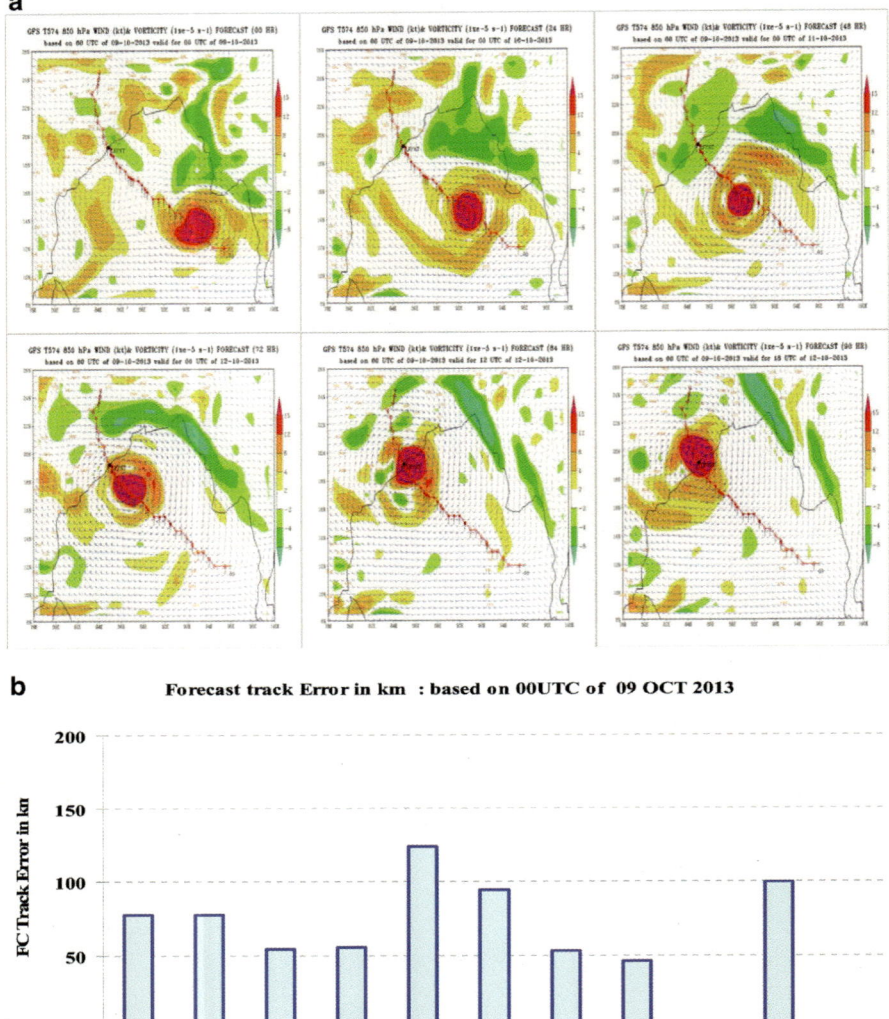

Fig. 2 GFS forecast of (**a**) wind and vorticity at 850 hPa and (**b**) track error in km based on initial condition of 00 UTC of 09 October 2013

on the initial condition of 9 Oct is found to be within the range of 120 km till 48 h of its forecast and the error increased to 150 km in 120 h as the forecast track showed northwesterly movement along with the observed track (Fig. 2).

The real time forecast using 9 October initial condition also shows landfall around 18 UTC of 12 Oct near Gopalpur in Orissa coast. However, the 96 h forecast

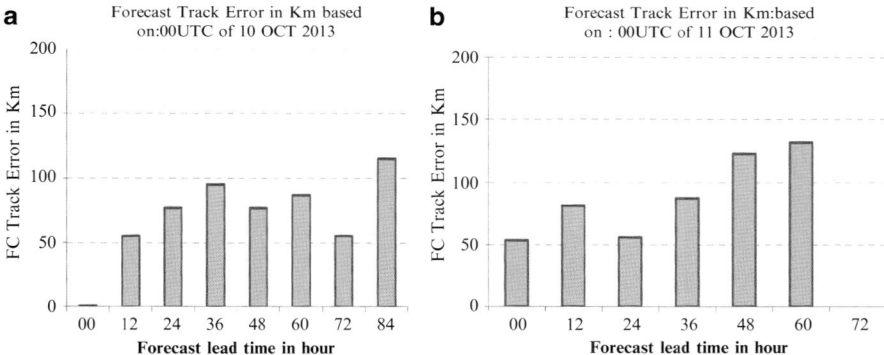

Fig. 3 GFS 24–72-h forecast track error in km based on initial condition of 00 UTC of (**a**) 10th and (**b**) 11th October 2013

position shows that the system was expected to hit the Orissa coast as it followed the northwestward movement like that of observed track. The forecast based on the initial condition of 10 Oct also shows northwestward movement of the cyclone and crossed the coast in its 84 h forecast position. The forecast and observed tracks showed northwestward movement of the system, although the forecast track showed slight north of the actual landfall point. The model forecast speed and landfall of the system is very close to that of observation. Based on the initial condition of 10 Oct, when the system was in the stage Deep Depression, the initial error is about 50 km (Fig. 3a) and the corresponding forecast errors also reduced substantially.

Similarly the 12 hourly forecast error based on 11 Oct initial condition as given in Fig. 3b is basically due to the movement of the system slightly to the north of the observed track although it had a very similar track parallel to that of the observed track. It is also indicated from Fig. 3 that the forecast tracks are improved with the initial conditions of 8–11 Oct with landfall error of range between 50 and 150 km. The mean initial error and 12 hourly mean forecast errors found from the GFS model runs is found to be less than 150. The track of the cyclone as obtained from the model simulations using different initial conditions are evaluated and compared with the best-fit track as estimated by IMD. Forecasts from all the initial conditions show that, in each case, the cyclone moves to the Orissa coast, whatever the initial condition is being chosen. The vector displacement error is also calculated at the landfall point.

4.2 Cyclone Intensity Forecast

GFS 10 m wind and MSLP for the intensity forecast based on 00 UTC of 08 October 2013 shows that the Depression near Andaman Sea would intensify into: (i) a DD at 00 UTC of 09 October 2013. (ii) cyclonic storm at 12 UTC of 09 October 2013 (iii) severe CS at 06 UTC of 10 October 2013, and (iv) very severe cyclonic storm at 06 UTC of 11 October 2013. Figure 4 represents the 12 hourly forecast of the

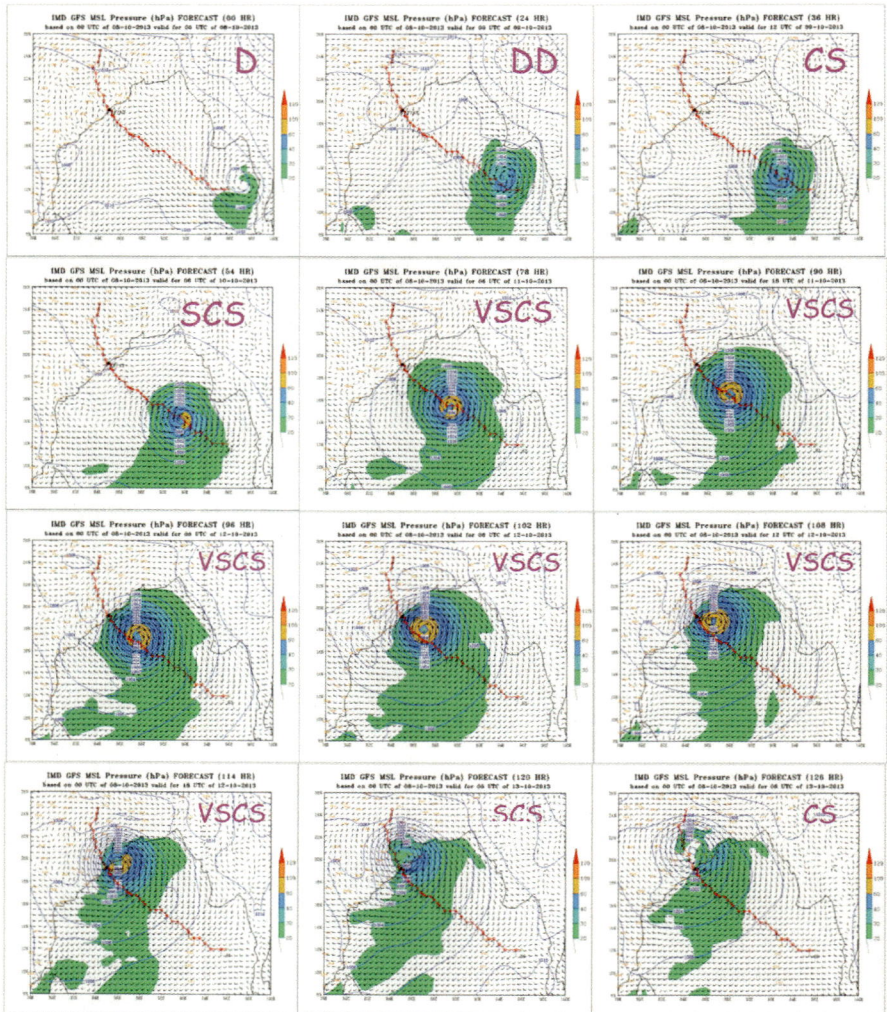

Fig. 4 GFS forecast of 10 m wind (kt) and MSLP (hPa) based on initial condition of 00 UTC of 08 October 2013; *D* Depression, *DD* Deep Depression, *CS* Cyclonic Storm, *SCS* Severe Cyclonic Storm, *VSCS* Very Severe Cyclonic Storm, *SuCS* Super Cyclonic Storm

MSLP and 10 m wind from GFS model simulation based on 8 Oct initial condition. On 8 Oct the system was only a depression with maximum wind speed of around 20–30 knots. Then it became cyclonic storm with wind speed reaching around 40 knots in the 36-h forecast. In the 54-h forecast the system became severe cyclonic storm (SCS) with wind speed reaching around 60 knots. Finally in the 78-h forecast, the system became very severe cyclonic storm (VSCS) with wind speed reaching in the range of 60–100 knots. GFS 84–114 h forecast shows that the intensity of VSCS was maintained till it made landfall on 17 UTC of 12 Oct 2013.

The central MSLP with the maximum sustainable wind of 100 kts is simulated in model forecast. The central MSLP with the maximum sustainable wind of 100 kts is simulated in 78–114 h forecast of GFS model based on 8 Oct initial condition very reasonably. Similarly, the GFS intensity forecast based on the initial condition of 9 Oct using 12 hourly forecasts of the mean MSLP and 10 m wind also gives the storm intensification reasonably accurate. From the results it may be inferred that simulation with GFS model wind at 10 m and MSLP forecast gives the storm intensification reasonably accurate with observation. It may also be noticed that GFS intensify the storm with less time delay, which is in reasonable accuracy with observation. GFS could predict the intensification, movement, and landfall and decaying of the system with reasonable accuracy.

4.3 Heavy Rainfall

The observed heavy rainfall associated with the Phailin cyclonic storm over eastcentral BOB from 03 UTC of 09th October to 12th October 2013 is shown in Fig. 5. The 24-h accumulated rainfall analysis at the resolution of 50 km is based on

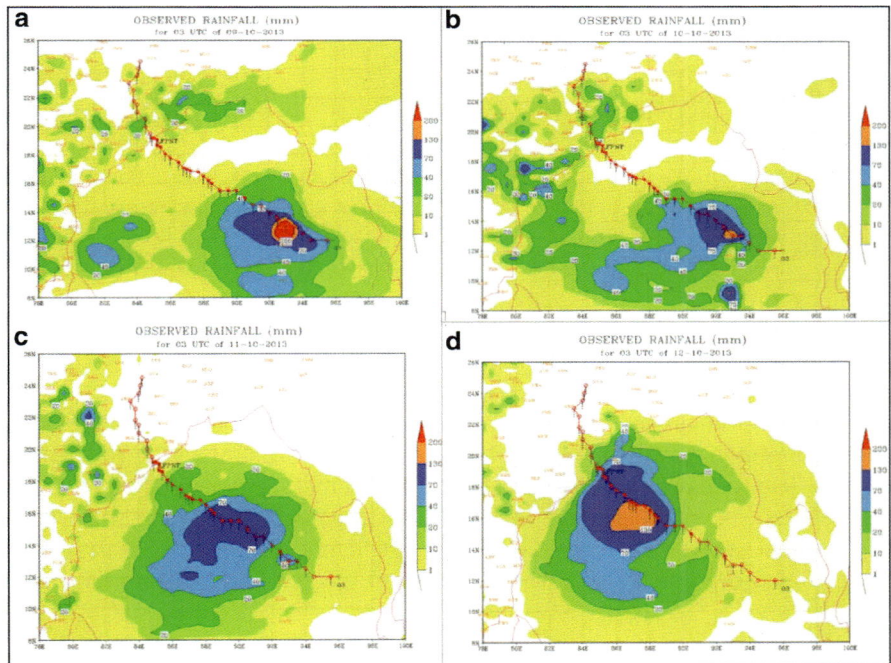

Fig. 5 Observed rainfall from rain gauge (land) and TRMM (sea) for (**a**) 09, (**b**) 10, (**c**) 11 and (**d**) 12th October 2013

the merged rainfall data combining gridded rain gauge observations prepared by IMD Pune for the land areas and Tropical Rainfall Measuring Mission (TRMM) 3B42RT data for the sea areas (Durai et al. 2010). It is seen from Fig. 5 that the heavy rainfall occurred over the sea to the southwest sector of the centre of the TC with a peak of >200 mm on 9, 10 and 12 October, but on 11th October it showed rainfall in the order of 70–130 mm only. In general, the rainfall estimate by TRMM could capture the magnitude and location of heavy rainfall associated with low pressure system reasonably well over the sea areas.

The spatial distributions of observed rainfall along with GFS forecast of wind at 850 hPa and rainfall valid for 13 Oct 2013 (top panel) and 14 October 2013 (bottom panel) in relation to movement of the system and occurrences of heavy rainfall are examined (Fig. 6). It ushered the landfall of the VSCS Phailin over Orissa coast and caused excess rainfall over these regions on 13 and 14 Oct 2013. Under the influence of this VSCS a heavy to very heavy rainfall occurred over coastal Orissa and adjoining areas during 13–14 October 2013. The case study selected is the exceptionally heavy rainfall due to Phailin on 13 and 14 October over Orissa.

On 13th October 2013 heavy rainfall was reported over Orissa and adjoining Jharkhand areas. Figure 6a shows the observed rainfall and 24–120-h rainfall forecast from GFS T574L64 valid for 13th over many parts of Orissa. The 24-h forecast could capture the spatial pattern of observed rainfall, but the magnitude is less than the actual. The location and magnitude of heavy rainfall on 13 Oct due to the landfall during 17 UTC of 12th Oct near Gopalpur is better captured by GFS T574 day-1 to day-5 forecast. The spatial distribution pattern suggested that the GFS T574 model forecasts, in general are better skillful in predicting heavy rainfall. Using the GFS T574 operational products, the location and intensity of heavy rainfall can be predicted up to 3 days in advance with an accuracy of spatial error less than 200 km, which can provide useful guidance for real time forecasting of heavy rainfall during monsoon depression over India.

The heavy rainfall on 14th October 2013 was reported over north Orissa and adjoining Jharkhand and Gangetic West Bengal areas as shown in Fig. 6b. The observed rainfall and 24–120-h rainfall forecast from GFS T574L64 valid for 14th Oct shows over northern parts of Orissa. The VSCS Phailin moved northwards after the landfall, so the very heavy rainfall occurred in the northern parts of Orissa on 14th. The 24–120-h forecast could capture the spatial pattern of observed rainfall, but the magnitude is more or less matching with the observed rainfall. The spatial distribution pattern suggested that the day to day GFS T574 model forecasts, in general, are better skillful in predicting heavy rainfall. GFS T574 model showed considerable skill in predicting the heavy rainfall due to cyclone. However, the accuracy in prediction of location and intensity fluctuates considerably.

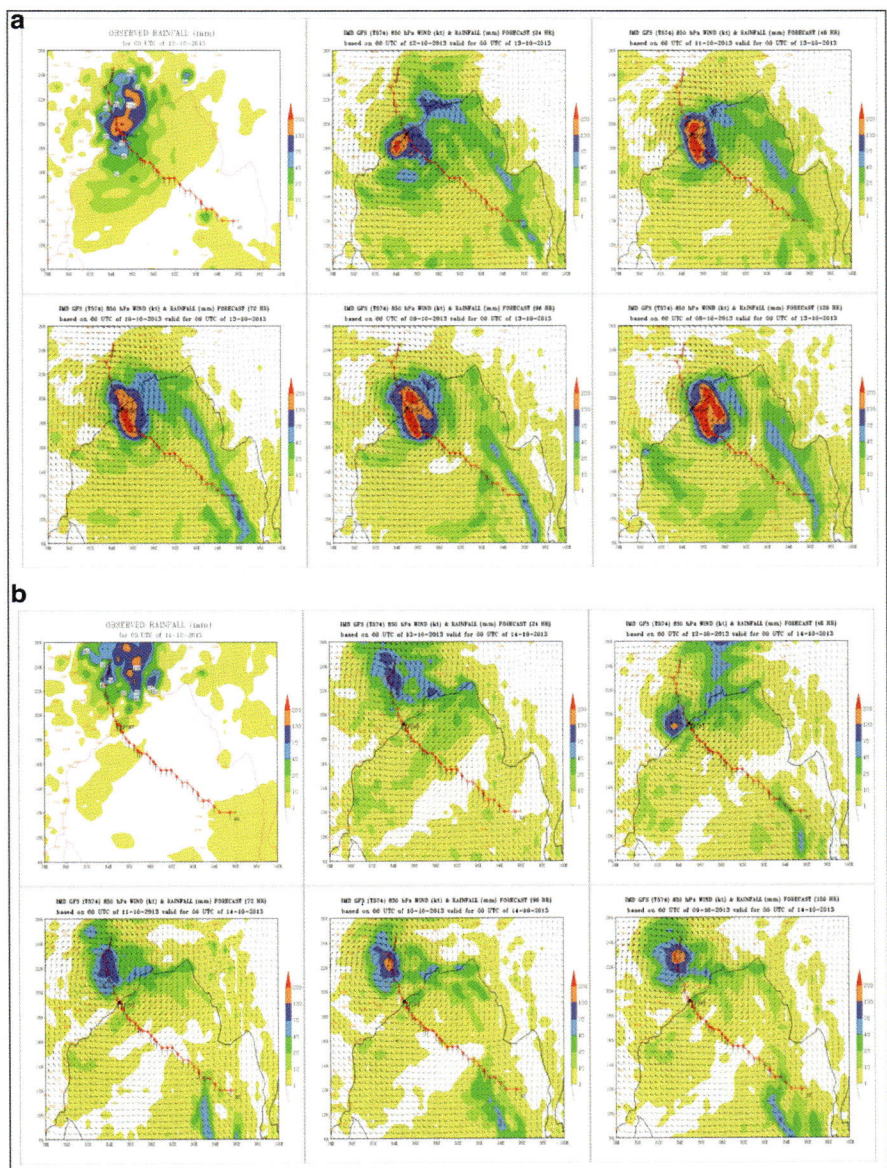

Fig. 6 (**a**) Observed rainfall and GFS T574 day-1 to day-5 forecasts of wind at 850 hPa and heavy rainfall on 13 Oct 2013 (*top panel*) and (**b**) 14 October 2013 (*bottom panel*)

Conclusion

NCEP-based GFS system has been in operational use at IMD New Delhi for daily medium range forecasts. This paper assesses the skill of GFS for the prediction of the intensity and track of very severe cyclonic storm (VSCS) 'Phailin' formed over Bay of Bengal (BOB). The GFS T574 MSLP and low level wind (10 m wind) forecast could capture the genesis location of depression formed over BoB (8th October 2013) up to 4–5 days in advance. Results demonstrate that GFS T574 provides skillful real-time forecasts of cyclone track and intensity over BoB. Using the GFS T574 operational products, the genesis location could be predicted up to 5–6 days in advance with an error less than 150 km, which provided useful guidance for real time forecasting of this TC over BoB. Due to inaccurate location of low-pressure systems by NWP models, in general, some mismatch prevails between the spatial distribution of forecast heavy rainfall and the observed one. Because of this double penalty, rainfall prediction skill deteriorates over Indian monsoon region.

The day-to-day GFS analysis is consistent with the observed cyclone track position of Phailin. The real time track forecast using GFS indicated that the system was expected to cross the Orissa coast near Gopalpur, around 18 UTC of 12 Oct 2013, with landfall error of 50–150 km and landfall time delay of 3–6 h. The movement and intensity of this system have been better captured by the GFS T574 wind and vorticity at 850 hPa and MSLP based on 8 Oct 2013 initial condition. The GFS forecast captured the formation and movement of 'Phailin' reasonably well, almost 120 h in advance with very less forecast error. In general, the high resolution GFS model (22 km) provided very useful guidance in terms of landfall point, landfall time, rapid intensification and decay after landfall. Further improvement in the forecast is expected with the possible inclusion of 3-Dimensional Hybrid Ensemble Kalmen Filter (Hamill et al. 2011) data assimilation and multiple physics in the GFS. However, the accuracy in prediction of location and intensity of cyclone fluctuates considerably.

Acknowledgements Authors are grateful to the Director General of Meteorology, India Meteorological Department for providing all facilities to carry out this research work. Acknowledgements are due to NCEP, USA for providing the source codes and support for the implementation of the upgraded Version GFS T574 at IMD. Authors acknowledge the technical support provided by NCMRWF for the implementation of the data decoders at IMD. Authors also acknowledge the use of TRMM products of NASA, USA in this work.

References

Durai VR, Roy Bhowmik SK (2013) Prediction of Indian summer monsoon in short to medium range time scale with high resolution global forecast system (GFS) T574 and T382. Clim Dyn, published online on 8 Aug 2013. doi:http://link.springer.com/article/10.1007%2Fs00382-013-1895-5

Durai VR, Roy Bhowmik SK, Mukhopadhaya B (2010) Evaluation of Indian summer monsoon rainfall features using TRMM and KALPANA-1 satellite derived precipitation and rain gauge observation. Mausam 61(3):317–336

Gray WM (1979) Hurricanes: their formation, structure, and likely role in the tropical circulation. In: Shaw DB (ed) Meteorology over the tropical oceans. Royal Meteorological Society, Bracknell, pp 155–218

Hamill TM, Whitaker JS, Fiorino M, Benjamin SG (2011) Global ensemble predictions of 2009's tropical cyclones initialized with an ensemble Kalman filter. Mon Weather Rev 139:668–688

Kleist DT, Parrish DF, Derber JC, Treadon R, Wu W-S, Lord S (2009) Introduction of the GSI into the NCEP Global Data Assimilation System. Weather Forecast 24:1691–1705

Parrish DF, Derber JC (1992) The National Meteorological Center's spectral statistical-interpolation analysis system. Mon Weather Rev 120:1747–1763

Rama Rao YV, Madhu Latha A, Suneetha P (2010) Evaluation of the WRF and Quasi-Lagrangian Model (QLM) for cyclone track prediction over Bay of Bengal and Arabian Sea. In: Indian Ocean tropical cyclones and climate change, Part 3. Springer

Saha S et al (2010) The NCEP climate forecast system reanalysis. Bull Am Meteorol Soc 91:1015–1057. doi:http://dx.doi.org/10.1175/2010BAMS3001.1

Part III
Heavy Rains

Observational Analysis of Heavy Rainfall During Southwest Monsoon over India

Pulak Guhathakurta

1 Introduction

The climate of south Asia is largely dominated by monsoon circulation. Most of the rainfall received in India is during the 4 months of southwest monsoon season. In recent years, India has faced frequent and severe floods that caused havoc in terms of economic loss and loss of human lives. The devastating floods are occurring almost every year but the places are not same for every occurrence. Most of these floods are categorized as flash floods which are generally associated with heavy precipitation. Heavy precipitation with cloud burst also caused disasters, particularly in northern states of the country. It may be mentioned that the information on the changes in extreme weather events is more important than the changes in mean pattern for better disaster management and mitigation. There is also high temporal variability of monsoon rainfall. This causes the extreme years with high monsoon rainfall departure from the long period mean value, the positive departure causing flood and negative departure causing drought. The variability of monsoon rainfall has been studied by many climate scientists and they have also drawn several conclusions. However the study of variability of rainfall is different from the studies of other climate parameters.

Using the data for the period 1901–2003, Guhathakurta and Rajeevan (2008) have shown that the all-India rainfall has no long-term trend neither for the monsoon rainfall nor for any monsoon months individually. However conclusions have been drawn regarding weakening of monsoon circulation over India using the data for the period 1951–2003/2004 (Naidu et al. 2011; Bawiskar 2009; Chung and Ramanathan 2006; Ramesh and Goswami 2007; Joseph and Simon 2005). As mentioned, the variability (both temporal and spatial) of rainfall is more than other

P. Guhathakurta (✉)
India Meteorological Department, Pune 411005, India
e-mail: pguhathakurta@rediffmail.com

climate parameters and the epochal variability of Indian monsoon rainfall has already been reported by several authors (Pant and Rupa Kumar 1997; Guhathakurta and Rajeevan 2008). Guhathakurta et al. (2011b) have considered annual frequency of rainfall events of different intensities and have identified zone of increasing trends in flood risk. The present study aims to analyze some of the extreme rainfall indices using reliable, consistent and sufficient number of raingauge station data. Since the country is divided in 36 meteorological sub-divisions, to know the distribution of properties of changes in frequency of heavy rainfall event spatially or more specifically for each of the 36 meteorological sub-divisions, we have calculated number of percentage of total stations having decreasing/increasing trends to the number of percentage of total stations for each sub-divisions. The trend analysis is done on the time series of frequencies considering three different periods viz. 1901–1950, 1951–2011 and 1901–2011. The consideration of frequency of station in percentage is due to unequal number of stations in three periods.

Thus it is required to present the latest status of trends in monsoon rainfall over India as a whole as well as in regional scales using the data for the period of 1901–2011.

2 Data Used and Methodology

For analysis of frequency of rain days (days with rainfall ≥ 0.1 mm), rainy days (days with rainfall amount 2.5 mm or more but less than 64.5 cm) and frequency of heavy rainfall (including very heavy and extremely heavy) days (days with rainfall amount 64.5 mm or more) we have considered daily rainfall data of more than 7,000 raingauge stations having good availability and also uniformly distributed over the country. For analysis of trend in mean rainfall we have considered new IMD districts, subdivision rainfall data series (Guhathakurta et al. 2011a). Our analysis contains three periods: (1) the complete period 1901–2011, (2) the first 50-year period 1901–1950 and (3) the remaining period after 1950 i.e. 1951–2011.

Both monotonic and non-monotonic (i.e. climate jump or change point) changes are investigated. Monotonic trends are generally gradual changes that are either increasing or decreasing with no reversal of direction. The least square linear fit is generally used to examine the existence of monotonic trend in the rainfall time series data and the statistic used for testing the significance is the Student's t-distribution.

The well known Mann Kendal non-parametric trend test is used to find the trend (if any) in the frequency of different rainfall events.

Non-monotonic or abrupt change point (climate jump) is the sudden changes in climate pattern which is also known as climate jump. The test used here is for change point detection developed by Pettitt (1979) which is a non-parametric test, which is useful for evaluating the occurrence of abrupt changes in climatic records

(Smadi and Zghoul 2006; Sneyers 1990; Tarhule and Woo 1998). One of the reasons for using this test is that it is more sensitive to breaks in the middle of the time series (Wijngaard et al. 2003). The statistic used for the Pettitt's test has been explained by Dhorde and Zarenistanak (2013), Kang and Yusof (2012) and many others. It is computed as follows:

The two samples $(x_1, x_2, \ldots, x_t)$ and $(x_{t+1}, x_{t+2}, \ldots, x_T)$ come from the same population. The test statistic $U_{t,T}$ is given by

$$U_{t,T} = \sum_{i=1}^{t} \sum_{j=t+1}^{T} \text{Sgn}(x_i - x_j)$$

The most significant change-point is found where the value $|U_{t,T}|$ is max: $K_T = \max|U_{t,T}|$ and the significant level associated with KT^+ or KT^- is determined approximately by

$$s = \exp\left(\frac{-6K_T^2}{T^3 + T^2}\right)$$

If ρ is smaller than the specific significance level, e.g., 0.05 in this study, the null hypothesis is rejected. In other words, if a significant change point exists, the time series is divided into two parts at the location of the change point.

In spatial scales, our analysis covers (1) the all-India area weighted rainfall and (2) regional scale covering 36 meteorological sub-divisions.

3 Variation of Rainfall in Regional Scale

3.1 Period 1901–1950

Figure 1 shows the results of trend analysis of June, July, August and September rainfall and also for the southwest monsoon rainfall for the period 1901–1950 for all the 36 meteorological sub-divisions. Not a single meteorological sub-division showed significant decaying trend in monsoon rainfall in this period. Eight sub-divisions showed significant increasing trends in the monsoon rainfall. Mostly regions of central India showed increasing trend. The sub-divisions Konkan & Goa, Madhya Maharashtra, west and east Madhya Pradesh, Chhattisgarh and Vidarbha from the central India, Nagaland, Manipur, Mizoram and Tripura from extreme northeastern parts and Himachal Pradesh from northern parts of India showed increasing trend. However there was large variation in monthly scale where three sub-divisions in June, eight in July, one in August and four sub-divisions in September showed increasing trend.

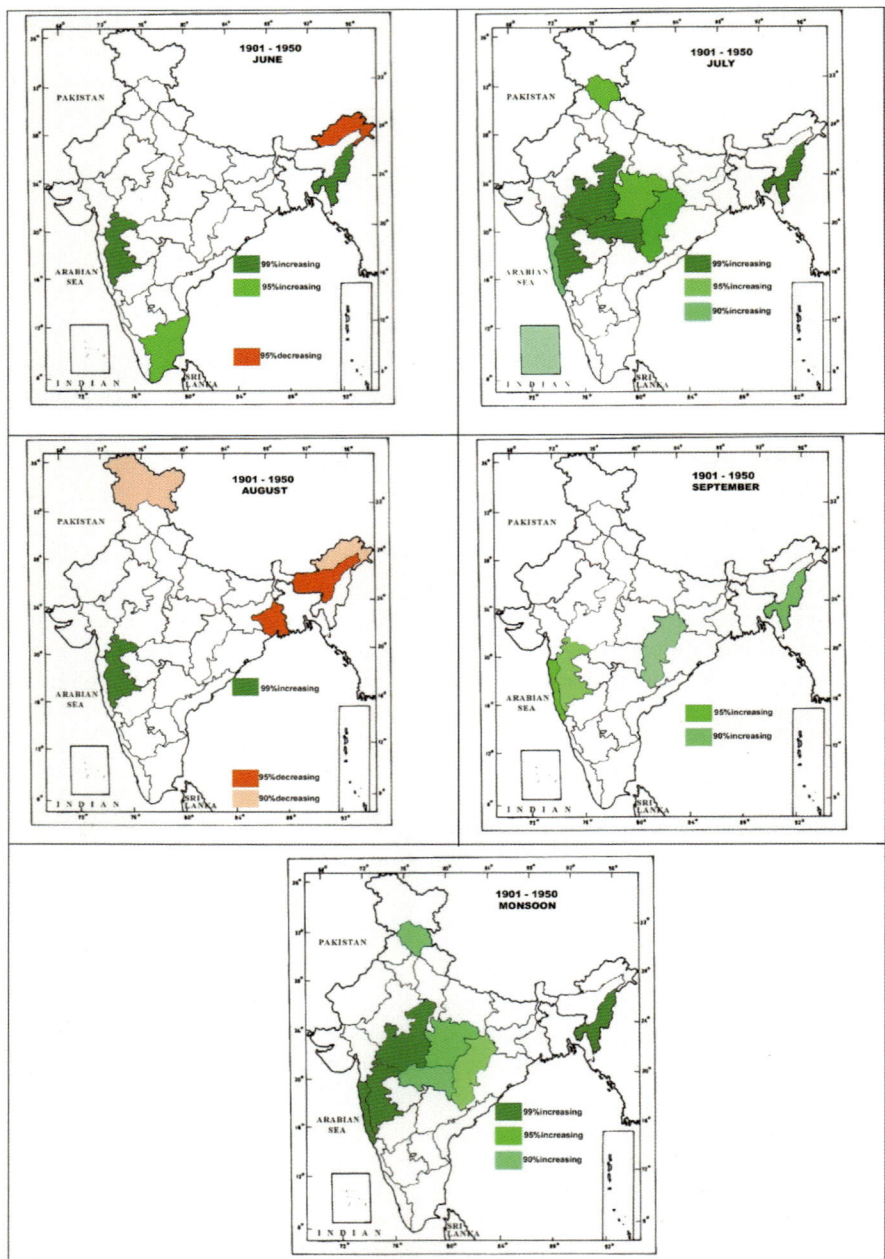

Fig. 1 Trends in the monthly and seasonal rainfall for the 36 meteorological sub-divisions of India for the period 1901–1950

3.2 Period 1951–2011

This period is dominated by decreasing trends in rainfall in many subdivisions (Fig. 2). Seven sub-divisions viz. Himachal Pradesh, west and east Uttar Pradesh, Chhattisgarh, sub-Himalayan West Bengal, Arunachal Pradesh and Nagaland, Manipur, Mizoram and Tripura showed significant decreasing trend in monsoon rainfall while Gangetic West Bengal showed significant increasing trend. In the monthly scale, only June rainfall has shown significant increasing trend in six sub-divisions with no increasing trend in subdivision rainfall for the other monsoon months.

3.3 Period 1901–2011

Figure 3 shows spatial variation of trend in rainfall for the monsoon months and also for the monsoon season for the period 1901–2011. Major changes are noticed in August rainfall where nine sub-divisions viz. south interior Karnataka, coastal Karnataka, Konkan & Goa, Madhya Maharashtra, Marathwada, Vidarbha, Telengana, coastal Andhra Pradesh and west Madhya Pradesh have reported significant increasing trends while eight subdivisions mostly from eastern, north eastern and northern parts of the country viz. Chhattisgarh, Bihar, Jharkhand, Arunachal Pradesh, Nagaland, Manipur, Mizoram & Tripura, east Uttar Pradesh, Uttarakhand and Himachal Pradesh have reported significant decreasing trends. Rainfall has decreased significantly in all the 4 months and as a result in monsoon season for the two sub-divisions from north eastern region viz. Arunachal Pradesh and Nagaland, Manipur, Mizoram & Tripura. Monsoon rainfall has increased significantly in five sub-divisions viz. south interior Karnataka, coastal Karnataka, Konkan & Goa, Jammu & Kashmir, and Gangetic West Bengal. Chhattisgarh, Jharkhand, Uttarakhand, Himachal Pradesh, Arunachal Pradesh, Nagaland, Manipur, Mizoram & Tripura, sub Himalayan West Bengal, Kerala, east Uttar Pradesh and east Madhya Pradesh have shown significant decreasing trends in monsoon rainfall during the period 1901–2011.

4 Annual 1-Day Maximum Rainfall and Return Period Analysis

From the daily rainfall, annual 1-day maximum rainfall series are prepared for each station. Table 1 gives the highest ever recorded 1-day point rainfall of more than 90 cm over India. We have 19 occurrences of more than 90 cm point rainfall. In this table we have shaded the cases which had occurred since the year 1970. Out of 19 cases 11 cases happened after 1970 indicating increasing frequency of occurrence of extreme rainfall events in the recent years. Also there are only five occurrences of more than 90 cm 1 day point rainfall in India other than northeastern states.

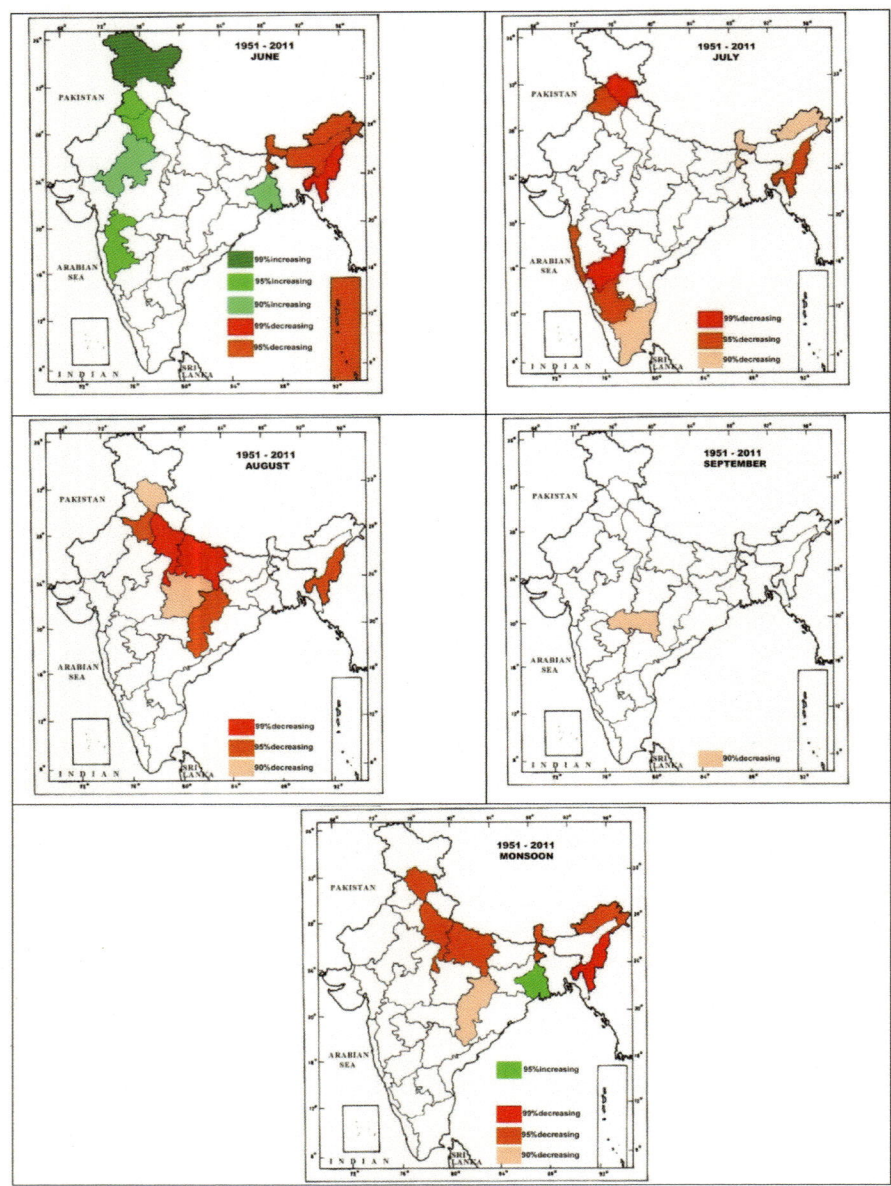

Fig. 2 Trends in the monthly and seasonal rainfall for the 36 meteorological sub-divisions of India for the period 1951–2011

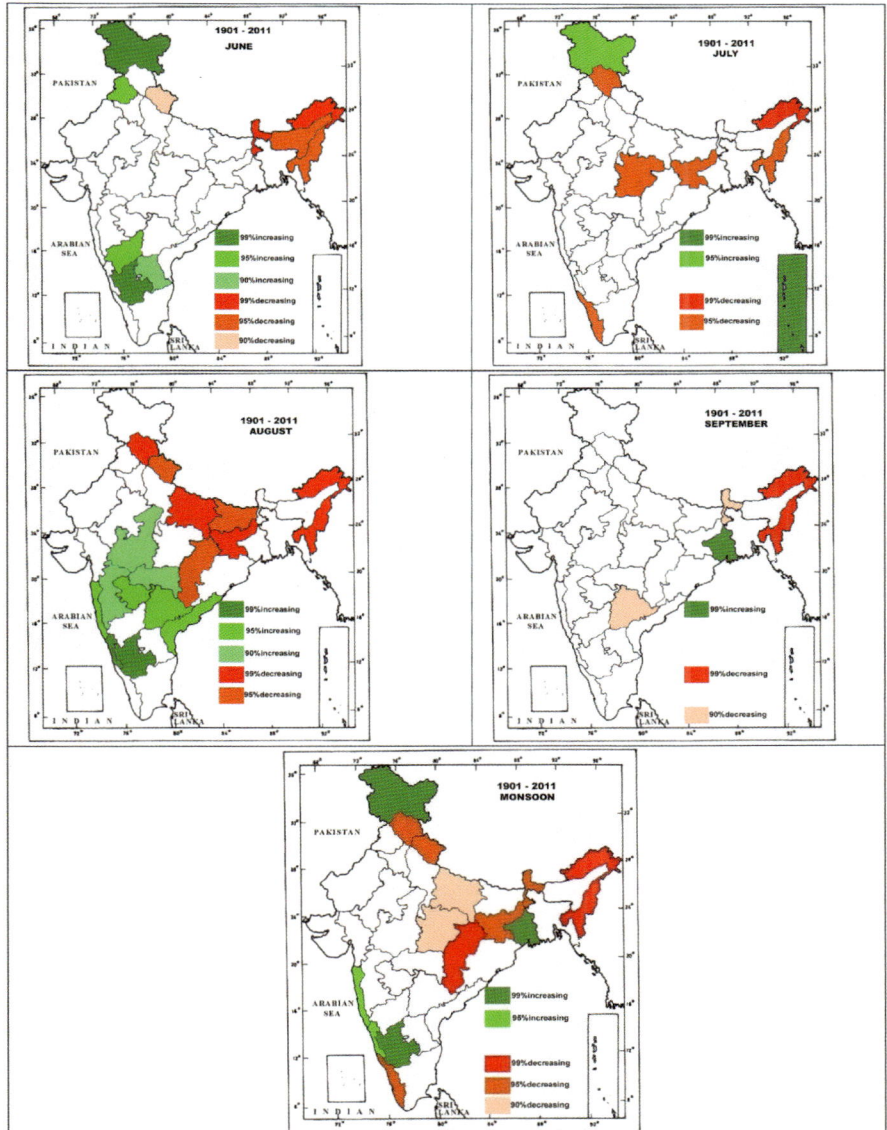

Fig. 3 Trends in the monthly and seasonal rainfall for the 36 meteorological sub-divisions of India for the period 1901–2011

Table 1 List of stations having highest 1-day point rainfall of more than 90 cm

	Station	State	1-day rainfall in cm	Date of occurrence
1	Cherrapunji Obsy	Meghalaya	156.3	16 Jun 1995
2	Amini Divi	Lakshadweep	116.8	6 May 2004
3	Cherrapunji	Meghalaya	103.6	14 Jun 1876
4	Ambernath	Maharashtra	101.0	27 Jul 2005
5	Cherrapunji	Meghalaya	99.8	12 Jul 1910
6	Mawsynram	Meghalaya	99.0	10 Jul 1952
7	Dharampur	Gujarat	98.7	2 Jul 1941
8	Cherrapunji	Meghalaya	98.5	13 Sep 1974
9	Mawsynram	Meghalaya	98.0	4 Aug 1982
10	Tamenlong	Manipur	98.0	10 Aug 1970
11	Cherrapunji	Meghalaya	97.4	5 Jun 1956
12	Mawsynram	Meghalaya	94.5	7 Jun 1966
13	Mumbai	Maharashtra	94.4	27 Jul 2005
14	Tamenlong	Manipur	94.0	28 Jul 1970
15	Cherrapunji	Meghalaya	93.0	15 Jun 1995
16	Guna	Madhya Pradesh	92.8	23 Aug 1982
17	Cherrapunji Obsy	Meghalaya	92.5	21 Jun 1934
18	Cherrapunji	Meghalaya	92.1	21 Jun 1934
19	Cherrapunji	Meghalaya	90.7	25 Jun 1970

We have done return period analysis of 1-day extreme rainfall series of 25 years return period for the period 1901–1950 and 1951–2005 to see the changes in flood risk. The extreme value (return value) x for a period of n years is determined by the following formula:

$$x = \psi - \beta \ln\left[-\ln(F)\right]$$

where ψ = average $- \gamma\beta$ (where γ is Euler's constant, approximately 0.557), $\beta = 0.78\sigma$ (where σ is the standard deviation) and $F = (n-1)/n$.

Spatial variation of extreme values of 25 years return period are clearly seen in Figs. 4 and 5 for the period of 1901–1950 and 1951–2005 respectively. Figure 6 brings out the changes in the flood risk between 1951 and 2005 and 1901–1950. The flood risk has increased in recent period (1951–2005) in most parts of the country except Chhattisgarh, parts of Odisha, Bihar, extreme west Rajasthan and west Madhya Pradesh. The increase is more in West Bengal, Assam & Meghalaya, Jharkhand, coastal Odisha, coastal Andhra Pradesh, Uttarakhand and adjoining areas and Kutch.

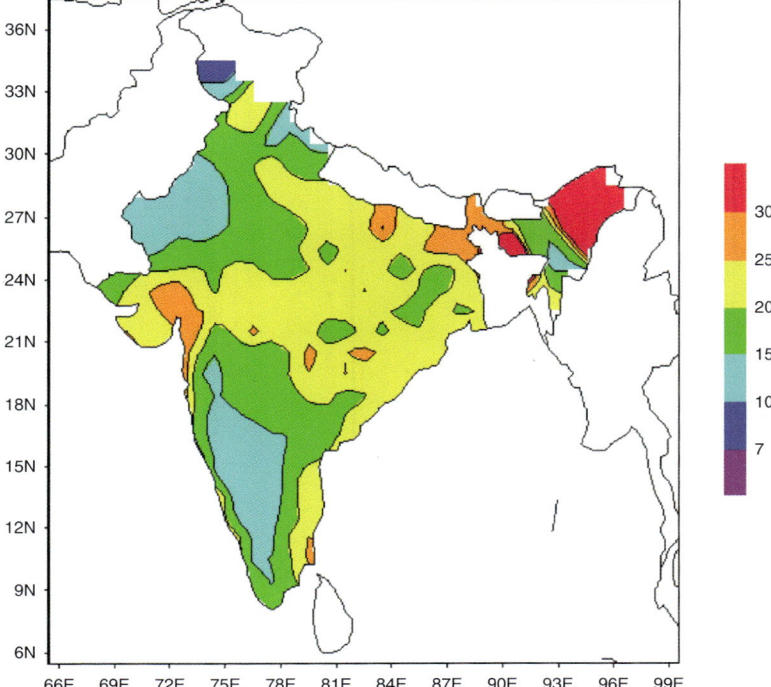

Fig. 4 Extreme values in cm for 25 years return period during 1901–1950

5 Changes in the Frequency of Rainfall of Different Categories

The daily rainfall has been classified in three categories (according to India Meteorological Department).

1. Rain day   Rainfall of amount 0.1 mm or more
2. Rainy day   Rainfall of amount greater than 2.4 mm but less than 64.4 mm
3. Heavy rainfall day   Rainfall of amount 64.5 mm or more

Heavy rainfall day here includes heavy, very heavy and extremely heavy categories.

The frequencies of all the three categories of rainfall stated above are computed from the daily rainfall data of all the stations for the four monsoon months and monsoon season. The trend analysis are then done on the time series of frequencies considering three different periods viz. 1901–1950, 1951–2011 and the complete

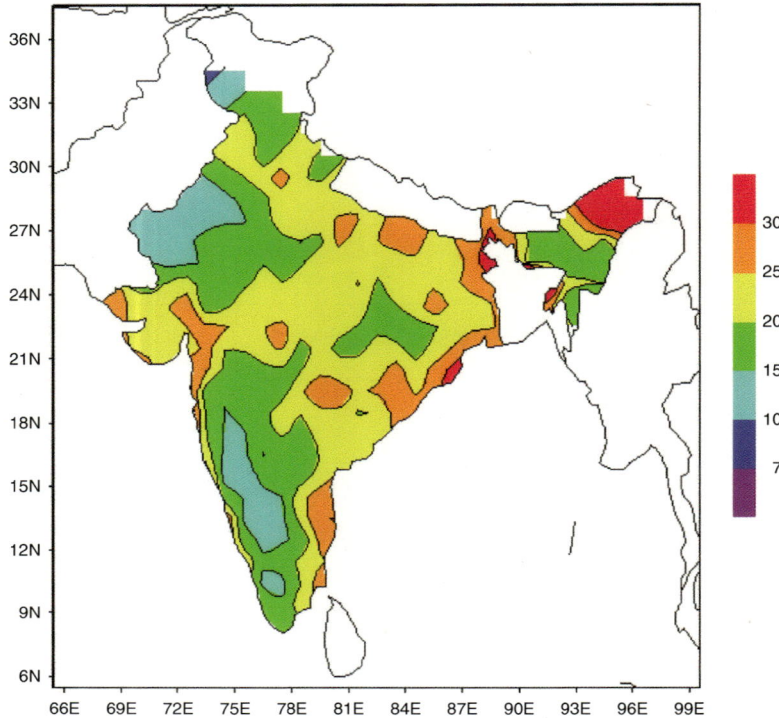

Fig. 5 Extreme values in cm for 25 years return period during 1951–2005

period 1901–2011 latest available. Since there are many missing years and the availability of the data for all the stations are not same we have all the stations which have more than 25 years for the first and second part of the period and all the stations which have more than 50 years of data for the complete period. To know the distribution of properties of trend spatially or more specifically for each of the 36 meteorological sub-divisions, we have calculated the number of percentage of total stations showing decreasing trend for each sub-divisions. The consideration of frequency of station in percentage is due to unequal number of stations in three periods (7,598 stations in period 1951–2011, 2,672 stations in the period 1901–1950 and 7,650 stations in the period 1901–2011).

Tables 2, 3 and 4 show the results of the trend analysis for the 36 met-subdivisions for the monsoon season during the three periods and for the frequency of rain day, rainy day and heavy rainfall day respectively. We can see that number of sub-divisions where +ve trend is more than the −ve trends are more in the period of 1901–1950 than the later period for all the three categories of rainfall. In the recent period (1951-latest) positive (increasing) trends in heavy rainfall frequency was more for the sub-divisions Jharkhand, Odisha, Gangetic and sub Himalayan West Bengal, Arunachal Pradesh and west Rajasthan whereas frequency of decreasing

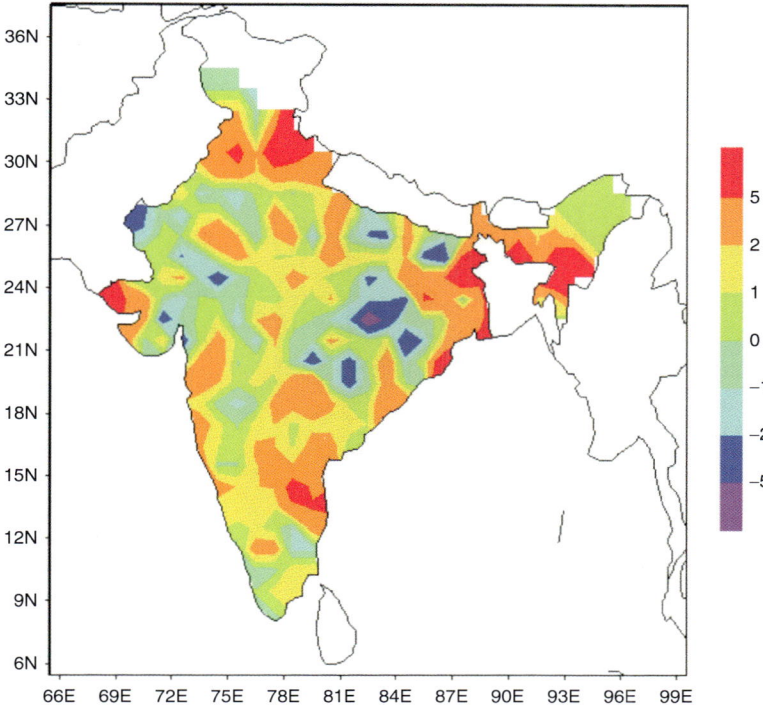

Fig. 6 Changes in flood risk (extreme values) in cm of 25 years return period during 1951–2005 from 1901 to 1950

trends in heavy rainfall are more in central India. In other words we can say the intensity of extreme rainfall events have decreased significantly in the central region during the period 1951-latest. Consideration of the complete period 1901-latest shows also that the intensity of extreme rainfall events have decreased significantly in the central region while it has been increased in peninsular India and coastal regions of eastern India and also eastern Rajasthan, Punjab, Haryana and Nagaland, Manipur, Mizoram and Tripura.

6 Change Point or Climate Jump in Southwest Monsoon Rainfall over India

The Pettitt test is used to detect the change point in southwest monsoon rainfall over 36 meteorological subdivisions of India. Figure 7 shows the change point (year) for each of the 36 met-subdivisions. The significant change point (95 % confidence level) is highlighted with red colour whereas text with black indicates change but

Table 2 Percentage frequency of number of stations in each subdivision having increasing (+ve) trend and decreasing (−ve) in rain days during southwest monsoon

Subd no.	No. of stns	% of stn 1901–1950		No. of stns	% of stns 1951–2011		No. of stns	% of stns 1901–2011	
		+Trend	−Trend		+Trend	−Trend		+Trend	−Trend
1	1	0	0	7	0	85.7	7	0	14.3
2	3	0	0	50	0	0	51	0	0
3	105	7.6	34.3	217	0.9	6	219	0.5	30.1
4	28	7.1	14.3	96	0	6.3	96	0	10.4
5	27	11.1	29.6	60	3.3	8.3	61	1.6	24.6
6	116	6	19.8	154	2.6	3.9	166	0.6	20.5
7	119	5.9	35.3	377	**12.7**	3.2	378	0.5	17.2
8	97	**19.6**	9.3	209	2.4	16.7	210	0	24.8
9	147	7.5	21.8	428	3.7	11.9	428	0.2	19.2
10	126	**32.5**	4	195	1	47.2	195	0.5	60
11	127	**22**	3.1	172	1.2	48.8	172	0	71.5
12	39	**20.5**	2.6	98	0	42.9	99	0	35.4
13	52	7.7	15.4	103	**6.8**	3.9	104	3.8	9.6
14	40	**15**	5	94	0	22.3	94	0	19.1
15	19	0	10.5	106	2.8	30.2	106	0.9	17
16	41	7.3	24.4	72	4.2	4.2	83	1.2	9.6
17	66	1.5	16.7	149	**5.4**	4	159	**8.2**	4.4
18	136	**10.3**	2.2	375	**10.7**	4.8	378	7.1	9
19	110	**32.7**	1.8	212	1.4	27.4	213	5.2	22.1
20	88	**25**	0	159	0	35.8	159	0	44.7
21	83	**4.8**	2.4	238	0	25.6	238	0	26.1
22	30	3.3	13.3	165	1.8	13.9	165	0.6	9.1
23	51	**15.7**	2	76	0	18.4	77	2.6	24.7
24	110	9.1	9.1	166	1.8	27.1	167	1.8	39.5
25	8	**12.5**	0	70	1.4	12.9	70	1.4	4.3
26	88	**35.2**	0	160	1.3	18.1	160	0	35.6
27	74	**23**	2.7	117	0	46.2	117	0.9	46.2
28	103	2.9	33	490	3.5	7.3	490	1.4	7.6
29	64	3.1	3.1	460	0.2	15.7	462	0.2	9.3
30	71	0	23.9	291	4.8	7.2	291	1.4	11.7
31	235	5.1	22.1	612	1.6	23	612	1.5	29.4
32	22	**36.4**	0	139	1.4	7.2	139	0	7.2
33	49	**16.3**	18.4	440	**7.7**	3.9	440	**4.1**	3.4
34	115	1.7	27.8	725	**23.3**	1.9	725	**5.9**	3.2
35	80	**16.3**	5	111	0	31.5	114	0.9	59.6
36	2	**50**	0	5	0	0	5	**20**	0

Table 3 Percentage frequency of number of stations in each subdivision having increasing (+ve) trend and decreasing (−ve) in rainy days during southwest monsoon

Subd no.	Stns	% of stn 1901–1950		Stns	% of stns 1951–2011		Stns	% of stns 1901–2011	
		+Trend	−Trend		+Trend	−Trend		+Trend	−Trend
1	1	0	0	7	0	85.7	7	0	14.3
2	3	0	0	50	0	0	51	0	0
3	105	1	21.9	217	1.4	5.1	219	0	25.6
4	28	7.1	7.1	96	1	3.1	96	1	7.3
5	27	0	11.1	60	3.3	6.7	61	0	18
6	116	4.3	15.5	154	0.6	3.2	166	0.6	19.3
7	119	0.8	24.4	377	**15.4**	2.4	378	0.3	9.3
8	97	**13.4**	8.2	209	3.3	9.6	210	0	19
9	147	5.4	17	428	2.8	8.9	428	0	16.6
10	126	**19**	3.2	195	0.5	35.9	195	0	57.4
11	127	**18.9**	0	172	0.6	39.5	172	0	54.1
12	39	**12.8**	0	98	0	42.9	99	0	34.3
13	52	1.9	9.6	103	1.9	5.8	104	1.9	4.8
14	40	**10**	2.5	94	0	22.3	94	0	14.9
15	19	**5.3**	0	106	0.9	30.2	106	0	17
16	41	0	19.5	72	1.4	1.4	83	1.2	9.6
17	66	0	6.1	148	2	4.1	158	2.5	3.2
18	136	**7.4**	0.7	375	3.7	4	378	1.9	3.2
19	110	**39.1**	0.9	212	0.9	19.3	213	3.3	10.8
20	88	**20.5**	0	159	0	27	159	0	31.4
21	83	0	0	238	0.4	3.4	238	0	3.4
22	30	0	0	165	0.6	3	165	0.6	0.6
23	51	2	0	76	5.3	5.3	77	0	3.9
24	110	**14.5**	0	166	2.4	7.2	167	**8.4**	4.2
25	8	0	0	70	0	11.4	70	0	2.9
26	88	**28.4**	0	160	0.6	11.9	160	0	12.5
27	74	**14.9**	1.4	117	0	48.7	117	0	47
28	103	3.9	8.7	490	1.4	6.1	490	1	2.4
29	64	**3.1**	0	460	0.2	10.2	462	0	5.2
30	71	0	2.8	291	**4.5**	2.4	291	1.7	2.4
31	235	0.9	10.2	611	1.5	25.2	611	1.3	24.2
32	22	**13.6**	0	139	0	1.4	139	0	0
33	49	**6.1**	4.1	440	2	2.5	440	**1.8**	0.5
34	115	3.5	6.1	725	**16.7**	0.4	725	**2.1**	1.7
35	80	3.8	3.8	111	0	25.2	114	0.9	49.1
36	2	**50**	0	5	0	0	5	**20**	0

Table 4 Percentage frequency of number of stations in each subdivision having increasing (+ve) trend and decreasing (−ve) in heavy rainfall days during southwest monsoon

Subd no.	Stns	% of stn 1901–1950		Stns	% of stns 1951–2011		Stns	% of stns 1901–2011	
		+Trend	−Trend		+Trend	−Trend		+Trend	−Trend
1	1	0	0	7	0	0	7	0	0
2	3	0	0	50	2	0	51	0	0
3	105	6.7	7.6	214	1.9	1.9	216	2.8	6.5
4	28	**14.3**	0	93	1.1	2.2	93	**2.2**	1.1
5	27	3.7	3.7	60	**6.7**	1.7	61	1.6	6.6
6	116	3.4	6	154	**1.9**	0.6	166	**2.4**	1.8
7	118	**7.6**	0.8	375	**11.5**	0.8	377	**4**	1.1
8	97	**6.2**	2.1	208	**8.2**	6.7	209	1.9	9.1
9	147	5.4	8.8	427	3.7	3.7	427	2.1	4.2
10	126	**5.6**	1.6	191	2.1	18.8	191	1	22
11	127	**7.1**	3.9	169	2.4	27.8	169	2.4	16.6
12	39	2.6	7.7	93	1.1	24.7	94	0	24.5
13	52	0	3.8	100	3	5	101	**8.9**	1
14	40	**5**	0	93	0	14	93	**3.2**	0
15	19	0	0	98	2	18.4	98	0	7.1
16	34	0	2.9	61	0	0	70	0	0
17	64	**3.1**	0	136	**2.2**	0	147	1.4	1.4
18	136	**2.2**	0.7	369	1.9	1.9	372	**2.7**	0.8
19	110	**22.7**	1.8	209	1.4	8.1	210	1	3.3
20	88	**8**	0	159	1.9	4.4	159	1.3	8.2
21	83	**6**	0	236	**2.1**	0.8	236	**5.1**	0
22	30	**10**	0	162	**1.2**	0.6	162	**1.2**	0
23	51	**11.8**	0	76	0	26.3	77	**16.9**	5.2
24	110	**11.8**	0	165	1.2	4.2	166	**4.8**	1.2
25	8	0	0	70	**2.9**	0	70	**4.3**	0
26	88	**2.3**	0	160	0	3.1	160	1.3	6.3
27	73	**6.8**	0	116	1.7	20.7	117	0	12.8
28	103	1	1.9	478	1.5	1.7	478	**1**	0.6
29	64	0	0	450	**1.3**	0.9	452	**1.8**	0
30	71	**1.4**	0	277	**0.4**	0	277	**0.4**	0
31	232	1.3	1.3	550	1.3	2.9	553	2.2	2.4
32	22	0	0	139	1.4	2.9	139	**0.7**	0
33	44	0	0	424	0.9	1.2	424	**0.5**	0
34	114	0	0.9	700	2.3	3.7	700	**1.3**	0.6
35	80	1.3	5	111	0	17.1	114	0.9	25.4
36	2	**50**	0	5	0	0	5	0	0

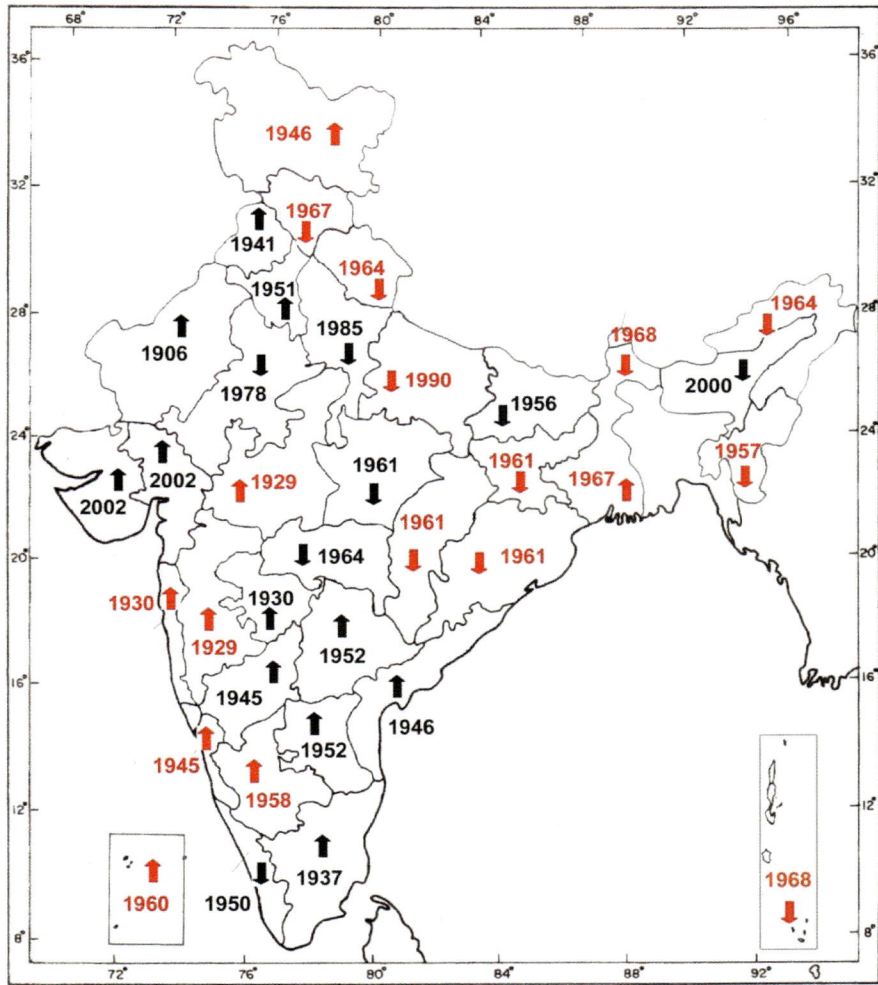

Fig. 7 The climate jump or change point (year) in southwest monsoon rainfall over 36 meteorological sub-divisions in India. *Red colour* indicates the change is significant (95 % confidence level). Upward and downward *arrows* represent upward and downward shift respectively

not significant. Also upward and downward arrows are used to indicate the upward shift or downward shift in southwest monsoon rainfall pattern.

Figure 7 indicates that there was a upward climate shift over Peninsular India around 1945–1955. Around 1960 there was downward shift over Chhattisgarh, east M.P., Vidarbha, Odisha, Jharkhand, Bihar, Sub Himalayan, West Bengal and most part of the north eastern region. It is necessary to study other climate features and their shift in light of the shift of rainfall pattern.

7 Rainfall Analysis for Neighbouring Countries: Bangladesh and Nepal

Using the same categories of rainfall i.e. dry day if rainfall is zero, rain day, rainy day and heavy rainfall day as stated earlier we have done trend analysis (Mann Kendal non-parametric test) of rainfall stations (Table 5) for the period 1947–1983 of Nepal and of Bangladesh (Table 6) for the period 1891–1946 and 1949–2011.

Figure 8 shows the results of trend analysis for Nepal. Significant increase in the frequency of dry day during SW monsoon over two stations in eastern parts of Nepal has been noticed. Even other four stations in this region have also reported increasing trend but not significant. Frequency of rain day has shown significant decreasing trends over three stations of east Nepal. There is no significant trend in the frequency of rainy days. But the frequency of heavy rainfall events has decreased in all the stations with significant trend in three stations during the period 1947–1983.

Comparative results of trend analysis of two periods viz. 1891–1946 and 1949–2011 of the frequency of rainfall events of different intensities in Bangladesh are shown in Fig. 9. Frequency of dry day has shown significant decreasing trends in eight out of 22 stations while only one station has significant increasing trend during the recent period. Almost similar pattern was present during 1891–1946. In general frequency of rain day and rainy day have shown increasing trends during both the periods. However frequency of heavy rainfall events has significantly decreased during the recent period which was not seen in the previous period. Thus for Bangladesh it can be concluded that southwest monsoon has become more favourable for agriculture and water sectors as dry days are becoming lesser and days with rainfall of moderate intensities are increasing.

Table 5 List of stations of Nepal considered for analysis

Station name	Latitude	Longitude
Amlekganj	27.28	85.00
Bairia	27.00	85.38
Bijapur	29.23	81.63
Chatra	26.82	87.15
Dhulikhel	27.62	85.55
Dumuhan	27.35	87.60
Ghumthang	27.87	85.87
Kathmandu	27.70	85.33
Munga	27.03	87.23
Sirha	26.65	86.22
Udaipur	26.93	86.52

Table 6 List of stations of Bangladesh considered for analysis

Station name	Latitude	Longitude
Barisal	22.75	90.37
Bogra	24.85	89.37
Chittagong	22.27	91.82
Comilla	23.43	91.18
Coxs Bazar	21.45	91.97
Dhaka	23.77	90.38
Dinajpur	25.65	88.68
Faridpur	23.60	89.85
Hatiya	22.43	91.10
Jessore	23.18	89.17
Khulna	22.78	89.53
M. Court	22.87	91.10
Mymensingh	24.72	90.43
Rajshahi	24.37	88.70
Rangamati	22.53	92.20
Rangpur	25.73	89.23
Sandwip	22.48	91.43
Satkhira	22.72	89.08
Sitakunda	22.58	91.70
Srimangal	24.30	91.73
Sylhet	24.90	91.88
Teknaf	20.87	92.30

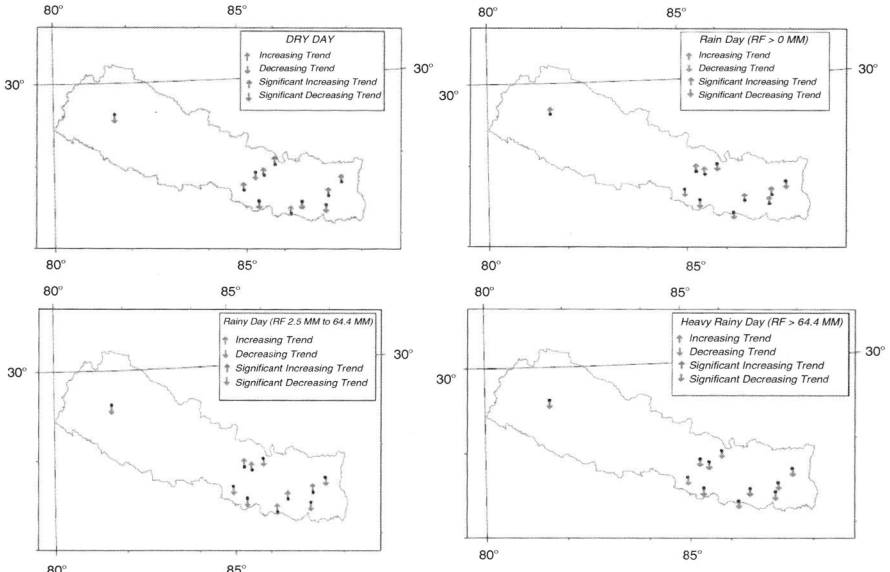

Fig. 8 Results of trend analysis of frequency of rainfall events of different intensities of Nepal

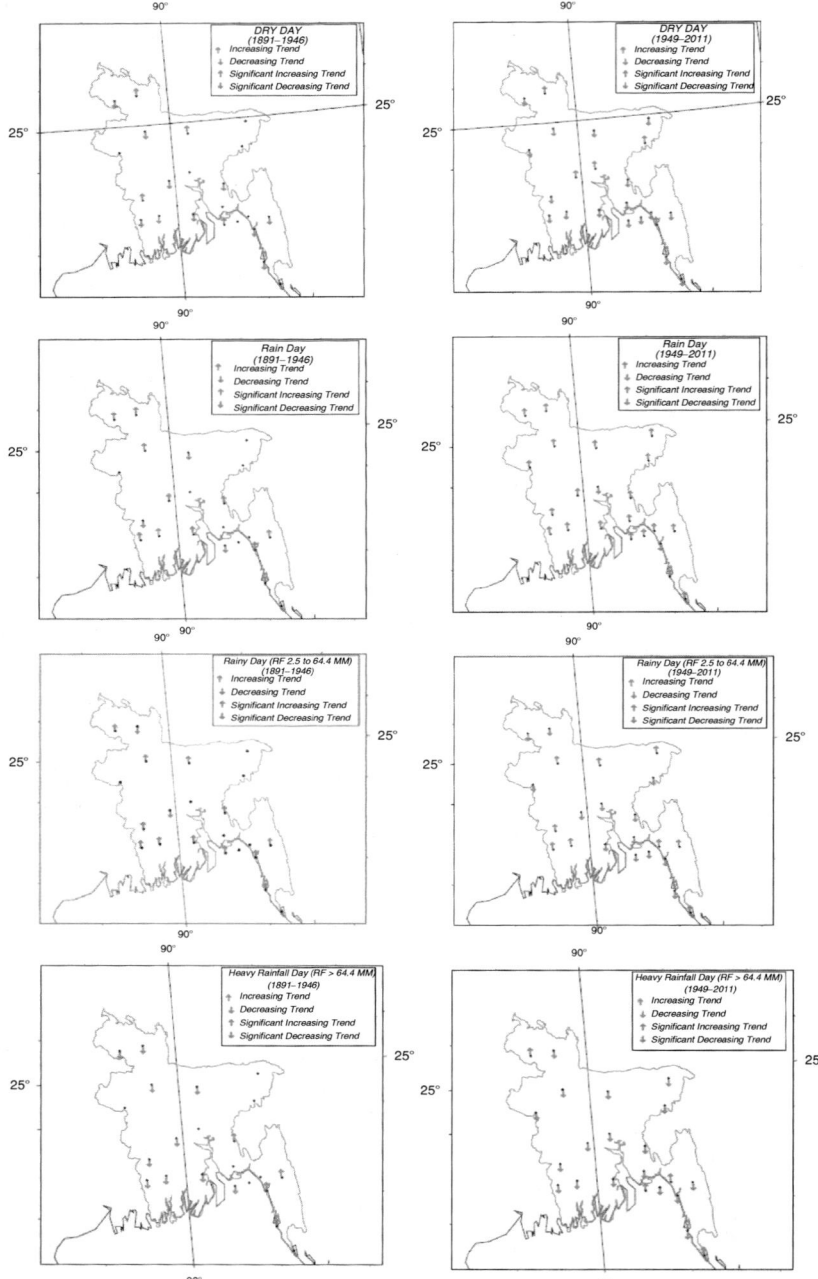

Fig. 9 Results of trend analysis of frequency of rainfall events of different intensities of Bangladesh for the periods 1891–1946 and 1949–2011

Conclusions

The rainfall data of longer series (1901–2011) were analyzed to study the temporal and spatial variability of rainfall pattern in India. The earlier study by Guhathakurta and Rajeevan (2008) analyzed the rainfall data for the period 1901–2003 and have shown the significant changes in the rainfall pattern in many meteorological sub-divisions of India. However during 2004–2011 country experienced two droughts and it necessitated to detect the changes in the rainfall pattern in the regional scale using the upto date data.

The trend analysis in the regional scales for the recent period 1951–2011 shows significant increasing trend in June rainfall only in six sub-divisions viz. Madhya Maharashtra; east Rajasthan; Punjab; Haryana, Delhi & Chandigarh; Jammu & Kashmir and Gangetic West Bengal while no sub-divisions reported increasing trend in other monsoon months. Seven sub-divisions viz. Himachal Pradesh, west and east Uttar Pradesh, Chhattisgarh, sub-Himalayan West Bengal, Arunachal Pradesh and Nagaland, Manipur, Mizoram & Tripura showed significant decreasing trend in monsoon rainfall while Gangetic West Bengal showed significant increasing trend in monsoon rainfall.

Analysis of the rainfall data for the complete period reveals that monsoon rainfall has increased significantly in five sub-divisions viz. south interior Karnataka, coastal Karnataka, Konkan & Goa, Jammu & Kashmir, and Gangetic West Bengal. Chhattisgarh, Jharkhand, Uttarakhand, Himachal Pradesh, Arunachal Pradesh, Nagaland, Manipur, Mizoram & Tripura, sub Himalayan West Bengal, Kerala, east Uttar Pradesh and east Madhya Pradesh have shown significant decreasing trends in monsoon rainfall. In August rainfall in nine sub-divisions viz. south interior Karnataka, coastal Karnataka, Konkan & Goa, Madhya Maharashtra, Marathwada, Vidarbha, Telengana, coastal Andhra Pradesh and west Madhya Pradesh have reported significant increasing trends while eight subdivisions mostly from eastern, north eastern and northern parts of the country viz. Chhattisgarh, Bihar, Jharkhand, Arunachal Pradesh, Nagaland, Manipur, Mizoram & Tripura, east Uttar Pradesh, Uttarakhand and Himachal Pradesh have reported significant decreasing trends. In July six subdivisions viz. Kerala, Jharkhand, east Madhya Pradesh, Himachal Pradesh, Arunachal Pradesh and Nagaland, Manipur, Mizoram & Tripura have reported decreasing trends while Jammu & Kashmir and Andaman & Nicobar Island have reported increasing trends.

Analysis of frequency of rainfall occurrence of different categories shows that frequencies of rain day have decreased in most parts of the country except west Rajasthan north and south interior Karnataka and Lakshadweep. Frequencies of rainy day (rainfall of moderate intensity in a day) have decreased in most parts of the country except Madhya Maharashtra, north and south interior Karnataka and Lakshadweep. In the large regions of the country i.e. 17 sub-divisions viz. Nagaland, Manipur, Mizoram & Tripura, Gangetic

(continued)

West Bengal, Odisha, Haryana, Delhi & Chandigarh, Punjab, east Rajasthan, Gujarat, Saurashtra & Kutch, Konkan & Goa, Madhya Maharashtra, Marathwada, coastal Andhra Pradesh, Rayalaseema, Telengana, coastal Karnataka, north and south interior Karnataka, increasing trends in the frequency of heavy rainfall events dominated during the period 1901–2011. In the recent period (1951–2011) only in the 11 subdivisions increasing trends in the frequency of heavy rainfall were dominant. However, in the central region there was no significant increase in the frequency of heavy rainfall events.

Changes in flood risk over India has been tested by comparing extreme values of 25 years return period between 1901 and 1950 and 1951–2005. The flood risk has increased in recent period (1951–2005) in most parts of the country except Chhattisgarh, parts of Odisha, Bihar, extreme west Rajasthan and west Madhya Pradesh. The increase is more in West Bengal, Assam & Meghalaya, Jharkhand, coastal Odisha, coastal Andhra Pradesh, Uttarakhand and adjoining areas and Kutch.

In Bangladesh dry days are becoming less and days with rainfall of moderate intensities are increasing. Frequency of heavy rainfall events (more than 64.5 mm in a day) has decreased in all the stations during 1949–2011. In Nepal also frequency of heavy rainfall events has decreased in all the stations during 1947–1983.

References

Bawiskar SM (2009) Weakening of lower tropospheric temperature gradient between Indian landmass and neighbouring oceans and its impact on Indian monsoon. J Earth Syst Sci 118(4):273–280

Chung CE, Ramanathan V (2006) Weakening of North Indian SST Gradients and the monsoon rainfall in India and the Sahel. J Clim 19:2036–2045

Dhorde AG, Zarenistanak M (2013) Three-way approach to test data homogeneity: an analysis of temperature and precipitation series over southwestern Islamic Republic of Iran. J Indian Geophys Union 17(3):233–242

Guhathakurta P, Rajeevan M (2008) Trends in rainfall pattern over India. Int J Climatol 28:1453–1469

Guhathakurta P, Koppar AL, Krishan U, Menon P (2011a) New rainfall series for the districts, meteorological sub-divisions and country as whole of India. National Climate Centre Research Report No. 2/2011, India Meteorological Department

Guhathakurta P, Sreejith OP, Menon PA (2011b) Impact of climate changes on extreme rainfall events and flood risk in India. J Earth Syst Sci 120(3):359–373

Joseph PV, Simon A (2005) Weakening trend of southwest monsoon current through Peninsular India from 1950 to present. Curr Sci 89(4):687–694

Kang HM, Yusof F (2012) Homogeneity tests on daily rainfall series in peninsular Malaysia. Int J Contemp Math Sci 7(1):9–22

Naidu CV, Muni Krishna K, Ramalingeswara Rao S, Bhanu Kumar OSRU, Durgalakshmi K, Ramakrishna SSVS (2011) Variations of Indian summer monsoon rainfall induce the weakening of easterly jet stream in the warming environment. Glob Planet Chang 75:21–30

Pant GB, Rupa Kumar K (1997) Climates of South Asia. Wiley, Chichester

Pettitt AN (1979) A non-parametric approach to the change point problem. J Appl Stat 28(2):126–135

Ramesh KV, Goswami P (2007) Reduction in Spatial and Temporal Extent of Monsoon rainfall. Geophys Res Lett 34: L 23704. doi:10.1029/2007GL031613

Smadi MM, Zghoul A (2006) A sudden change in rainfall characteristics in Amman, Jordan during the mid 1950s. Am J Environ Sci 2(3):84–91

Sneyers S (1990) On the statistical analysis of series of observations, Technical note no. 5 143. Secretariat of the World Meteorological Organization, Geneva

Tarhule A, Woo M (1998) Changes in rainfall characteristics in northern Nigeria. Int J Climatol 18:1261–1271

Wijngaard JB, Klein Tank M, Konnen GP (2003) Homogeneity of 20th century European daily temperature and precipitation series. Int J Climatol 23:679–692

WMO (2008) Guide to hydrological practices, vol 1, WMO no. 168. WMO, Geneva

Long Term Trends in the Extreme Rainfall Events over India

D.S. Pai and Latha Sridhar

1 Introduction

Rainfall and surface air temperature are the two key elements of climate that are commonly used as indicators of global climate change due to the availability of long time series of these elements from most parts of the world. Rainfall, a component of terrestrial hydrological cycle, determines the availability of water and the level of the soil moisture. During the last few decades, rainfall extremes have been found to be increasing around the world with distinct regional differences, and the increase is linked to the warming of the atmosphere which has taken place since pre-industrial times (IPCC 2007). Its immediate implication is the increased flood risk around the world. In India, there have been several studies on daily extreme rainfall events over the region based on station and grid point rainfall data (Sen Roy and Balling 2004; Joshi and Rajeevan 2006; Goswami et al. 2006; Rajeevan et al. 2008; Ghosh et al. 2009; Guhathakurta et al. 2011).

Goswami et al. (2006), using a daily $1° \times 1°$ gridded rainfall data set of Rajeevan et al. (2006) based on fixed network of 1,803 rain gauge stations, examined the trends in the extreme rainfall (ER) events over central India during the southwest monsoon season (June–September) for the period 1951–2000 by classifying the ER events into three categories; moderate rainfall ($\geq$5–100 mm) or MR events, heavy rainfall ($\geq$100 mm) or HR events and very heavy rainfall ($\geq$150 mm) or VHR events. The study of Goswami et al. (2006) observed significant rising trends in the frequency and the magnitude of HR and VHR events and significant decreasing trend in the frequency of MR events resulting in to insignificant trend in the mean rainfall over central India during the period. Rajeevan et al. (2008), on examining the long-term trends of ER events over central India using another daily $1° \times 1°$

D.S. Pai (✉) • L. Sridhar
India Meteorological Department, Pune 411005, India
e-mail: sivapai@hotmail.com

gridded rainfall data set prepared based on a fixed network of 1,380 rain gauge stations for the period 1901–2004, found a statistically significant long-term trend and significant inter-annual and inter-decadal variations in the number of VHR events over central India. Rajeevan et al. (2008) also found significant decreasing trend in the number of MR events and an association with the SST anomalies over the equatorial Indian Ocean and trends in the VHR events.

In this study, long term trends in the ER events during the southwest monsoon season have been examined over three broad geographical regions that are most prone to ER events using a new daily gridded rainfall data set of higher spatial resolution ($0.25° \times 0.25°$, latitude×longitude) covering a longer period of 110 years (1901–2010) over the Indian main land (Pai et al. 2013). These regions are Central India (CI), where the rainfall pattern is influenced by the north-south oscillation of monsoon trough, northeast India (NEI) where the topography plays an important role in deciding the rainfall pattern over the region and west coast (WC), which is on the windward side of the mountain range of Western Ghats.

The Sect. 2 provides details of the various data sets used in this study. The Sect. 3 describes various results of the study and the Sect. 4 presents summary and conclusions.

2 Data Used and Methodology

The main data used in this study is the new daily ($0.25° \times 0.25°$) gridded rainfall data developed by Pai et al. (2013). This grid point data was prepared using daily rainfall data from all the rain gauge stations over the country for the period 1901–2010 readily available in the archive of National Data Centre, IMD, Pune. The daily rainfall records from 6,955 rain gauge stations with varying availability periods were used, which is the highest number of stations used by any studies so far for preparing the grid point rainfall data over the country. Out of these 6,955 stations, 547 were IMD observatory stations, 494 were hydro-meteorology observatories and 74 are agromet observatories. The remaining are rainfall reporting stations maintained by the State Governments. Though the spatial density of the station points was not uniform throughout the country, there was good number of stations representing most areas of the country, particularly the three regions under consideration of this study. On an average, data from about 2,600 stations per year were available for the preparation of daily grid point data. However, the data density varied from year to year from about 1,450 in the first year (1901) to about 3,950 during the period 1991–1994. The data density was relatively higher ($\geq 3,100$ stations per day) from 1951 onwards except in the last 2–3 years when the density reduced to about 1,900 stations per day. More details of the development of this data set can be obtained from Pai et al. (2013).

For the analysis of the ER events over the country, three regions (CI, NEI and WC) most prone for ER events during the southwest monsoon season (June–September) were identified. The daily grid point ER events (and its three categories separately) aggregated over each of the three regions were then used to examine the

3 Results

3.1 Trends in the Frequency of the Extreme Rainfall

Figure 1 depicts the mean of the grid point frequency of ER events over Indian mainland per season averaged over entire data period (1901–2010). As seen in Fig. 1, we can clearly delineate three regions on the map that received mean number of ER events ≥30 per season. The spatial domains of WC, CI and NEI enclose 199, 2,034 and 398 grids respectively. From Fig. 1, we can see that the grid points within

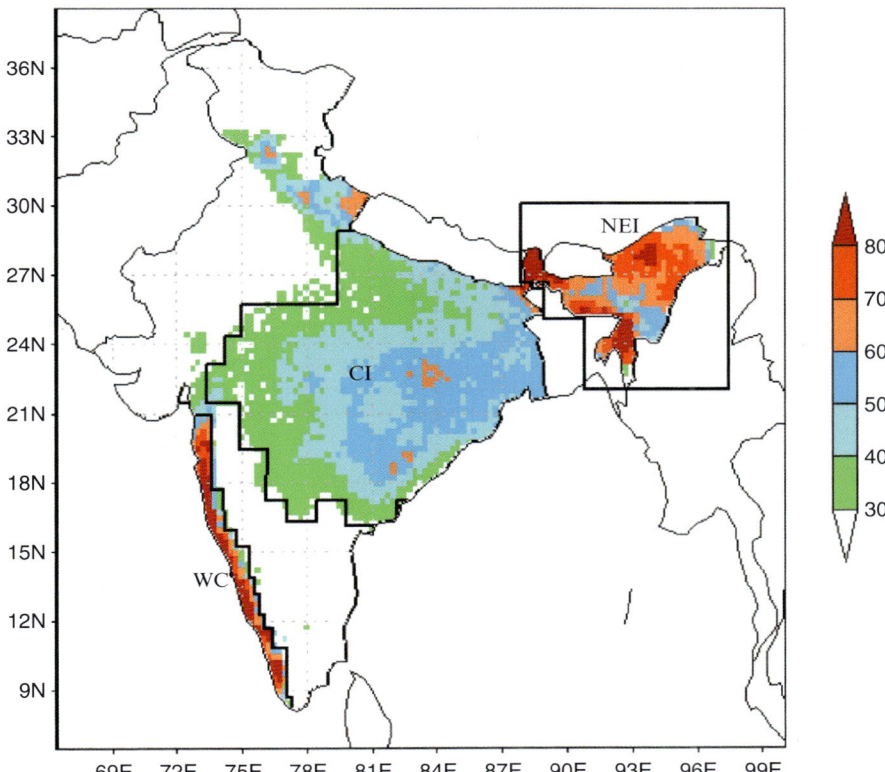

Fig. 1 The spatial distribution of mean and frequency of ER events over India during the SW monsoon season computed for the period 1901–2010. Unit is number/season

CI received about 30–60 days/season of ER events and most of the grid points within the other two regions (WC and NEI) received ≥60 days/season of ER events. Over WC and NEI, the frequency of ER events was much higher over most of the areas. It has been observed that during the data period, ER events occurred during, 35, 54 and 56 %, of the total possible grid point days (number of grid points × number of days (122)) per season respectively over CI, NEI and WC. The highest grid point rainfalls recorded over CI, NEI and WC during the entire data period were 763.38 mm, 939.45 mm and 820.86 mm respectively.

The trends of area weighted rainfall and aggregated grid point ER events over CI, NEI and WC during the southwest monsoon season are given in Tables 1, 2 and 3 respectively. The linear trends computed for the two halves (1901–1955 and 1956–2010) of the data period are also presented in Table 1. The trends that were significant at 95 % (99 %) are shown with a single asterisk or * (double asterisk or **). It was observed that the area weighted rainfall and ER events of all the three regions are highly correlated. The C.C. between the area weighted rainfall and number of

Table 1 Trends in the rainfall over Central India during the southwest monsoon season (June–September)

		Period		
Trends/decade in	Rainfall categories (mm)	1901–2010	1901–1955	1956–2010
Area averaged rainfall (mm/day)	≥0	−0.03	0.15*	−0.09
Aggregated number of grid point events over the region (number)	≥5	−519.7	1664.6*	−1781.7*
	5–100	−545.6	1648.5*	−1863.6*
	100–150	12.6	15.6	44.2*
	≥150	13.2**	0.6	37.6**
Average of 25 heaviest rainfall events (mm/day)	≥5	7.05**	0.86	14.49**

The trends that are significant at 95 % (99 %) are shown with * (**)

Table 2 Trends in the rainfall over Northeast India during the southwest monsoon season (June–September)

		Period		
Trends/decade in	Rainfall categories (mm)	1901–2010	1901–1955	1956–2010
Area averaged rainfall (mm/day)	≥0	−0.11*	0.31**	−0.27
Aggregated number of grid point events over the region (number)	≥5	−179.9**	395.3**	6.2
	5–100	−196.3**	356.3**	41.7
	100–150	4.1	23.5*	−35.7**
	≥150	12.3*	15.5*	0.2
Average of 25 heaviest rainfall events (mm/day)	≥5	17.60**	5.97	20.93*

The trends that are significant at 95 % (99 %) are shown with * (**)

Table 3 Trends in the rainfall over West Coast during the southwest monsoon season (June–September)

Trends/decade in	Rainfall categories (mm)	Period		
		1901–2010	1901–1955	1956–2010
Area averaged rainfall (mm/day)	≥0	0.05	0.45*	−0.27
Aggregated number of grid point events over the region (number)	≥5	−60.8	205.6*	−177.7
	5–100	−72.4*	178.6	−174.3
	100–150	6.7*	19.2*	−6.5
	≥150	5.0**	7.8	3.0
Average of 25 heaviest rainfall events (mm/day)	≥5	4.49**	6.74	6.12*

The trends that are significant at 95 % (99 %) are shown with * (**)

ER events during the data period 1901–2010 for CI, NEI and WC were 0.94, 0.84 and 0.80 respectively. It has been seen that, over CI, about 53 % of the rainfall events were ER events with MR events contributing 52 %, and HR and VHR together contributing only 1 %. Over NEI, about 66 % of rainfall events were ER events (MR = 65 %, HR and VHR = 1 %) and over WC, about 67 % of the rainfall events (rainfall >0) were ER events (MR = 64 %, HR and VHR = 3 %). Thus in all the three regions, most of the ER events were MR events and the HR and HVR events were significantly less compared to MR events.

As seen in Table 1, no significant linear trends were observed in the seasonal rainfall averaged over CI during the data period as during the second half of the data period. However, a positive linear trend of 0.15 mm/day per decade significant at 95 % level was observed in the first half (1901–1955). The long-term linear trend in the number of ER events/season during the total period showed decrease of 519.7 events/decade which is not statistically significant. However, increasing trend of 1664.6 events/decade during the first half and decreasing trend of 1781.7 events/decade during the second half of the data period both significant at 95 % level were observed. Figure 2 shows the interannual variation of category-wise number of ER (MR, HR and VHR) events over CI respectively for the period 1901–2010 with trend lines fitted for the total and two halves of the data period. Among the various categories of ER events over CI, during the entire data period, MR events showed decreasing trend (545.6 events per decade), HR events showed increasing trend (12.6 events/decade) and VHR events showed increasing trend (13.2 events per decade) all at 95 % significant level. During first half, significant increasing trend (1648.5 events/decade at 95 % significant level) was observed only in the MR events. On the other hand, during the second half, significant decreasing trend (1863.6 events/decade at 95 % level) was observed in the MR events, while increasing trends were observed in the HR (44.2 events per decade at 95 % level) and VHR events (37.6 events per decade at 1 % level).

Over NEI, an increasing trend of 0.34 mm/day/decade significant at 99 % level during the first half and a decreasing insignificant trend (0.27 mm/day/decade)

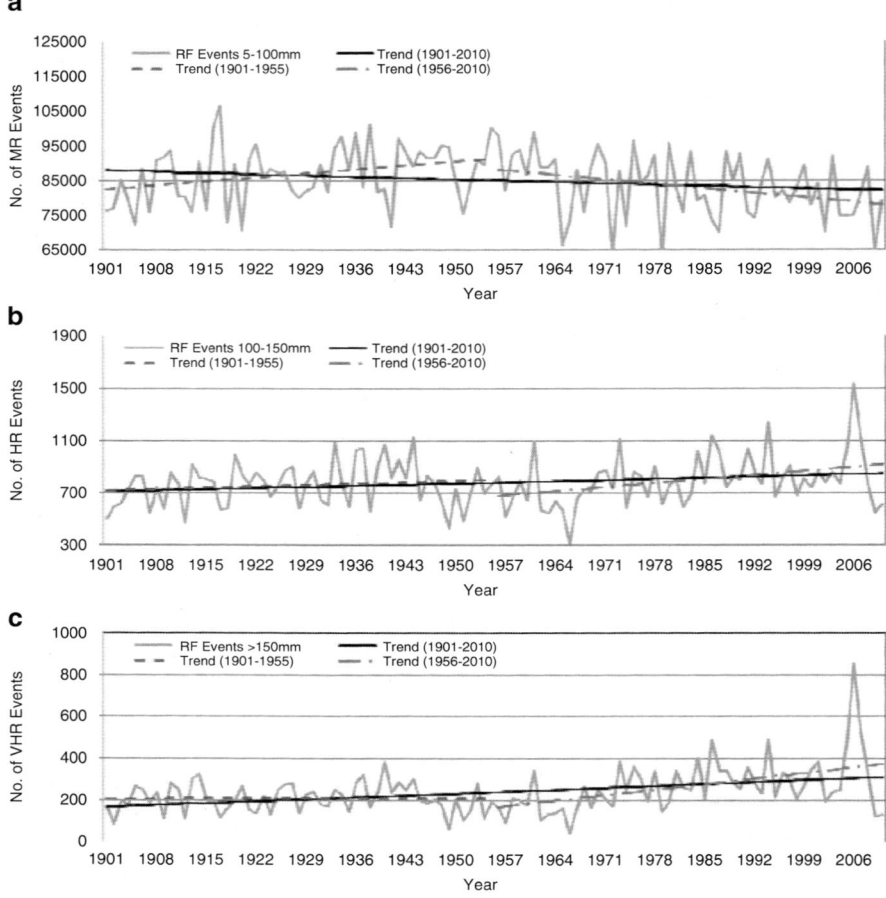

Fig. 2 Interannual variation of category-wise number of extreme rainfall events: (**a**) MR, (**b**) HR and (**c**) VHR events over Central India (CI)

during the second half were observed in the area averaged monsoon season rainfall over NEI. The net result was decreasing trend of 0.11 mm/day/decade significant at 95 % during the total data period.

The long-term linear trend in the number of ER events over NEI during the total period showed decrease of 179.9 events per decade significant at 99 % level (Table 2). Interestingly, positive trends of 395.3 events per decade (significant at 99 % level) and 6.2 events per decade were observed during the first and second halves of the period. The interannual variation of category-wise number of extreme rainfall events over NEI for the entire data period with trend lines fitted for various data periods are shown in Fig. 3. As seen in Fig. 3 and Table 2, during the total data

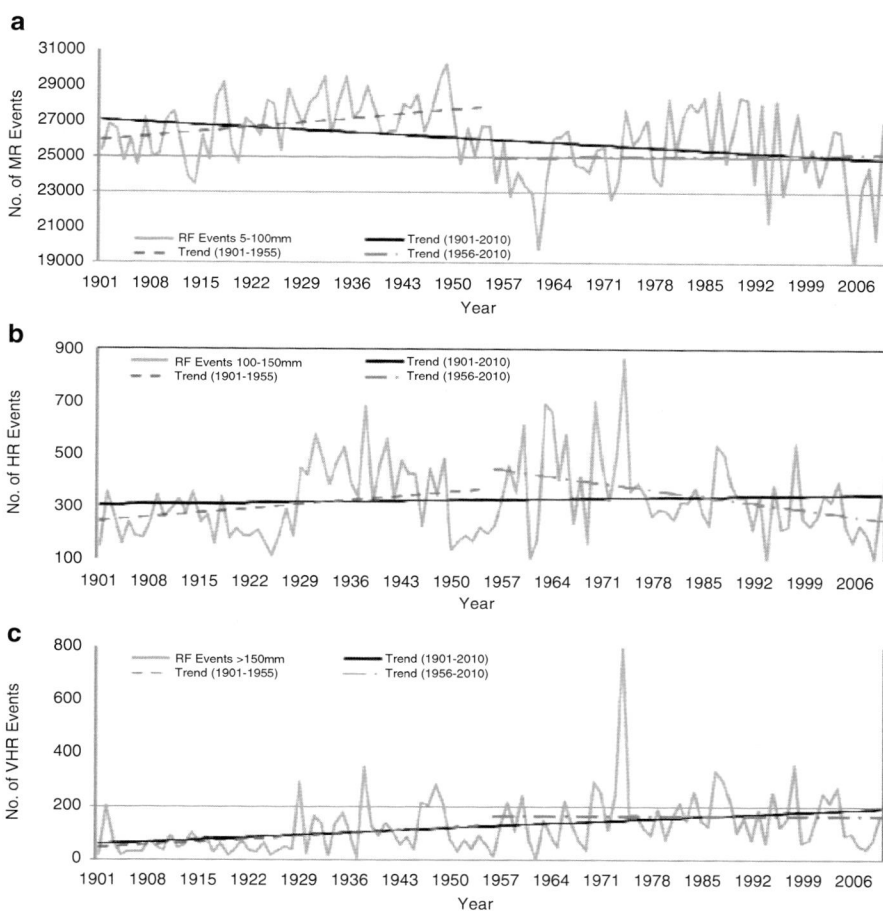

Fig. 3 Interannual variation of category-wise number of extreme rainfall events: (**a**) MR, (**b**) HR and (**c**) VHR events over Northeast India (NEI)

period, there is decreasing trend in the MR events (196.3 events/decade at 1 % significant level), and increasing trends in HR events (4.1 events/decade) and VHR events (12.3 events/decade at 95 % significant level). During the first half, significant increasing trends were observed in all the three categories of ER events. On the other hand, during the second half (1956–2010), significant trend (decreasing) was observed only in the HR events (35.7 events/decade at 99 % level).

Over the WC, no significant long-term linear trends were observed in the area weighted season rainfall during the entire data period as well as during the second half of the data period. However, during the first half, increasing trend of 0.45 mm/day/decade significant at 95 % level was observed. Figure 4 shows the interannual

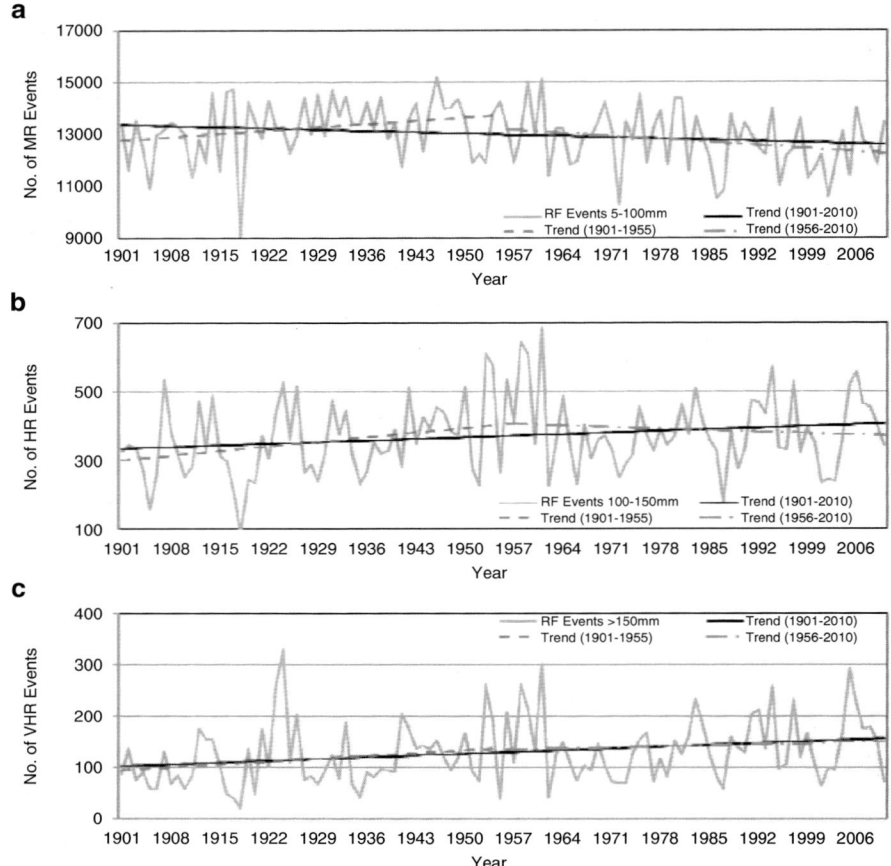

Fig. 4 Interannual variation of category-wise number of extreme rainfall events: (**a**) MR, (**b**) HR and (**c**) VHR events over West Coast (WC)

variation of category-wise number of ER events over WC for the entire data period with trend lines fitted for various data periods. As seen in Table 3 and Fig. 4, significant increasing trend in the ER events (205.6 events per decade at 95 % level) was observed during the first half and significant decreasing trend in the MR events (72.4 events per decade at 95 % level) was observed during the total period. Similarly significant increasing trends were observed in the HR events (6.7 events per decade at 95 % level) and VHR events (5.0 events per decade at 99 % level) during the total period. A significant increasing trend in the HR events (19.2 events per decade at 95 % level) during the first half was also observed.

3.2 Long-Term Trends in the Intensity of ER Events

The change in the intensity of ER events are as important as the change in the frequency of ER events in measuring the risk potential of ER events. Therefore, the long-term trends in the intensity of the ER events were examined for the complete data period as well as the two halves of the data period. For this purpose, 25 heaviest grid point ER events during each of the monsoon season for the total data period for each of the three geographical regions under this study were identified. For each region, average of the heaviest 25 rainfall amounts was then computed to represent as the intensity of ER events in that season for that region. Figure 5a–c

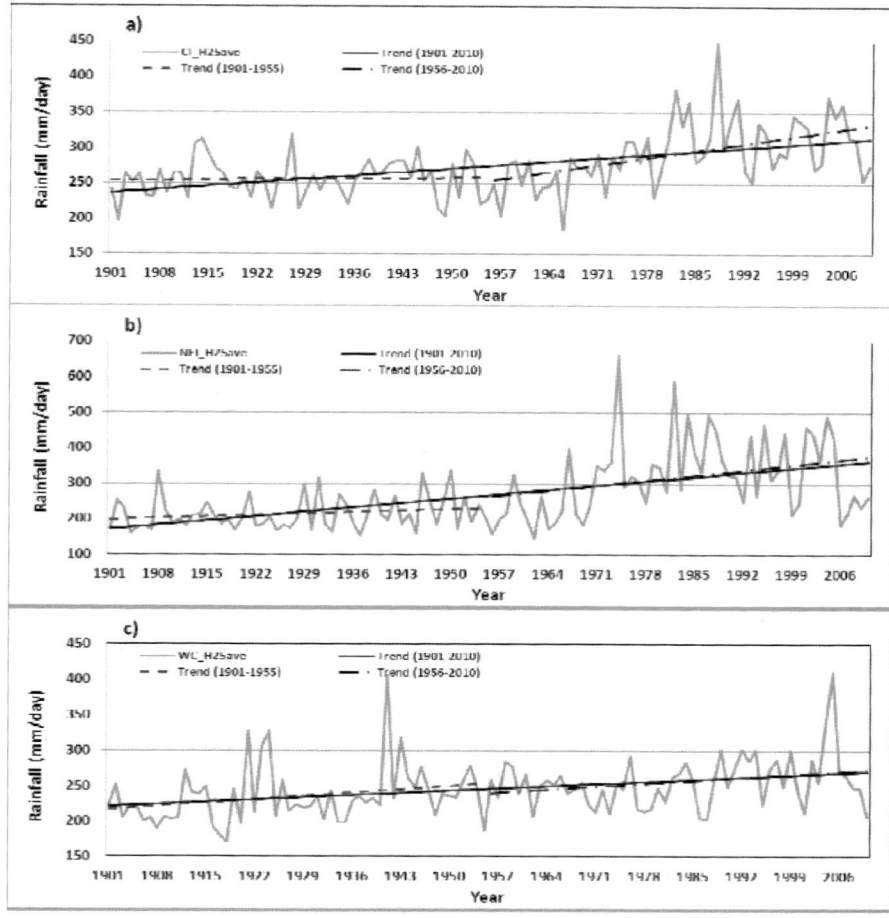

Fig. 5 Interannual variation of the average rainfall of heaviest 25 daily rainfall events over: (**a**) Central India, (**b**) Northeast India and (**c**) West Coast

show the interannual variation of the intensity of the ER events over CI, NEI and WC for the entire data period with linear trend lines fitted for the total period as well as its two halves. The long-term trends in the intensity of ER events computed for various periods in respect of CI, NEI and WC are also given in Tables 1, 2 and 3.

From Fig. 5 and Tables 1, 2 and 3, it can be seen that over all the three regions and during all the three data periods, the intensity of the ER events showed increasing trends. The trends in the intensity of rainfall during the total data period (1901–2010) over all the three regions were significant at 99 % level. On the other hand, over all the three regions, the trends in the intensity of ER events during the first half (1901–1955) were not significant. During the second half (1956–2010), the trend in the intensity of ER events over CI was significant at 99 % level and that over NEI and WC were significant at 95 % level.

Conclusions

Significant long-term trends in the number and intensity of the ER events were observed in all the three geographical regions under study (CI, NEI and WC). Whereas the intensity of ER events (average of 25 highest rainfall events) over all the three regions showed significant increasing trends during the second half and the total period, the signs and magnitude of the trends in the frequency of the ER events differed in these regions. Over CI, the signs of the trends in the frequency of ER events during the total period as well during the two halves of the period were same as that of MR. Significant increasing trends in the first half and decreasing trends in the second half of the data period resulted in relatively weaker decreasing trends in both ER and MR events during the total period. In tune with this, slight decreasing trend was observed in the area averaged rainfall over CI during the total period. However, the most important point was the significant increasing trends observed in the frequency of HR and VHR events during the second half (1956–2010). Thus the results on long-term trends in the ER events reported in this study are consistent with the results of Goswami et al. (2006) and Rajeevan et al. (2008).

Over NEI, significant increasing trends were observed in the frequency of ER events and that of all the three categories of ER events during the first half (1901–1955) of the data period. As a result, highly significant (at 99 % level) increasing trends were observed in the frequency of ER events as well as in the area weighted season rainfall in the first half. On the other hand, during the second half, only significant trend was observed in the HR events (decreasing at 99 % level). During the total period (1901–2010), significant (at 99 % level) decreasing trends were observed in the MR as well as the ER events. It may be noted that the trends in the MR and ER events during the first half were significant at 99 % level but the trends during the second half were low and

(continued)

insignificant. Significant decreasing trend was also observed in the area weighted rainfall during the total period.

Over WC, the increasing trends in the frequency of all the three categories of ER events during the first half were significant trend only. In HR events insignificant increasing trend in the ER events and in the area weighted season rainfall over the region was seen. In the second half, no significant trends were observed in all the categories of ER events. On the other hand, during the entire data period (1901–2010), significant trends were observed in all the categories of MR events. However, like in the case of CI, significant increasing trends in the HR and VHR events and significant decreasing trends in the MR events resulted in insignificant decreasing trend in ER events and no trend in the area weighted seasonal rainfall over the region.

From the trend analysis of frequency and intensity of the ER events over CI, NEI and WC, it can be concluded that the disaster potential over CI during the recent years have increased due to significant increasing trends in the frequency (areal coverage) of HR and VHR events as well as the intensity over CI during the recent half of the data period. On the other hand, the disaster potential over NEI and WC has increased in terms of the intensity of the rainfall events.

Acknowledgements We are thankful to Mr. S.S. Krishnaiah, ADGM (Research) and Mr. B. Mukhopadyay, DDGM (Climatology) for their encouragement, guidance and support during various stages of this work.

References

Ghosh S, Luniya V, Gupta A (2009) Trend analysis of Indian summer monsoon rainfall at different spatial scales. Atmos Sci Lett 10:285–290

Goswami BN et al (2006) Increasing trend of extreme rain events over India in a warming environment. Science 314:1442–1445

Guhathakurta P, Shreejith OP, Menon PA (2011) Impact of climate change on extreme rainfall events and flood risk in India. J Earth Syst Sci 120(3):359–373

IPCC (2007) Summary for Policymakers. In: Metz B, Davidson OR, Bosch PR, Dave R, Meyer LA (eds) Climate change 2007: mitigation. Contribution of Working Group III to the Fourth Assessment Report of the Intergovernmental Panel on Climate Change. Cambridge University Press, Cambridge

Joshi U, Rajeevan M (2006) Trends in precipitation extremes over India, Tech. rep. 3. National Climate Centre, IMD, Pune

Pai DS, Sridhar L, Rajeevan M, Sreejith OP, Satbhai NS, Mukhopadhyay B (2013) Development and analysis of a new high spatial resolution (0.25 × 0.25 deg.) long period (1901–2010) daily gridded rainfall data set over India, NCC RR no. 1/2013. IMD, Pune

Rajeevan M, Bhate J, Kale JD, Lal B (2006) High resolution daily gridded rainfall data for the Indian region: analysis of break and active monsoon spells. Curr Sci 91(3):296–306

Rajeevan M, Bhate J, Jaswal AK (2008) Correction to "Analysis of variability and trends of extreme rainfall events over India using 104 years of gridded daily rainfall data". Geophys Res Lett 35, L23701. doi:10.1029/2008GL036105

Sen Roy S, Balling RC Jr (2004) Trends in extreme daily precipitation indices in India. J Clim 24(4):457–466

Diagnostic Study of Heavy Rainfall Events in Monsoon Season over Northern Part of Bangladesh Using NWP Technique

Md. Abdul Mannan and Md. Mahbub Alam

1 Introduction

Bangladesh is a play-ground of natural calamities due to its geographical position, i.e., conical shaped Bay of Bengal in the south, the Himalayas in the north and passing of tropic of cancer through it. Bangladesh experiences tropical cyclones, storm surges, monsoon depressions, floods, droughts, nor'westers, tornadoes, heavy rainfall etc. causing heavy loss of lives and damages to properties. Life-giving rain comes during southwest monsoon and accounts for over 75 % of the total annual rainfall. The high lands situated to the east and north of Bangladesh play their due role to this monsoonal rain and the rainfall patterns in Bangladesh are governed by southwest monsoon (Mannan et al. 2013a, b). But most of the heavy rainfall events occur in monsoon months and the northeastern, eastern and southeastern regions of the country are most susceptible for this meteorological phenomenon. The socio-economic sectors affected by heavy rainfall are: agriculture, urban/town planning and construction, energy, water resource management, fisheries, forestry, human health and social services, disaster management, transportation, tourism, sports, etc.

In the tropical and subtropical coastal regions of Asia, especially adjacent to prominent terrain features, episodes of heavy rainfall exceeding 100 mm/day occur rather frequently and events of 300 mm/day or more are occasionally observed (Kharin et al. 2005; Chen et al. 2007; Chokngamwong and Chiu 2008). While tropi-

---

Md.A. Mannan (✉)
SAARC Meteorological Research Centre (SMRC), Dhaka, Bangladesh

Bangladesh Meteorological Department (BMD), Dhaka, Bangladesh
e-mail: mannan_u2003@yahoo.co.in

Md.M. Alam
SAARC Meteorological Research Centre (SMRC), Dhaka, Bangladesh

Department of Physics, Khulna University of Engineering & Technology (KUET), Khulna, Bangladesh

cal cyclones account for many of these events, others occur in conjunction with monsoon horizontal wind regimes (Chen et al. 2007). Despite the synoptic-scale character of the monsoon, numerous local effects ultimately conspire to determine the location of heavy rainfall. Climatology points to the maxima in mesoscale convective system (MCS) stratiform over coastal regions, within which most of the heaviest widespread rain events occur across much of southern Asia during the monsoon (Romatschke et al. 2010). Because of a general lack of mesoscale observations of lower-tropospheric wind, temperature, and water vapour, it is challenging to document the mesoscale processes in such regions, but such processes determine the all-important location of heavy rainfall and its societal consequences.

Dhar and Nandargi (1993) have found that severe rainstorms i.e., heavy rainfall over Indian region do not occur uniformly and most of the rainstorms are caused due to low pressure systems (LPS) which include low, depression, deep depression and cyclonic storm. Orography plays a significant role in intensity and distribution of heavy rainfall. Smith (1979) has suggested the three independent mechanisms of orographic rainfall, viz. (i) large scale slope precipitation due to orographically forced vertical motion or convection triggered by smooth orographic ascent bringing the air to saturation resulting in precipitation, (ii) rainfall from the pre-existing clouds is partially evaporated before hitting low ground and (iii) the rainfall due to orographic control of the formation of cumulonimbus clouds in a conditionally unstable air mass. However, Dubey and Balakrishnan (1992) have studied the frequency distribution of heavy rainfall days over different parts of India and the causative systems of these heavy rainfall events. Rakhecha and Pisharoty (1996) have studied point and spatial distribution of heavy rainfall over India during monsoon season. Dhar and Mhaiskar (1973) have studied point and spatial distribution of rainfall over India in association with depression/storm and found the occurrence of intense rainfall to the south of the monsoon trough extending from the centre of depression/storm. Heavy rainfall in particular over Bangladesh is predominantly determined by the interaction of basic monsoon flow with the orography (Mannan et al. 2013a, b). In addition, induced low pressure system within monsoon trough is one of the major causes for low level convection and heavy rainfall over Bangladesh and adjoining areas (Mannan et al. 2013a, b).

According to Bangladesh Meteorological Department (BMD), heavy rainfalls (HRs) were recorded over northwestern part of Bangladesh and its surrounding areas on 19 September 2012. The prediction processes of these events are still inadequate and demands further study for its improvement. The present study comprehensively examines the environmental conditions of HR that occurred on 19 September 2012 over northwestern part of Bangladesh. Simulated parameters are investigated. It is believed that the present study greatly contributes to the understanding and forecasting of the environmental conditions of HR over northern part of Bangladesh.

1.1 Description of the Event

On 19 September 2012 widespread HR occurred over the extreme northwestern part of Bangladesh. At this time the maximum amounts of rainfall of 131, 118 and 101 mm were recorded respectively at Rangpur, Dinajpur and Sayedpur. Light

Diagnostic Study of Heavy Rainfall Events in Monsoon Season 243

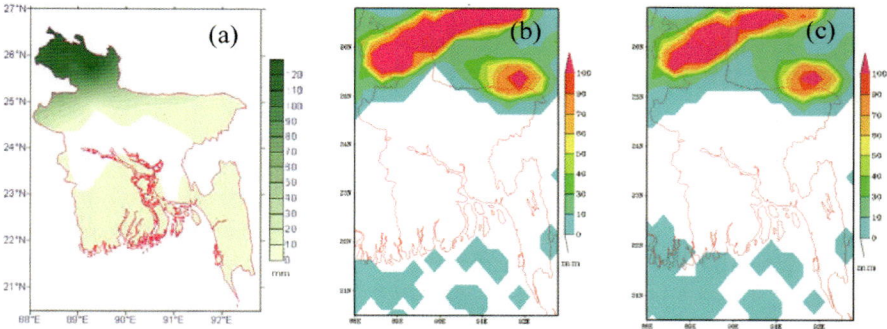

Fig. 1 Distribution of (**a**) observed, (**b**) TRMMV6 and (**c**) TRMMV7 rainfall over Bangladesh during 19 September 2012

Table 1 Classification of rainfall category in Bangladesh

Category of rainfall	Intensity
Trace	Non measurable amount/24 h
Light rainfall	< or = 10 mm/24 h
Moderate rainfall	11–22 mm/24 h
Moderately heavy rainfall	23–43 mm/24 h
Heavy rainfall	44–88 mm/24 h
Very heavy rainfall	>88 mm/24 h

amounts of rainfall were also recorded over southern coastal part and other regions of northern part of Bangladesh. TRMM products of version 6 and 7 (hereafter referred as TRMMV6 and TRMMV7) had also the signature of high amounts of rainfall over the same area (Fig. 1).

2 Methodology

A number of raingauge stations are set in the northern part of Bangladesh under BMD. But the stations of Bogra, Dinajpur, Mymensing and Rangpur are having longer periodic records of rainfall. In this regard, daily rainfall data of these stations during 1 June–30 September for the period of 1951–2012 are collected from BMD and used in this study for calculating the frequency of monthly HR over these locations as well as over the northern part of Bangladesh following the criteria given in Table 1 (Mannan and Karmakar 2008). Trends of the frequency of moderately HR, HR and very HR are calculated in the time scales of 1950–2012, 1981–2012 and 2001–2012 and analyzed.

WRF ARW model (version 3.2.1) with the grid resolution of 9 km is used to diagnose the event of 19 September 2012 using Ferrier (FR), Kessler (KS), Lin et al. (LN), WRF Single-Moment 5 Class (WSM5), WRF Single-Moment 6 Class (WSM6) and Thompson Graupel (TH) microphysics (MPs) schemes with the combination of Betts-Miller-Janjic (BMJ), Grell-Devenyi ensemble (GD), Kain-Fritsch

(KF) and New Grell (NG) cumulus scheme (CPs). Therefore, the 24 combinations of MPs and CPs are: BMJFR, BMJKS, BMJLN, BMJTH, BMJWSM5, BMJWSM6, GDFR, GDKS, GDLN, GDTH, GDWSM5, GDWSM6, KFFR, KFKS, KFLN, KFTH, KFWSM5, KFWSM6, NGFR, NGKS, NGLN, NGTH, NGWSM5 and NGWSM6. The coverage area of the model domain is 12–30°N and 80–100°E. The topographic data used in the model is obtained from USGS land covers data set. NCEP data have been provided at every 6 h as initial and boundary conditions. The model has been run with 19 sigma levels in the vertical direction from the ground to the 100 hPa level to simulate the event. The parameters of sea level pressure (SLP), relative humidity (RH), wind at 10 m, u and v-components of upper wind, vorticity, convergence and divergence, CAPE, convective rain and non-convective rain are extracted on hourly basis for analyzing by using GrADS. To furnish the simulated result, the forecast skills of Probability of Detection (PoD), False Alarm Ratio (FAR), Percentage Correct (PEC), Frequency Bias (FBI) and Critical Success Index (CSI) are calculated for the HR (which includes moderately HR, HR and very HR) at raingauge location for all the combinations of CPs and MPs.

3 Results and Discussion

3.1 HR Frequency in Monsoon Season over Northern Part of Bangladesh

3.1.1 Temporal Variation of the Frequency of Moderately HR

The frequency and trend of moderately HR for different time scales is given in Table 2. Frequency of moderately HR is the highest at Mymensingh and lowest at Dinajpur but it is 10.7 in the northern part of Bangladesh. The trends of the frequency of moderately HR at Mymensingh and Rangpur in the time scales of 1950–2012, 1981–2012 and 2001–2012 are positive. It is positive at Bogra in the scale of 1950–2012 but negative for other time scales. It is positive at Dinajpur during 2001–2012 and negative for other time scales. Accordingly, the trend of the frequency of moderately HR is negative during 2001–2012 but positive for other time scales over northern part of Bangladesh (Fig. 2).

Table 2 Frequency and trend of moderately HR over northern part of Bangladesh

Station	Average frequency	Trend of moderately HR		
		1950–2012	1981–2012	2001–2012
Bogra	10.5	+0.004	−0.046	−0.615
Dinajpur	10.3	−0.009	−0.047	+0.018
Mymensingh	11.8	+0.073	+0.163	+0.199
Rangpur	10.5	+0.061	+0.004	+0.094
N-Bangladesh	10.7	+0.033	+0.018	−0.076

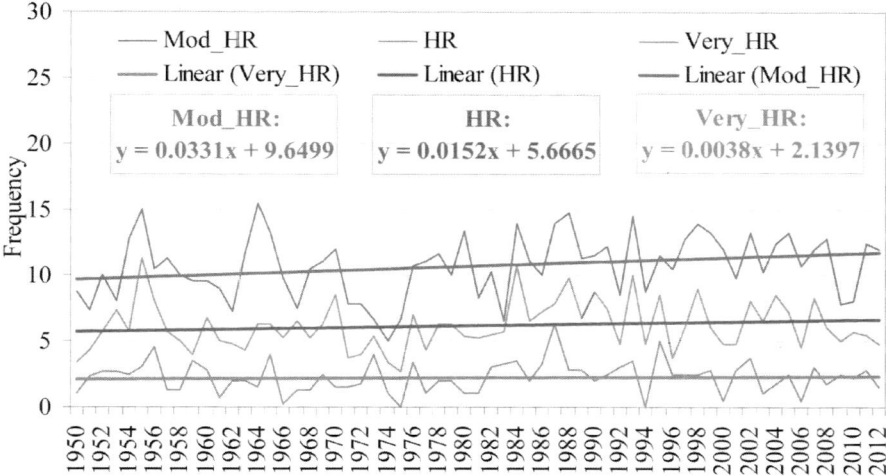

Fig. 2 Temporal variation of the frequency of moderately heavy (Mod HR), heavy (HR) and very heavy (Very HR) rainfall over northern part of Bangladesh during 1950–2012

Table 3 Frequency and trend of HR over northern part of Bangladesh

Station	Average frequency	Trend of moderately HR		
		1950–2012	1981–2012	2001–2012
Bogra	5.5	+0.012	−0.069	−0.136
Dinajpur	6.7	+0.006	−0.066	−0.458
Mymensingh	6.2	−0.018	−0.093	+0.171
Rangpur	6.5	+0.060	+0.024	−0.21
N-Bangladesh	6.2	+0.015	−0.051	−0.158

3.1.2 Temporal Variation of the Frequency of HR

The frequency and trend of HR of different time scales are given in Table 3. Frequency of HR is the highest at Dinajpur and lowest at Bogra but it is 6.2 in the northern part of Bangladesh. The trends of the frequency of HR are positive at Bogra and Dinajpur during the time scale of 1950–2012 but negative during 1981–2012 and 2001–2012. It is negative at Mymensingh in the scale of 1950–2012 and 1981–2012 but positive during 2001–2012. At Rangpur, it is positive in the scales of 1950–2012 and 1981–2012 but negative during 2001–2012. Accordingly, the trend of HR is positive during 1950–2012 and negative for other time scales over northern part of Bangladesh (Fig. 2).

3.1.3 Temporal Variation of the Frequency of Very HR

The frequency and trend of very HR for different time scales are given in Table 4. Frequency of very HR is the highest at Rangpur and lowest at Bogra but it is 2.3 in the northern part of Bangladesh. The trends of the frequency of very HR

Table 4 Frequency and trend of very HR over northern part of Bangladesh

Station	Average frequency	Trend of moderately HR		
		1950–2012	1981–2012	2001–2012
Bogra	1.7	+0.002	−0.006	+0.056
Dinajpur	2.5	+0.015	−0.025	−0.108
Mymensingh	2.1	−0.014	−0.038	+0.017
Rangpur	2.9	0.009	−0.078	−0.106
N-Bangladesh	2.3	0.004	−0.037	−0.035

are positive at Bogra and Dinajpur during 1950–2012 and negative for the duration of 1981–2012. It is negative at Mymensingh in the scales of 1950–2012 and 1981–2012 but positive for 2001–2012. At Rangpur, it is positive for the period of 1950–2012; however negative for other time scales. Accordingly, the trend of very HR is positive during 1950–2012; however negative for other two time scales (Fig. 2).

3.2 Synoptic Conditions During 19 September 2012 over Bangladesh and Adjoining Areas

Monsoon trough extended from Uttar Pradesh to Assam across Bihar, sub-Himalayan West Bengal and northern parts of Bangladesh with its extension over northwest Bay of Bengal. A distinct low pressure area persisted within the monsoon trough over northwestern part of Bangladesh (Figs. 3, 4 and 5).

3.3 Simulated Parameters

3.3.1 Wind Flow and Divergence

A strong convergence was generated over Bihar and adjoining sub-Himalayan West Bengal of India at 0300 UTC of 19 September 2012 which then became stronger and moved east-southeastwards to northwestern parts of Bangladesh following the surface wind flow and reached northwestern part of Bangladesh at 1200 UTC. It then intensified and moved further in the same direction. A strong wind convergence associated with this system extended upto 700 hPa level and moved along the surface flow. As a result, strong convergence persisted upto 700 hPa but divergence persisted above this level over the same area during 0900–1200 UTC over northwestern part of Bangladesh and adjoining areas. Similar situation was observed at Bogra, Ishurdi, Mymensingh, Rangpur and Sayedpur where high amounts of rainfall are recorded (Fig. 6).

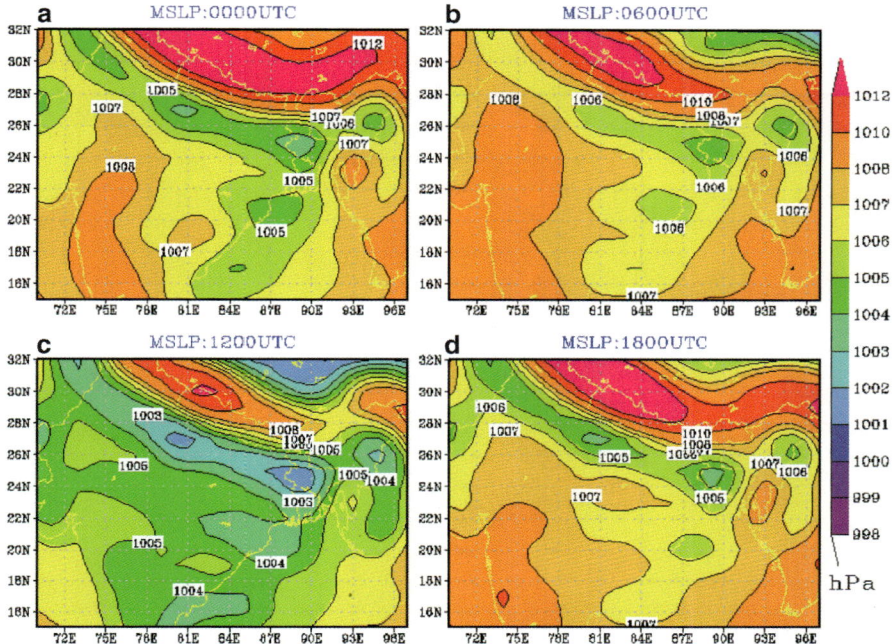

Fig. 3 Distribution of SLP over Bangladesh and adjoining areas at (**a**) 0000, (**b**) 0600, (**c**) 1200 and (**d**) 1800 UTC during 19 September 2012 derived from NCEP data

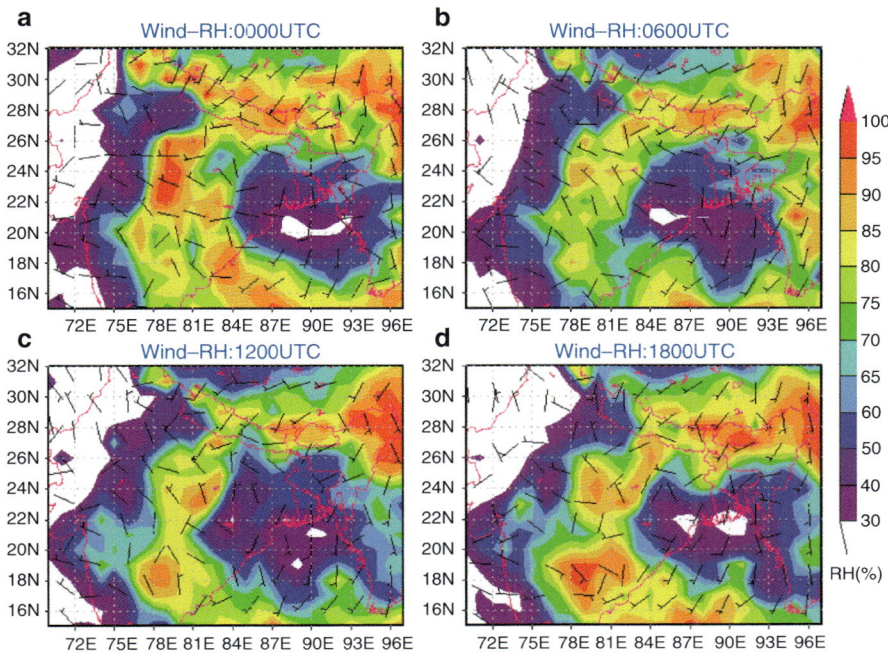

Fig. 4 Distribution of RH over Bangladesh and adjoining areas at (**a**) 0000, (**b**) 0600, (**c**) 1200 and (**d**) 1800 UTC during 19 September 2012 derived from NCEP data

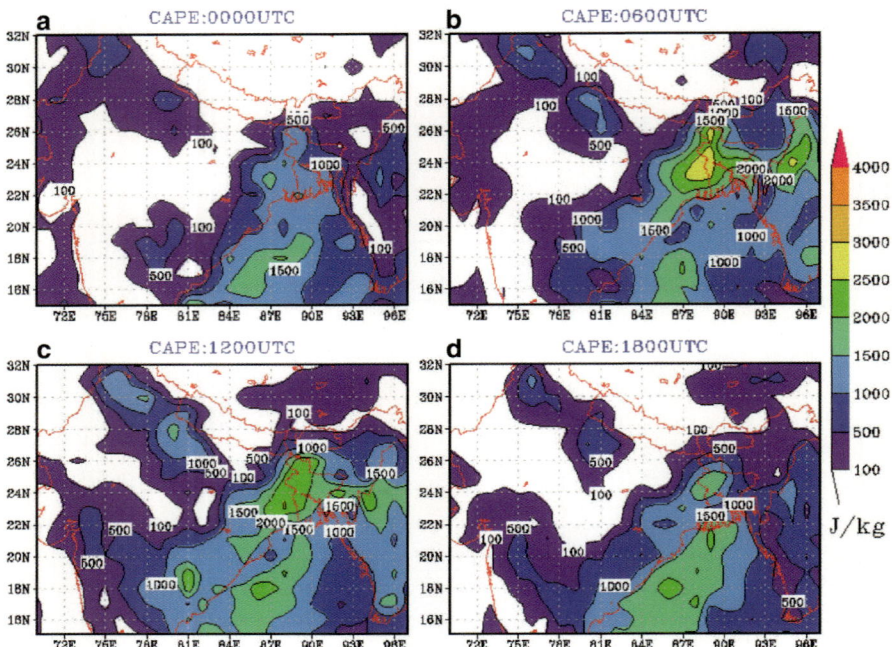

Fig. 5 Distribution of CAPE over Bangladesh and adjoining areas at (**a**) 0000, (**b**) 0600, (**c**) 1200 and (**d**) 1800 UTC during 19 September 2012 derived from NCEP data

3.3.2 Stream Flow and Vorticity

A vortex was generated over Bihar and adjoining sub-Himalayan West Bengal at 0300 UTC of 19 September 2012 at surface level which then intensified and moved east-southeastwards to northwestern part of Bangladesh at 1200 UTC (Fig. 7). It then intensified further and moved in the same direction. A strong confluence associated with this system extended upto 700 hPa level and also moved following the surface vortex. Under this situation, strong positive vorticity persisted upto 700 hPa but negative vorticity persisted above this level during 0900–1200 UTC over northwestern part of Bangladesh. Similar situation was observed at Bogra, Ishurdi, Mymensingh, Rangpur and Sayedpur which are located in the northern part of Bangladesh (Fig. 8).

3.3.3 CAPE

Strong CAPE field persisted in the morning of 19 September 2012 over northwestern part of Bangladesh and adjoining area of Bangladesh and India with the coverage over other parts of Bangladesh. With the progress of time the CAPE field became stronger over the same area. In support of wind field and moisture advection

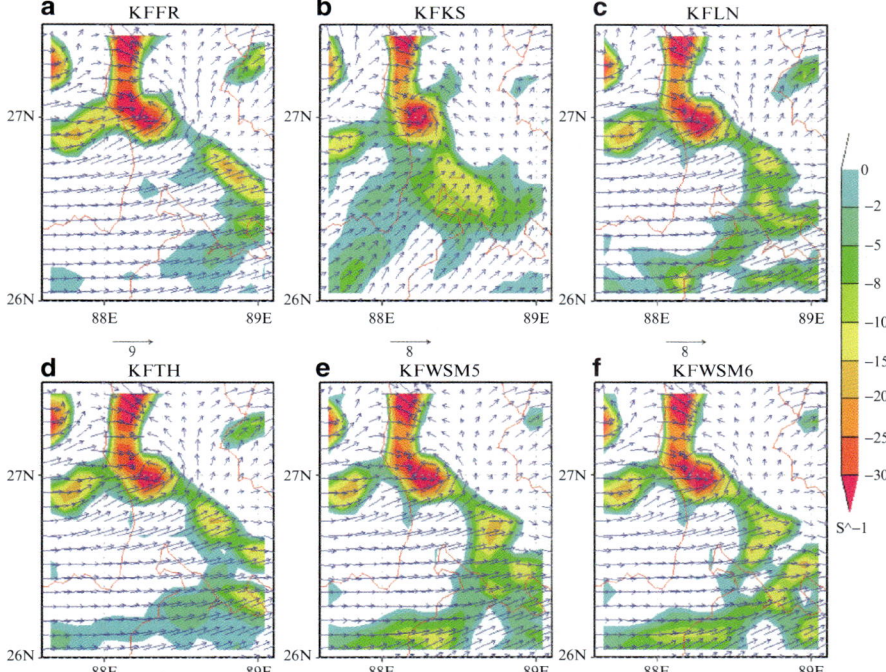

Fig. 6 Divergence and wind at 10 m for (**a**) KFFR, (**b**) KFKS, (**c**) KFLN, (**d**) KFTH, (**e**) KFWSM5 and (**f**) KFWSM6 over northwestern part of Bangladesh and adjoining areas at 1000 UTC on 19 September 2012

a strong CAPE appeared over West Bengal of India with its extension over northwestern part of Bangladesh. Simulated CAPE field for BMJ with the selected MPs was not so strong but for GD, KF and NG with the MPs was very strong for generation of instability, moisture convergence and convection over the northern part of Bangladesh (Fig. 9). Strong CAPE with temporal variation in the lower levels was also found at the stations located over northern part of Bangladesh.

Simulated maximum CAPE at Bogra, Dinajpur, Mymensingh, Rangpur and Sayedpur locations for the combinations of CPs and MPs are given in Table 5. The magnitude of maximum CAPE varying from 1,359 to 3,305 J/kg was found at Mymensingh and Bogra. The CAPE at Dinajpur varied from 1,834–1,975, 1,822–2,634, 1,818–2,179 and 1,820–2,760 J/kg respectively for BMJ, GD, KF and NG. The CAPE at Rangpur was 1,789–2,013, 1,808–2,975, 1,923–2,474 and 1,862–2,535 J/kg respectively for BMJ, GD, KF and NG. Similarly, the CAPE at Sayedpur was 1,836–1,983, 1,806–2,742, 1,804–2,453 and 1,833–2,812 J/kg respectively for BMJ, GD, KF and NG. Likewise, the CAPE at Bogra was also in ranges of 1,818–2,108, 2,354–3,305, 2,319–2,817 and 2,364–3,180 J/kg for BMJ, GD, KF and NG. All of these CAPEs are above the CAPE thresholds for high convection and indicate most unstable atmosphere over the heavy rainfall recorded area over Bangladesh.

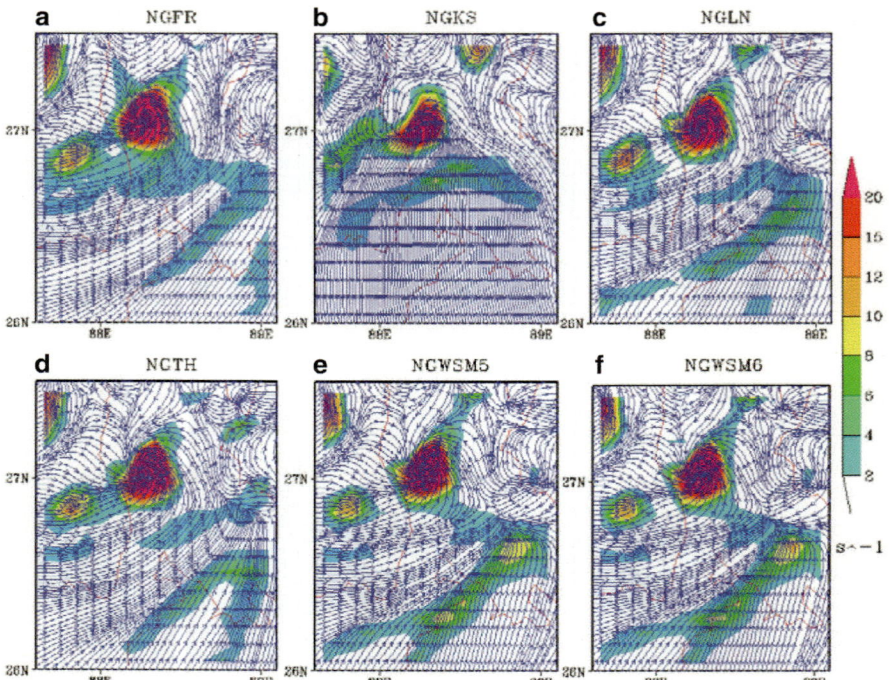

Fig. 7 Stream flow and vorticity at 10 m for (**a**) NGFR, (**b**) NGKS, (**c**) NGLN, (**d**) NGTH, (**e**) NGWSM5 and (**f**) NGWSM6 over northwestern part of Bangladesh and adjoining areas at 1200 UTC on 19 September 2012

3.4 Rainfall

Simulation for BMJ combinations depicted that the signatures of high rainfall was over the southern parts of Bhutan and adjoining area of India. In addition, signature of rainfall was observed over northern extreme part of Bangladesh for all BMJ combinations but the amounts were low. The amounts of maximum rainfall for BMJFR, BMJKS, BMJLN, BMJTH, BMJWSM5 and BMJWSM6 over Bangladesh territory and adjoining areas were 77.4, 12.9, 92.2, 58.7, 38.9 and 40.6 mm respectively. For GD combinations, the signatures of very high rainfall were over the southern part of Bhutan and adjoining areas of India. In addition, signatures of high rainfall were found over Sherpur region of Bangladesh and adjoining Meghalaya of India which is about 100 km away from Rangpur where high amounts of rainfalls were recorded (Fig. 10). The maximum amounts of rainfall at this place were 169.1, 169.1, 129.0, 179.6, 139.5 and 130.4 mm respectively for GDFR, GDKS, GDLN, GDTH, GDWSM5 and GDWSM6. The patterns of the rainfall signatures for KF and NG with the combinations of different MPs were similar to GD combinations but the amounts of rainfall were different (Figs. 11 and 12). The amounts of

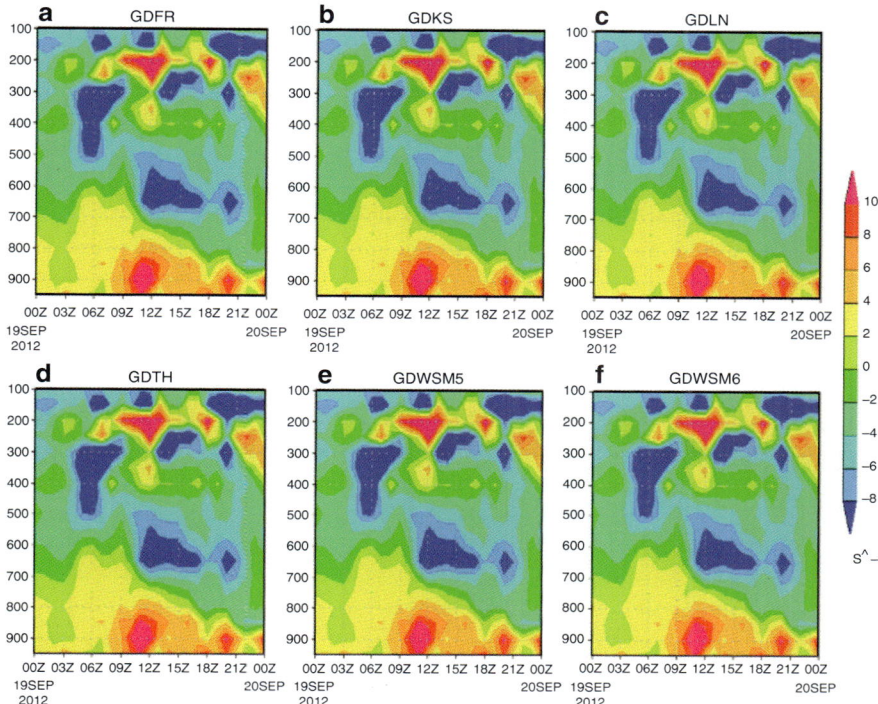

Fig. 8 Vorticity for (**a**) GDFR, (**b**) GDKS, (**c**) GDLN, (**d**) GDTH, (**e**) GDWSM5 and (**f**) GDWSM6 at Bogra during 19 September 2012

maximum rainfall at this place were 116.9, 110.9, 136.5, 115.6, 133.0 and 139.0 mm for KFFR, KFKS, KFLN, KFTH, KFWSM5 and KFWSM6. Similarly, the amounts of maximum rainfall for NGFR, NGKS, NGLN, NGTH, NGWSM5 and NGWSM6 were 173.2, 94.3, 147.0, 153.4, 138.0 and 135.8 mm respectively.

3.5 *Skill Scores of Heavy Rainfall Prediction*

Model simulated isolated moderately HR and HRs over Bangladesh for the combinations of GD, KF and NG with the selected MPs. But BMJ combination could not simulate HR rainfall at raingauge locations. Simulated maximum rainfall at Bogra and Ishudri were 82.7 and 57.2 mm for NGKS and NGTH respectively whereas observed rainfall was nil. Similarly, simulated maximum rainfall at Dinajpur, Rangpur and Sayedpur was 22.2, 41.0 and 42.6 mm for GDWSM5, GDLN and KFFR respectively which was much lower than observed. HR was simulated at Bogra for GDKS, NGFR and NGKS but the amounts were lower than observed.

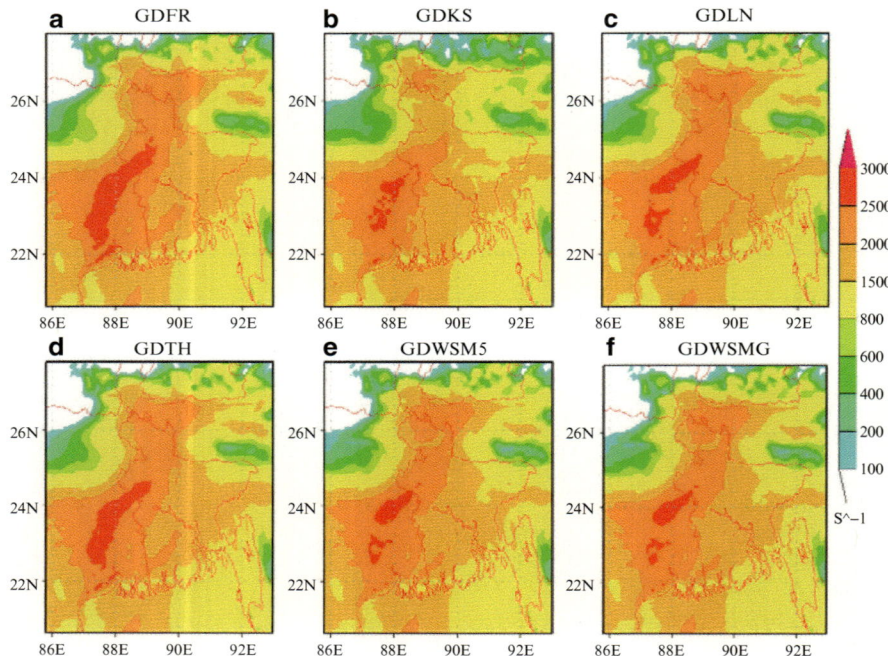

Fig. 9 CAPE field for (**a**) GDFR, (**b**) GDKS, (**c**) GDLN, (**d**) GDTH, (**e**) GDWSM5 and (**f**) GDWSM6 over northwestern part of Bangladesh and adjoining areas at 0700 UTC on 19 September 2012

Model could not simulate heavy rainfall at Dinajpur, Rangpur and Sayedpur where the recorded rainfalls were 118, 131 and 101 mm respectively. Accordingly, the forecast skills in terms of PoD, FAR, PEC, FBI and CSI were very low. The magnitudes of these skill scores for all the combinations of CPs and MPs are given in Table 6. Table 6 depicts that the PoD was maximum for KFFR and KFKS; FAR was minimum for BMJFR, BMJKS, BMJLN, BMJTH, BMJWSM5, BMJWSM6, KFFR and KFKS; PEC was highest for KFFR and KFKS; FBI was maximum for GDFR and GDWSM5; and CSI was highest for KFFR and KFKS (Fig. 13).

Table 5 Simulated CAPE for different combination of CPs and MPs at raingauge locations over northern part of Bangladesh during 19 April 2013

Combination of CPs and MPs	Maximum CAPE (J/kg)				
	Bogra	Dinajpur	Mymensingh	Rangpur	Sayedpur
BMJFR	2,108	1,834	1,653	1,930	1,929
BMJKS	1,928	1,840	1,514	1,789	1,836
BMJLN	1,935	1,891	1,588	2,013	1,956
BMJTH	1,971	1,906	1,702	1,952	1,794
MBJWSM5	2,006	1,901	1,359	1,980	1,983
BMJWSM6	1,818	1,975	1,602	1,979	1,979
GDFR	3,305	2,486	2,413	2,975	2,643
GDKS	2,354	1,822	1,852	1,808	1,806
GDLN	3,034	2,575	2,102	2,519	2,579
GDTH	3,138	2,609	2,276	2,817	2,742
GDWSM5	2,769	2,634	2,126	2,679	2,538
GDWSM6	2,581	2,625	2,136	2,710	2,691
KFFR	2,522	2,138	2,420	2,429	2,453
KFKS	2,319	1,818	2,095	1,923	1,804
KFLN	2,743	2,124	2,361	2,381	2,202
KFTH	2,746	2,193	2,585	2,474	2,365
KFWSM5	2,596	2,077	2,272	2,287	2,224
KFWSM6	2,817	2,075	2,351	2,438	2,226
NGFR	3,180	2,760	2,615	2,535	2,650
NGKS	2,364	1,820	1,710	1,862	1,833
NGLN	3,142	2,562	2,113	2,398	2,586
NGTH	3,150	2,893	2,025	2,473	2,812
NGWSM5	2,968	2,454	2,066	2,458	2,536
NGWSM6	2,963	2,472	2,068	2,449	2,530

Conclusion

The frequency of HR shows increasing trend for the three categories of moderately HR, HR and very HR over northern part of Bangladesh during the time scales of 1950–2012. It shows negative trend during 1981–2012 and 2001–2012 over the same area. Model simulated the synoptic situation associated with the HR event of 19 September 2012 which was very similar to NCEP observation. Simulated rainfalls for BMJ had no signature over northwestern part of Bangladesh. GD, KF and NG combinations had signature of heavy rainfall but the location was about 100 km away from the heavy rainfall recorded area. Model simulated HR at some places of Bangladesh where HRs were not recorded. It could not simulate heavy rainfall at Dinajpur, Rangpur and Sayedpur where high amounts of rainfall were recorded. Accordingly, the forecast skills were very poor. Extensive research is required to make use of the simulated result in HR prediction over northern part of Bangladesh.

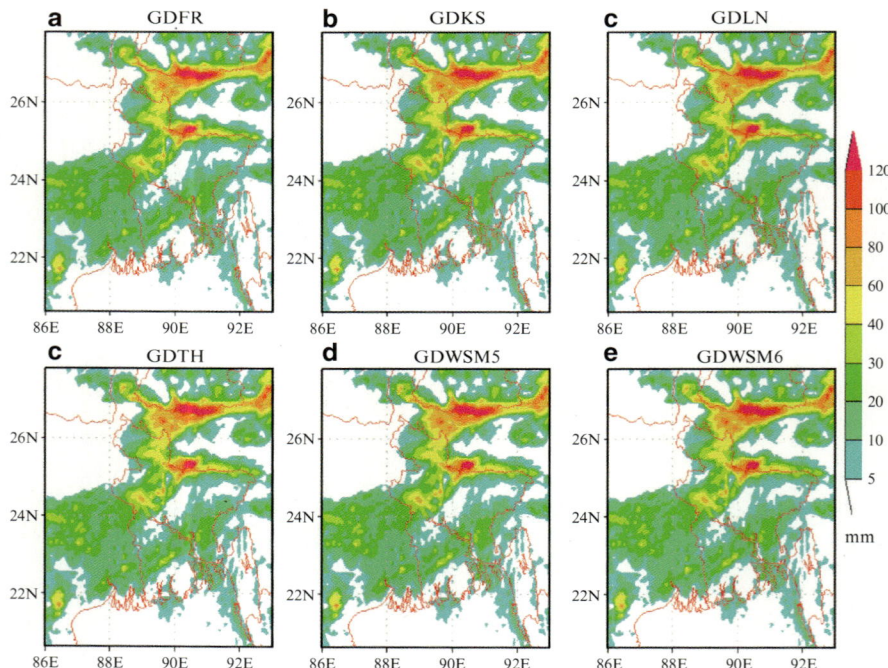

Fig. 10 Rainfall for (**a**) GDFR, (**b**) GDKS, (**c**) GDLN, (**d**) GDTH, (**e**) GDWSM5 and (**f**) GDWSM6 over Bangladesh on 19 September 2012

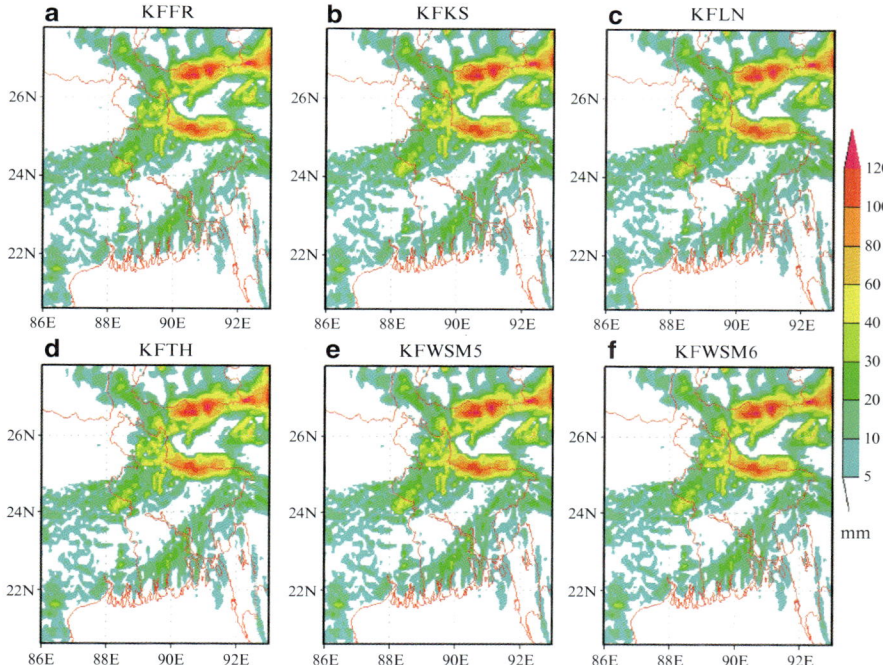

Fig. 11 Rainfall for (**a**) KFFR, (**b**) KFKS, (**c**) KFLN, (**d**) KFTH, (**e**) KFWSM5 and (**f**) KFWSM6 over Bangladesh on 19 September 2012

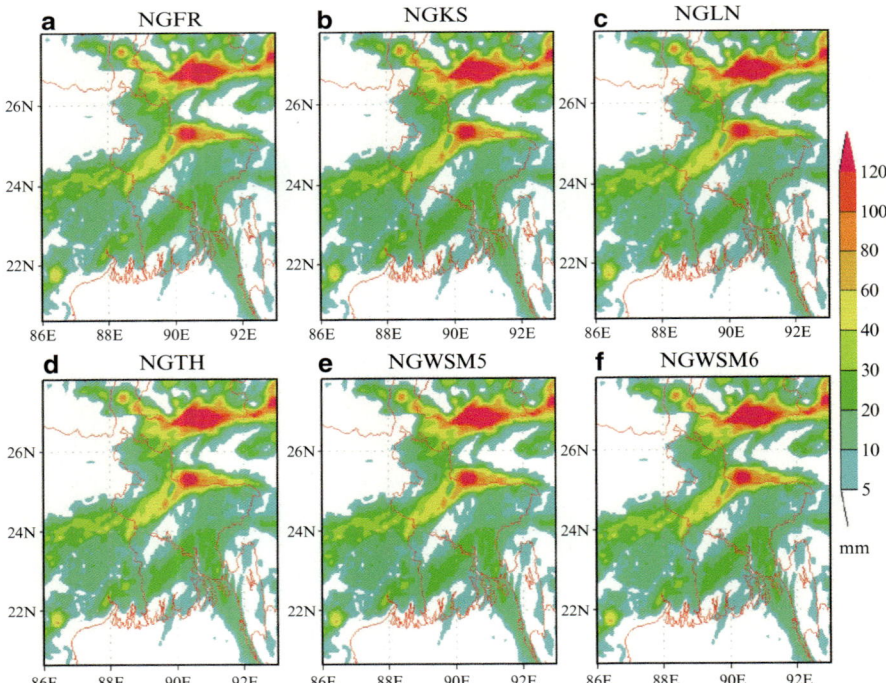

Fig. 12 Rainfall for (**a**) NGFR, (**b**) NGKS, (**c**) NGLN, (**d**) NGTH, (**e**) NGWSM5 and (**f**) NGWSM6 over Bangladesh on 19 September 2012

Table 6 Skill score for heavy rainfall prediction at raingauge locations for different combination of CPs and MPs over Bangladesh on 19 September 2012

Combination of CPs and MPs	Skill scores for moderately Heavy Rainfall				
	PoD	FAR	PEC	FBI	CSI
BMJFR	0.47	0.11	0.44	0.53	0.44
BMJKS	0.47	0.11	0.44	0.53	0.44
BMJLN	0.47	0.11	0.44	0.53	0.44
BMJTH	0.47	0.11	0.44	0.53	0.44
BMJWSM5	0.47	0.11	0.44	0.53	0.44
BMJWSM6	0.47	0.11	0.44	0.53	0.44
GDFR	0.44	0.29	0.40	0.61	0.37
GDKS	0.42	0.29	0.37	0.59	0.36
GDLN	0.45	0.23	0.41	0.58	0.40
GDTH	0.45	0.23	0.41	0.58	0.40
GDWSM5	0.44	0.29	0.40	0.61	0.37
GDWSM6	0.45	0.23	0.41	0.58	0.40
KFFR	0.48	0.11	0.47	0.55	0.46
KFKS	0.48	0.11	0.47	0.55	0.46
KFLN	0.48	0.14	0.46	0.56	0.44
FTH	0.47	0.14	0.44	0.55	0.43
KFWSM5	0.47	0.17	0.44	0.56	0.43
KFWSM6	0.48	0.14	0.46	0.56	0.44
NGFR	0.43	0.26	0.39	0.58	0.38
NGKS	0.44	0.26	0.40	0.56	0.40
NGLN	0.44	0.23	0.4	0.57	0.39
NGTH	0.43	0.26	0.39	0.58	0.38
NGWSM5	0.44	0.23	0.39	0.56	0.39
NGWSM6	0.44	0.23	0.39	0.56	0.39
TRMMV6	0.49	0.11	0.49	0.56	0.46
TRMMV7	0.49	0.11	0.49	0.56	0.46

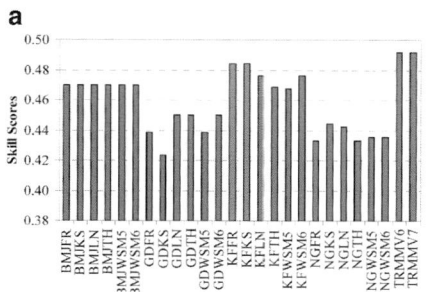
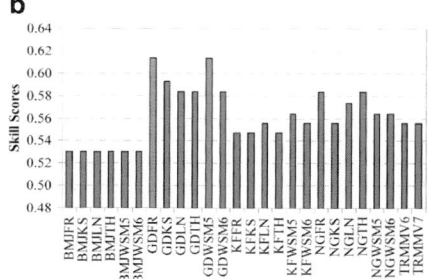

Fig. 13 Skill scores of (**a**) PoD and (**b**) FBI for moderately HR over Bangladesh during 19 September 2012

References

Chen C-S, Chen Y-L, Liu C-L, Lin P-L, Chen W-C (2007) The statistics of heavy rainfall occurrences in Taiwan. Weather Forecast 22:981–1002

Chokngamwong R, Chiu LS (2008) Thailand daily rainfall and comparison with TRMM products. J Hydrometeorol 9:256–266

Ciesielski PE et al (2010) Quality-controlled upper-air sounding dataset for TiMREX/SoWMEX: development and corrections. J Atmos Ocean Technol 27:1802–1821

Dhar ON, Mhaiskar PR (1973) Areal and point distribution of rainfall associated with depressions/storms on the day of crossing the east coast of India. Indian J Met Geophys 24:271–278

Dhar ON, Nandargi S (1993) The zones of severe rainstorm activity over India. Int J Climatol 13:301–311

Dubey DP, Balakrishnan TK (1992) A study of heavy to very heavy rainfall over MP for the period 1977 to 1987. Mausam 43:326–329

Kharin VV, Zwiers FW, Zhang X (2005) Inter-comparison of near-surface temperature and precipitation extremes in AMIP-2 simulations, re-analyses, and observations. J Clim 18:5201–5223

Mannan A, Chowdhury AM, Karmakar S, Islam N (2013a) Application of NWP model and its validation in prediction of heavy rainfall in Bangladesh. J Eng Sci 4(1):127–140

Mannan A, Chowdhury AM, Karmakar S, Islam N (2013b) Application of NWP model in prediction of heavy rainfall in Bangladesh. Elsevier Procedia Eng 56:667–675. doi:10.1016/j.proeng.2013.03.176

Rakhecha PR, Pisharoty PR (1996) Heavy rainfall during monsoon season: point and spatial distribution. Curr Sci 71:179–186

Romatschke U, Medina S, Houze RA Jr (2010) Regional, seasonal, and diurnal variations of extreme convection in the South Asian region. J Clim 23:419–439

Smith RB (1979) The influence of mountains on the atmosphere. Adv Geophys 21:187–230

Analysis of Increasing Heavy Rainfall Activity over Western India, Particularly Gujarat State, in the Past Decade

Manorama Mohanty, Kamaljit Ray, and Kalyan Chakravarthy

1 Introduction

The study of spatial and temporal variations of rainfall during monsoon has been studied by many scientists. India receives about 80 % of total annual rain during the southwest summer monsoon season from June to September. The number of moderate rain days and low rain days averaged over whole of India have significantly decreased whereas the number of heavy rain days is found to have increased (Dash et al. 2009). However, the trends in heavy rainfall events are not uniform over central India. For example, Mohapatra and Mohanty (2005) have found that there is no significant trend in frequency of very heavy rainfall in recent years over Orissa. By using a daily rainfall data set, Goswami et al. (2006) have shown (i) significant rising trends in the frequency and the magnitude of extreme rain events and (ii) a significant decreasing trend in the frequency of moderate events over central India during the monsoon season from 1951 to 2000.

Most of the studies during the last four decades have clearly pointed out that the monsoon rainfall is trendless, particularly on an all-India scale (Mooley and Parthasarthy 1984). Extreme rainfall studies in different regions of the country have been made by many scientists. Heavy rainfall studies over Bombay and Kerala state have been made by Prasad and Agarwal (1996) and Saseendran et al. (1995), respectively. Desai et al. (1996) studied very heavy rainfall over northwestern parts of India including Punjab, Himachal Pradesh and Haryana. Case studies of very heavy rainfall over Kolkata have been made by Banerjee et al. (1967) and Dhar and Ramachandran (1970). However the studies on heavy rainfall activity

M. Mohanty
Meteorological Centre, Ahmedabad, India

K. Ray (✉) • K. Chakravarthy
India Meteorological Department, Lodhi Road, New Delhi, India
e-mail: kamaljit_ray@rediffmail.com

over Gujarat state are limited. Hence the objective of this study is to analyze heavy rainfall activity over Gujarat state and some case studies of extremely heavy rainfall over Saurashtra and Kutch in the month of September in the past few years which have contributed significantly to the total seasonal rainfall over this subdivision.

2 Data and Methodology

Gujarat state is located in the extreme western part of India. It is situated roughly between 20° 15′ N and 24° 30′ N. It has Sind (Pakistan) and Rajasthan to the north, Maharashtra to the south, Madhya Pradesh to the east, the Thar desert in the north-east. The mighty Arabian Sea envelops the state from the south and south-west. The state experiences a tropical dry climate. Based on meteorological parameters, the state is divided into two subdivisions such as (i) Saurashtra-Kutch, the sea coast land and (ii) Gujarat region, the northeast corner of the Indian west coast which is the mainland of the state. The bulk of the rainfall activity over this state occurs during the months of July and August under the influence of synoptic scale systems.

It has tropical climate namely sub-humid, arid and semi-arid spread over different regions of the state. North Gujarat region comprising Kutch, part of Banaskantha, Mehsana and north western part of Saurashtra has arid climate while the south Gujarat has sub-humid climate and the rest of the state, has semi-arid climate. Temperature varies from 6 to 45 °C. Annual rainfall varies from 250 mm in the north west to more than 1,500 mm in south Gujarat. Out of 225 talukas, 56 talukas are drought prone.

Sub-division wise data has been compiled from DRMS (District-wise rain monitoring scheme) stations in various districts for the period 1974–2013. The annual averages during monsoon months and season as a whole are calculated and analyzed. The percentage departure of monsoon rainfall from long period average have been calculated and analyzed. The percentage departure is defined as [(actual rainfall − normal rainfall)/normal rainfall]×100. The frequency of heavy, very heavy and extremely heavy rainfall events have been found out based on the data of 40 years (1974–2013). For this study heavy rainfall event with daily 24-h cumulative rainfall recorded at 0,830 h IST amounting to 65–124 mm, very heavy rainfall amounting to 125–244 mm and extremely heavy rainfall amounting to 245 mm or more are considered. The decadal variations of frequency of heavy rains, very heavy rains and extremely heavy rains reported over selected ten stations (Departmental observatory only) of Gujarat state have also been calculated and analyzed.

3 Result and Discussion

The yearly average rainfall and the anomalies over two sub-divisions are presented in Sect. 3.1. The decadal variations in average seasonal rainfall are presented in Sect. 3.2. The extreme monthly and seasonal rainfall is analyzed in Sect. 3.3.

The contribution of heavy rainfall to the total monthly and seasonal rainfall are analyzed and presented in Sect. 3.4. The spatial distribution of rainfall are analyzed and discussed in Sect. 3.5. Three case studies of extreme rainfall events are also presented in this Section.

3.1 Yearly Average Rainfall and the Anomalies over Two Sub-divisions

The monthly and seasonal rainfall time series for Gujarat region (IMD Sub-division) is shown in Fig. 1 for the period 1974–2013. The analysis indicates an increasing trend in July and September months. The total seasonal rainfall also shows an increasing tendency. Figure 2 shows the time series for Saurashtra and Kutch where the trend was increasing in the months of July and September. The seasonal rainfall was more or less same in the past 40 years for Saurashtra and Kutch. The percentage departure of seasonal rainfall from the long period average based on the period 1974–2013 have also been analyzed for Gujarat region and Saurashtra-Kutch and shown in Figs. 3 and 4.

The following categories of rainfall departures—normal: percentage departure −19 % to +19 %; excess: percentage departure +20 % and deficient: percentage departure −20 % are utilised for Indian monsoon rainfall. Out of the total 40 years considered, seasonal rainfall was excess for 15 years, it was deficient for 12 years and normal for 13 years over Gujarat region. In Saurashtra and Kutch, the rainfall was excess for 13 years, deficient for 13 years and it was normal for 14 years. The years with positive and negative anomaly were equal in the past 40 years. Particularly in Saurashtra and Kutch the anomaly was continuously positive after 2002 (except for 2004 and 2012). The seasonal rainfall over these two subdivisions was found to be highly variable with large inter-annual variations.

3.2 Decadal Variations in Average Seasonal Rainfall

Figure 5 shows decadal averages of seasonal rainfall for the period 1974–2013. The decadal seasonal average rainfall for Gujarat region shows a continuous increasing trend in the past four decades with an increase from an average rainfall of 486 mm in the decade 1974–1983 to an average rainfall of 1,026 mm in the decade 2004–2013. Similarly in the last three decades the average seasonal rainfall over Saurashtra and Kutch was also found to have increased from 378 to 674 mm. The average seasonal rainfall was found to have increased appreciably in the last decade for both subdivisions. The analysis was also performed individually for IMD stations located in two subdivisions. The rainfall data for 40 years of nine departmental observatories was considered for studying the rainfall activity over four stations

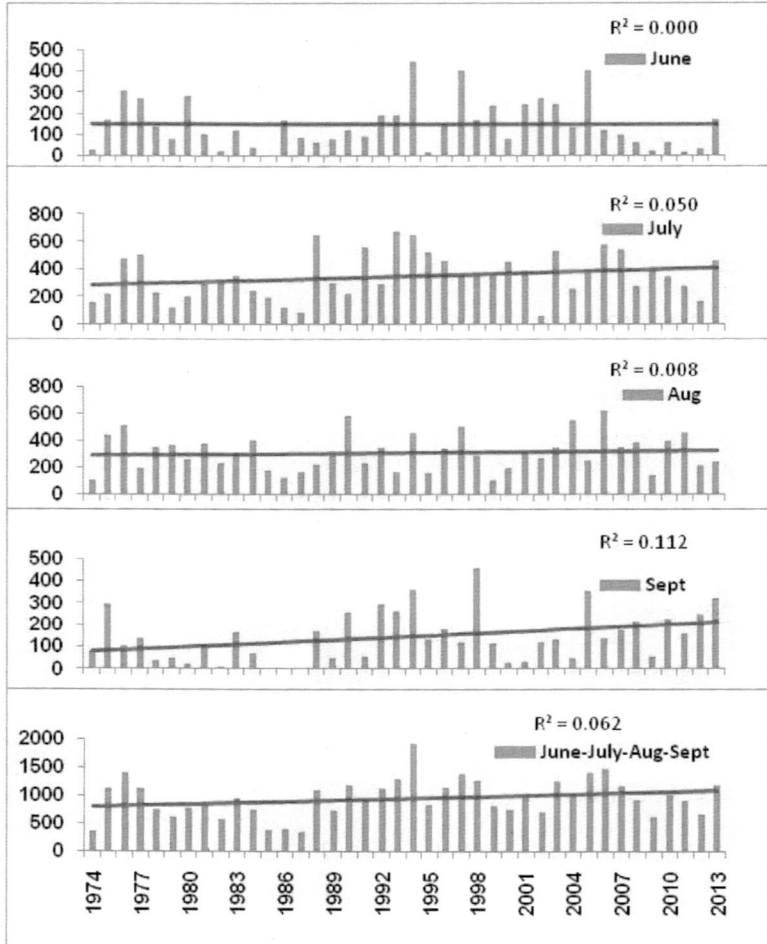

Fig. 1 The time series of monthly and seasonal rainfall during the monsoon season from June to September for Gujarat region for the period 1974–2013

(Ahmedabad, Deesa, Baroda, Surat) in Gujarat region and five stations in Saurastra-Kutch (Porbandar, Rajkot, Bhavnagar, Veraval, Bhuj). Figure 6 shows the decadal average seasonal rainfall for various IMD stations in Gujarat region and Fig. 7 for stations in Saurashtra and Kutch. It can be seen that the average seasonal rainfall is highest in the last decade for all stations. Ray et al. (2009) also analyzed seasonal rainfall for all observatories for the 40 years period (1969–2008) in Gujarat and found that 30 years moving averages of seasonal rainfall showed an increasing trend for all the stations during the above period.

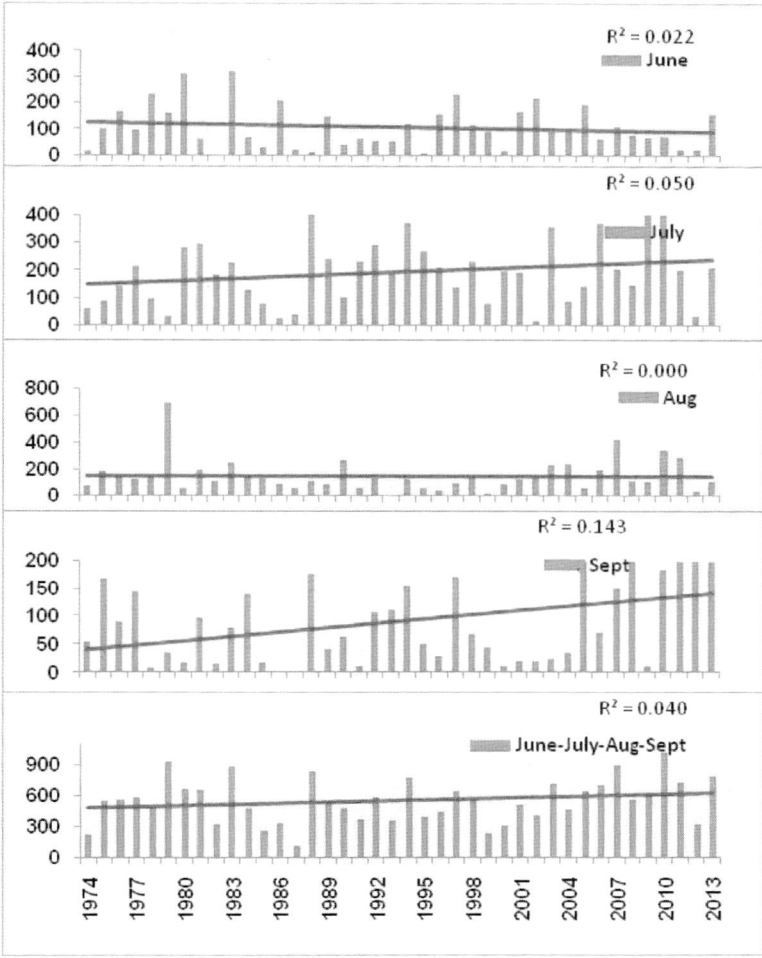

Fig. 2 The time series of monthly and seasonal rainfall during the monsoon season from June to September for Saurashtra-Kutch for the period 1974–2013

3.3 Extreme Monthly and Seasonal Rainfall

The decadal variations of frequency of heavy rains, very heavy rains and extremely heavy rains reported over four stations of Gujarat region were also studied. Tables 1 and 2 give the decadal variation of frequency of heavy/very heavy/extremely heavy rainfall over various IMD stations in two subdivisions. It is observed that the frequency of heavy rains (≥ 65 mm) has increased significantly in last decade in all the stations. The rise was appreciable in all stations in south Gujarat and coastal Saurashtra.

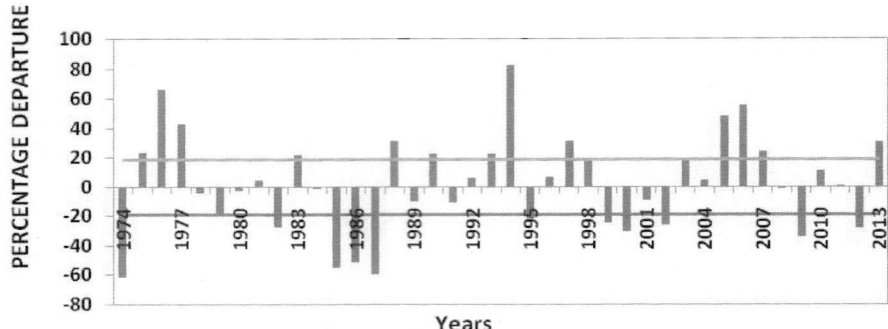

Fig. 3 Time series analysis of percentage departure of seasonal rainfall from long period average for Gujarat region during 1974–2013

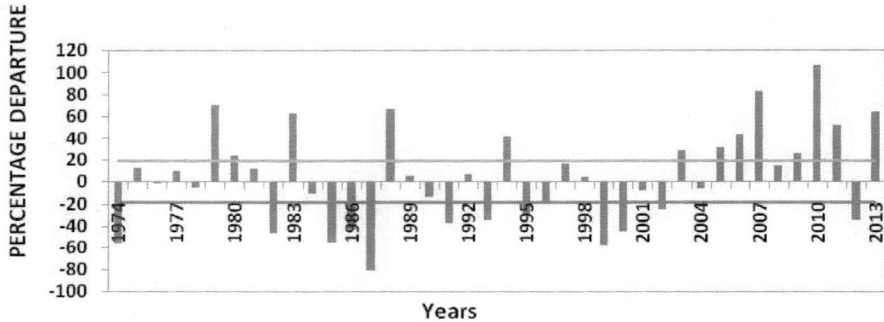

Fig. 4 Time series analysis of percentage departure of seasonal rainfall from long period average for Saurashtra-Kutch region during 1974–2013

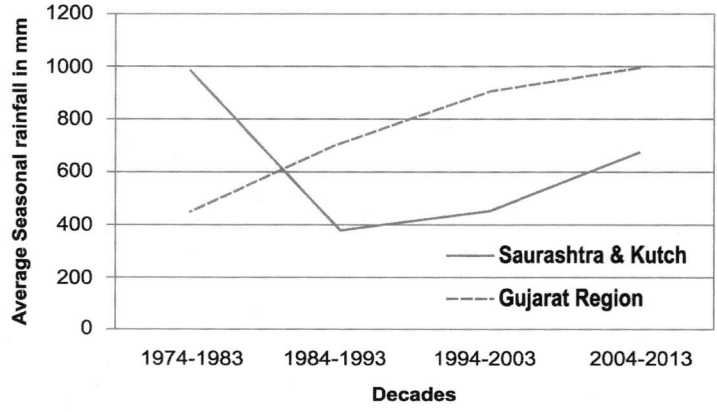

Fig. 5 Decadal variations of average seasonal rainfall for Saurashtra & Kutch and Gujarat region

Analysis of Increasing Heavy Rainfall Activity over Western India

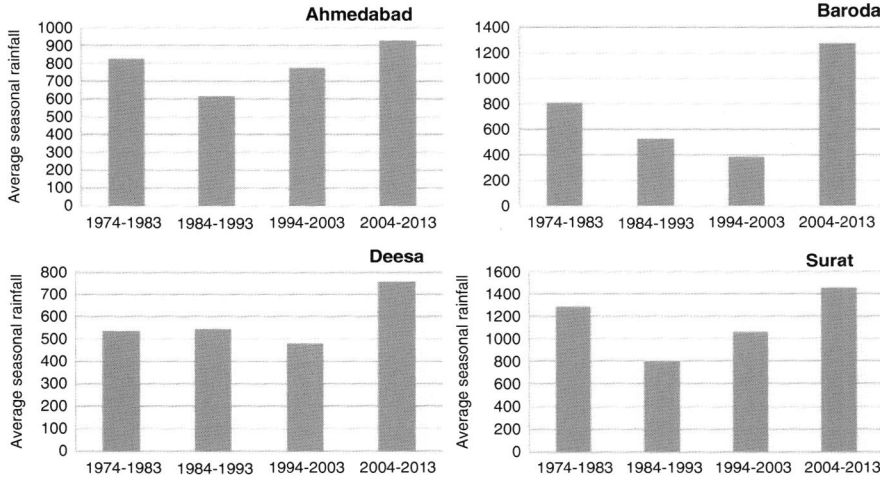

Fig. 6 Decadal variations of average seasonal rainfall for stations for Gujarat region

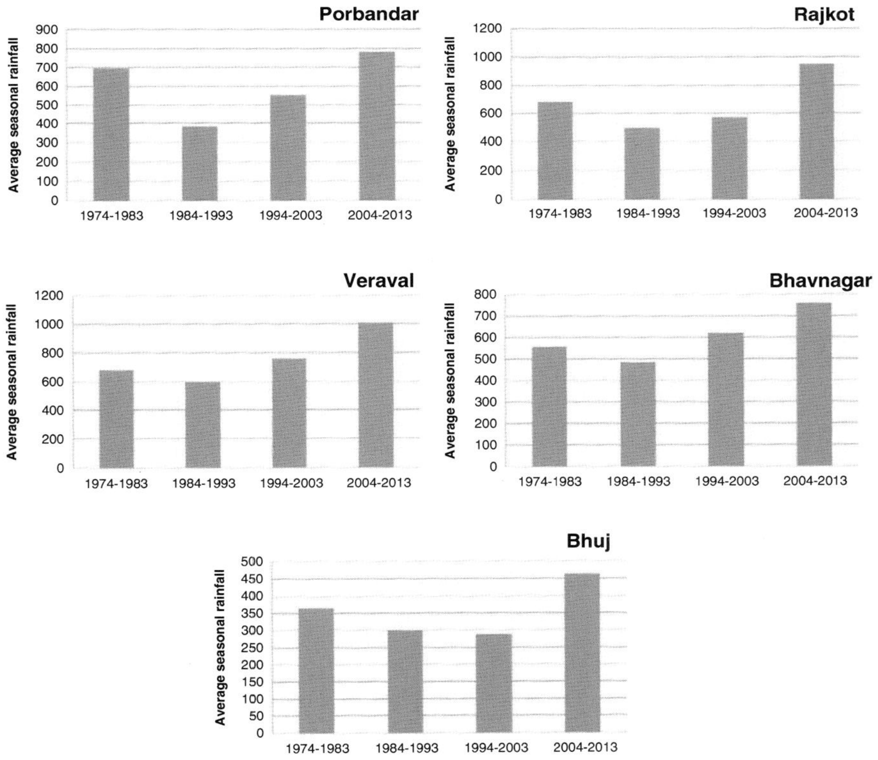

Fig. 7 Decadal variations of average seasonal rainfall for stations in Saurashtra and Kutch

Table 1 Decadal variation of frequency of heavy/very heavy/extremely heavy rainfall over the stations in Gujarat region

No. of events	1974–1983	1984–1993	1994–2003	2004–2013
Ahmedabad				
Ex-heavy rainfall	1	2	3	2
Very heavy rainfall	8	5	5	1
Heavy rainfall	18	14	17	**30**
Total	27	21	25	**33**
Deesa				
Ex-heavy rainfall	2	0	0	3
Very heavy rainfall	2	9	4	1
Heavy rainfall	19	7	19	**25**
Total	23	16	23	**29**
Baroda				
Ex-heavy rainfall	0	1	0	4
Very heavy rainfall	4	1	4	7
Heavy rainfall	24	10	18	**35**
Total	28	12	22	**46**
Surat				
Ex-heavy rainfall	3	6	3	4
Very heavy rainfall	12	7	9	**18**
Heavy rainfall	36	29	33	**46**
Total	51	42	45	**68**

The results also indicate that higher numbers of heavy rainfall events were recorded in the first and last decades as compared to the second and third decades. This coincided with the increase of seasonal rainfall over these stations in the last decade indicating an important role played by heavy rains in contributing to the seasonal rainfall.

3.4 Contribution of Heavy Rainfall to the Total Monthly and Seasonal Rainfall

Table 3 shows the percentage contribution of heavy rainfall to the total rainfall in various months for Gujarat region. It was found that of the total rainfall in the months of July and August, more than 60 % was contributed by heavy rains in all 5 years. The total September rain also had a contribution of more than 30 % from heavy rains except in the year 2011 when the heavy rains contributed only 19 %. The contribution of heavy rains towards total seasonal rainfall was higher in Saurashtra and Kutch subdivision (Table 4). More than 40 % of the rains in June are due to heavy spells only as can be seen from Table 4. In the year 2012 the percentage of heavy rains was less in all months, specifically in July and August,

Analysis of Increasing Heavy Rainfall Activity over Western India   267

Table 2 Decadal variation of frequency of heavy/very heavy/extremely heavy rainfall over the stations in Saurashtra and Kutch

No. of events	1974–1983	1984–1993	1994–2003	2004–2013
Bhuj				
Ex-heavy rainfall	1	0	0	0
Very heavy rainfall	3	2	3	**5**
Heavy rainfall	8	8	6	**11**
Total	12	10	9	**16**
Porbandar				
Ex-heavy rainfall	0	1	0	4
Very heavy rainfall	4	1	4	**7**
Heavy rainfall	24	10	18	**35**
Total	28	12	22	**46**
Bhavnagar				
Ex-heavy rainfall	4	1	1	3
Very heavy rainfall	4	1	8	**7**
Heavy rainfall	18	13	15	**30**
Total	26	15	24	**40**
Rajkot				
Ex-heavy rainfall	2	0	1	1
Very heavy rainfall	3	2	0	**6**
Heavy rainfall	5	16	13	12
Total	10	18	14	19
Veraval				
Ex-heavy rainfall	0	2	4	5
Very heavy rainfall	7	5	5	11
Heavy rainfall	19	21	25	26
Total	26	28	34	**42**

Table 3 Percentage of rainfall contributed by heavy rainfall (>=64.5 mm) to the total amount of rainfall received over Gujarat region in the past 5 years

Gujarat Region	2009	2010	2011	2012	2013
June	6.8	12.3	0	6.07	29.9
July	37.9	34.1	32.3	25.5	36.8
August	30.5	34.9	33.8	25.2	26.7
September	32.5	38	19	36.74	58.4

Table 4 Percentage of rainfall contributed by heavy rainfall (>=64.5 mm) to the total amount of rainfall received over Saurashtra and Kutch in the past 5 years

Saurashtra and Kutch	2009	2010	2011	2012	2013
June	45.3	23.2	44.3	21.2	44
July	58.1	46.6	31.3	6.7	27.6
August	22.1	46.9	34.2	1.5	18.7
September	0	26.3	42.7	25.63	61.4

which led to deficient rains over Saurashtra and Kutch. There was no spell of heavy rains in Gujarat region in June 2011 but Saurashtra and Kutch got heavy rains in association with the depression that formed over the northeast Arabian Sea, near Lat. 20.0°N/Long. 71.5°E on 11 June, and crossed Saurashtra coast on 12 June. Except for the year 2013, for all other years the monsoon onset over Gujarat region was in the first week of July while the onset over the adjoining subdivision (Saurashtra and Kutch) was in the second week of June. As we can see from Tables 3 and 4, heavy rains in the month of June were very less in Gujarat region while in Saurashtra and Kutch the formation of low pressure system/depression in Arabian Sea in the month of June caused heavy rains, thus facilitating the monsoon onset over this coastal subdivision.

3.5 Spatial Distribution of Rainfall

The spatial distribution of rains over two subdivisions was also analyzed during the past 5 years. Daily spatial distribution of rainfall based on more than 200 rain gauge data over Gujarat State was calculated for each year during the period 2009–2013 and results are shown in Tables 5 and 6.

The analysis for Gujarat region indicated the following:

1. Years with excess rainfall (2010, 2011, 2013) had more number of days with fairly widespread and widespread distribution of rainfall as compared to years with deficient rainfall (2009, 2012).
2. There were on an average 109 rainy days in the whole monsoon season. The average days with fairly widespread and widespread rainfall activity were around 40.
3. 75 % of days with fairly widespread and widespread distribution of rainfall were in the months of July and August.

Similarly the analysis for Saurashtra and Kutch region indicated the following:

1. Years with excess rainfall (2009, 2010, 2011, 2013) had more number of days with fairly widespread and widespread distribution of rainfall as compared to years with deficient rainfall (2012).
2. There were on an average 98 rainy days in the whole monsoon season. The average days with fairly widespread and widespread rainfall activity were around 22.
3. The contribution of fairly widespread and widespread days was highest in the month of July (50 %) followed by September (23 %).

If we analyze last 5 years data, then it was seen that the withdrawal of monsoon is delayed from Gujarat due to formation of low pressure in Bay of Bengal or Arabian Sea in the month of September, which is giving extremely heavy rains to the state. Three such case studies in the last decade are discussed below.

Table 5 Spatial distribution of rainfall over Gujarat region

June					July					August					September					June–September				
	ISO	SCA	FWS	W.S		ISO	SCA	FWS	W.S		ISO	SCA	F.W.S	W.S		ISO	SCA	F.W.S	W.S		ISO	SCA	FWS	W.S
2009	12	0	0	0	2009	7	10	4	10	2009	23	2	1	2	2009	10	4	1	0	2009	52	16	6	12
2010	22	2	2	0	2010	10	9	6	6	2010	5	14	4	7	2010	10	10	5	2	2010	47	35	17	15
2011	24	0	0	0	2011	12	12	3	4	2011	6	8	4	13	2011	12	10	1	5	2011	54	30	8	22
2012	21	3	0	0	2012	14	12	3	2	2012	15	1	9	3	2012	15	7	6	3	2012	65	23	18	8
2013	17	7	1	3	2013	4	11	5	11	2013	13	8	4	6	2013	15	7	2	4	2013	49	33	12	24
AVG	19	6	4	2	AVG	6	9	10	8	AVG	9	9	8	5	AVG	12	6	4	3	AVG	53	27	25	16

Table 6 Spatial distribution of rainfall over Saurashtra and Kutch

June					July					August					September					June–September				
	ISO	SCA	FWS	WS		ISO	SCA	FWS	WS		ISO	SCA	FWS	WS		ISO	SCA	FWS	WS		ISO	SCA	FWS	WS
2009	5	4	1	0	2009	8	5	10	5	2009	12	1	0	1	2009	12	1	0	0	2009	37	11	11	6
2010	21	1	3	0	2010	7	7	13	3	2010	14	8	3	5	2010	14	7	3	1	2010	56	23	22	9
2011	14	0	0	0	2011	12	8	4	4	2011	12	7	6	4	2011	18	4	4	2	2011	56	19	14	10
2012	17	2	0	0	2012	23	2	1	0	2012	24	6	0	0	2012	14	6	7	2	2012	78	16	8	2
2013	20	4	2	3	2013	8	9	11	2	2013	14	8	2	1	2013	19	3	2	3	2013	61	24	17	9
AVG	15	2	1	0.6	AVG	12	6	8	3	AVG	15	6	2	2	AVG	15	4	3	2	AVG	58	19	15	7

3.5.1 Case Study 1

24 September-28 September, 2013: During monsoon 2013, the longest spell with heavy to extremely heavy rains occurred during the period 22–29 September 2013 over Gujarat State. The highest rainfall recorded each day from 22 to 29 was as follows:

Date	Station	Rainfall (cm)
22 September,	Quant (Dist. Baroda)	16
23 September,	Sagbara (Dist. Narmada)	36
24 September,	Quant (Dist. Baroda)	43
25 September,	Umerpada (Dist. Surat)	37
26 September,	Rajkot	39
27 September.	Khambhalia (Dist. Jamnagar)	45
28 September	Bachau (Dist. Kutch)	31
29 September,	Dantiwada (Dist. Banaskantha)	23

Figure 8 shows the isohyets on individual days based on the data received from 220 raingauge stations. The extremely heavy rains belt shifted from east to northwest across Gujarat state during the period 24–28 September. Very heavy to extremely heavy rains were recorded on 23 and 24 September in south Gujarat region due to the low pressure area lying over west Madhya Pradesh and adjoining Gujarat region. As the low pressure area moved westwards and lay over Saurashtra and Kutch and adjoining west Rajasthan, the very heavy to extremely heavy rainfall belt moved to Saurashtra and Kutch on 25 and 26 September. The low pressure area further narrowed down to Kutch and adjoining extreme west Rajasthan, giving extremely heavy rains to Kutch and adjoining north Gujarat region. The system maintained its intensity due to associated upper air cyclonic circulation extending upto mid tropospheric levels and also due to interaction with the mid tropospheric westerly trough extending upto Gujarat, lying along 70.0°E.

3.5.2 Case Study 2

3–7 September, 2011: In the year 2011 also extremely heavy rains led to flood-like situation over Saurashtra and Kutch from 3 to 7 September. Gujarat received heavy to very heavy rainfall with isolated extremely heavy falls during 1st week of September 2011. Rainfall activity was higher in Saurashtra and Kutch as compared to Gujarat region. Monsoon in Saurashtra and Kutch was active (observed rainfall upto four times of normal) on 3, 4 and 5 September. Intensity of rainfall increased and was highest on 6 and 7 September. Monsoon was vigorous (observed rainfall more than four times of normal) and some of the stations also received extremely heavy rainfall (more than 25 cm) on these 2 days (Figs. 9 and 10).

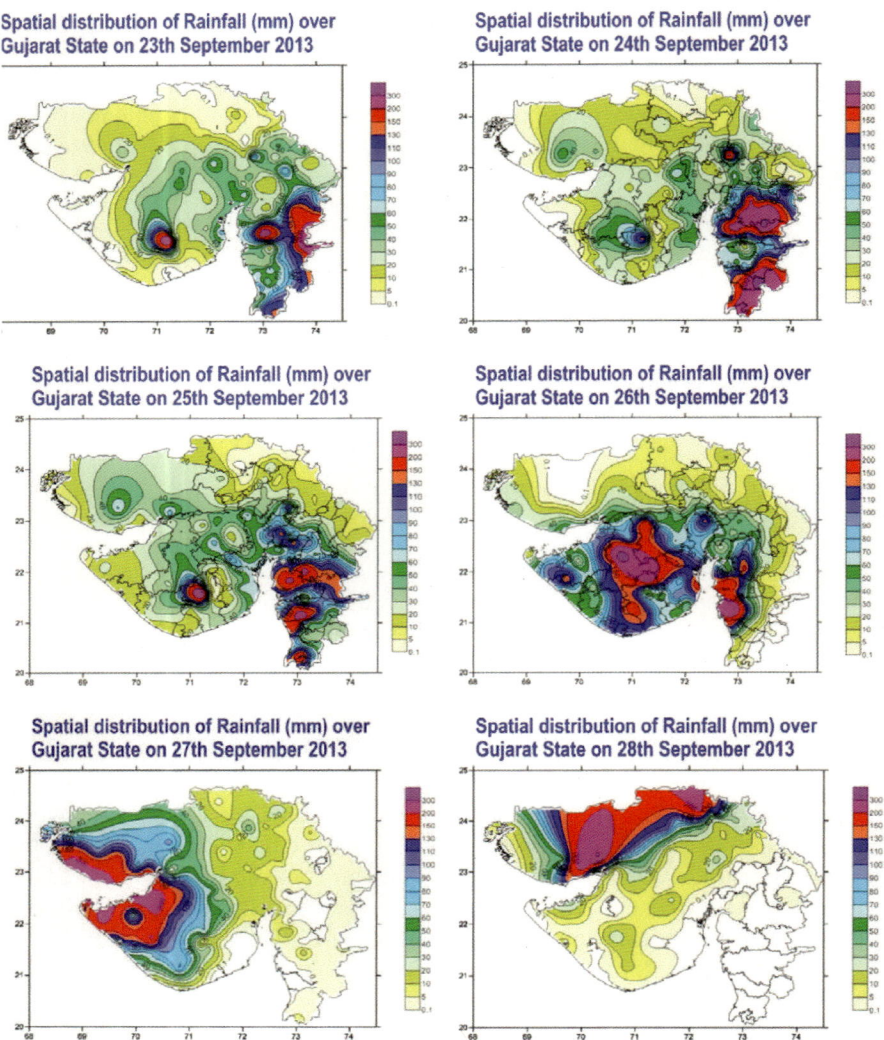

Fig. 8 Isohytel analysis over Gujarat during the period 23–28 September, 2013

Considering the synoptic conditions, a low pressure area formed over northwest Bay of Bengal on 30 August, it moved across Odisha, Chhattisgarh and Madhya Pradesh and lay over southwest Rajasthan and adjoining Gujarat on 4 September and over Gujarat and neighbourhood on 5 and 6 September where it became well marked (Fig. 11). It lay over southwest Rajasthan and adjoining northwest Gujarat and southeast Pakistan and thereafter moved over to southeast Pakistan and adjoining southwest Rajasthan and Kutch area. During the above period its associated upper air cyclonic circulations extended upto mid-tropospheric levels tilting southwards with height.

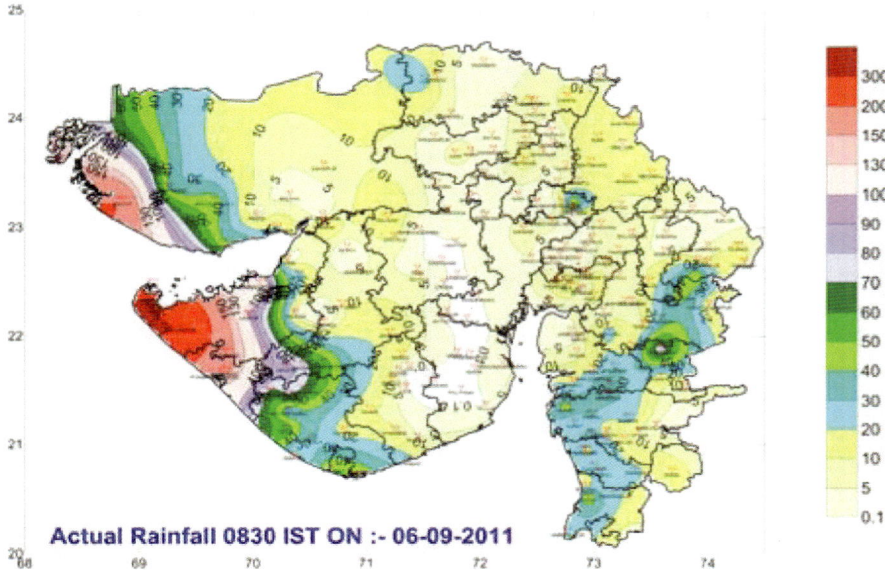

Fig. 9 Isohytel analysis of 24-h cumulative rainfall on 6 September

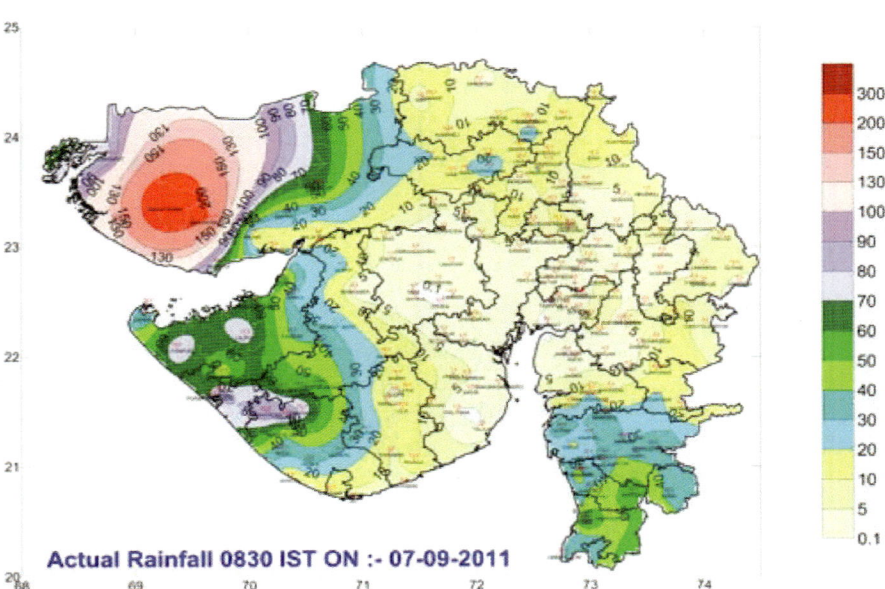

Fig. 10 Isohytel analysis of 24-h cumulative rainfall on 7 September

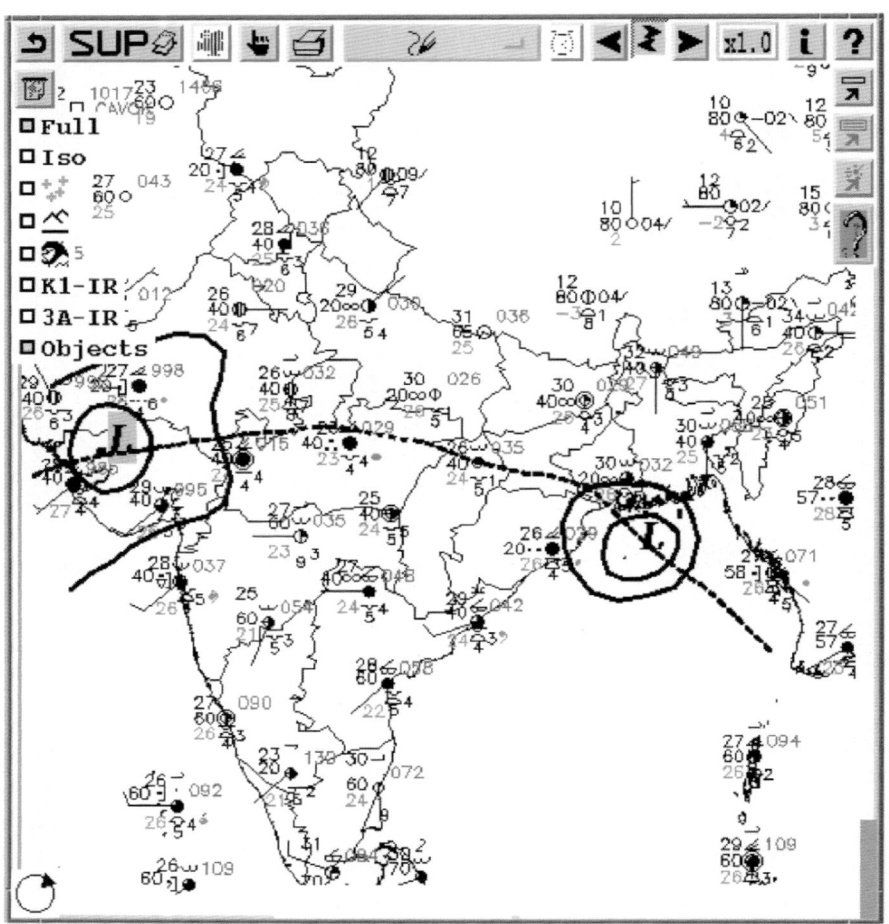

Fig. 11 Surface isobaric analysis of 6th September 2011

During this period, there was also a trough in mid-tropospheric westerlies lying between 65 and 70.0°E. Due to combined effect of this trough and low pressure areas, lower level southwesterly winds strengthened over northeast Arabian Sea enhancing moisture incursion over Gujarat. The monsoon trough also ran much south of its normal position and passed through Gujarat from 1 to 6 September.

3.5.3 Case Study 3

15–20 September, 2008: A deep depression formed over Bay of Bengal on 15 September. The system crossed Odisha coast on 16 September, it weakened into a depression and lay over central Madhya Pradesh near Satna on 19 September.

This system gave active to vigorous rainfall activity over Gujarat state from 15 to 20 September. Rainfall of the order of 504 mm at Dhanduka (Ahmedabad), 265 mm at Ranpur (dist. Ahmedabad), 478 mm at Lakhtar (dist. Surendernagar), 372 mm at Wadhwan (Dist. Surendernagar), 302 mm at Limbdi (Dist. Surendernagar), 258 mm at Kharaghoda and Dhrangadhria (Dist. Surendernagar) was recorded on 18 September. The total mean rainfall for the month of September was 212.3 mm (excess) as against normal of 159.6 mm for Gujarat region and 277 mm (excess) as against normal of 71.6 mm for Saurashtra and Kutch. The percentage departure rainfall from normal for Gujarat region and Saurashtra & Kutch subdivision was excess in the month of September. The total cumulative rainfall till 12 September was 741.8 mm for Gujarat region and 346.1 mm for Saurashtra and Kutch i.e. −14 % and −23 % departures from normal respectively. The deep depression that formed from 15 to 19 September changed the scenario of Gujarat rainfall from deficient to normal. The percentage rainfall departures for the entire season changed to −2 % for Gujarat region and +15 % for Saurashtra and Kutch.

Conclusions
(i) The average seasonal rainfall was found to have increased appreciably in the decade 2004–2013 for both sub-divisions of Gujarat state.
(ii) The frequency of heavy rains (>65 mm) has increased significantly in the last decade (2004–2013) in all observatories of Gujarat state.
(iii) Low pressures/depressions forming in Arabian Sea in the month of June and the low pressure/depressions forming in the month of September were contributing significantly to the total seasonal rainfall of Gujarat, particularly for Saurashtra and Kutch in the past decade.
(iv) The increasing trend in the heavy rainfall events in the last decade was also found to be due to the increase in cyclonic activity over Arabian Sea at the beginning and end of the monsoon season.

Although we have discussed three cases to support the above trend, more number of years can be considered for better results. Heavy rains were found to contribute significantly to the total seasonal rainfall.

References

Banerjee P, Mukhopadhyay PK, Handa BK, Chatterjee SD (1967) Abnormal rainfall over Calcutta on 10th October, 1965. Indian J Met Geophys 18:291
Dash SK, Kulkarni MA, Mchanty UC, Prasad K (2009) Change in the characteristics of rain events in India. J Geophys Res 114:1–12
Desai DS, Thade NB, Huprikar MG (1996) Very heavy rainfall over Punjab, Himachal Pradesh and Haryana during 24-27 September 1988 – case study. Mausam 47(3):269–274
Dhar ON, Ramachandran G (1970) Short duration analysis of Kolkata rain. Indian J Met Geophys 21:93–102

Goswami BN, Venugopal V, Sengupta D, Madhusoodanan MS, Xavier PK (2006) Increasing trend of extreme rain events over India in a warming environment. Science 314:1442–1445

Mohapatra M, Mohanty UC (2005) Some characteristics of very heavy rainfall over Orissa during summer monsoon season. J Earth Syst Sci 114:17–36

Mooley DA, Parthasarthy B (1984) Fluctuations in all India summer monsoon rainfall during 1871–1978. Clim Chang 6:287–301

Prasad T, Agarwal AL (1996) A day of exceptionally heavy rainfall over Bombay. Mausam 47(4):425–428

Ray K, Mohanty M, Chincholikar JR (2009) Climate variability over Gujarat, India. In: ISPRS archives XXXVIII-8/W3 workshop proceedings: impact of climate change on Agriculture, pp 38–43

Saseendran SA, Singh KK, Bahadur J, Dhar ON (1995) 1 to 10 days extreme rainfall studies for Kerala State. Mausam 46(2):175–180

Simulation of Heavy Rainfall Event over Gujarat During September 2013

S.I. Laskar, S.D. Kotal, S.K. Bhattacharya, and S.K. Roy Bhowmik

1 Introduction

Heavy rainfall events are known to occur over the Indian subcontinent during southwest monsoon season under the influence of off-shore troughs, off-shore vortices, depressions over the Bay of Bengal and Arabian Sea and Mid-Tropospheric Cyclones (MTC). Rainfall amounts of 100–300 mm in a day at and around the weather systems along the west coast of India and other parts of the country are common during southwest monsoon season. These rainfall events are caused by organized Mesoscale Convective Systems (MCSs) embedded in large scale synoptic systems (Benson and Rao 1987; Sikka and Gadgil 1980). Extreme rainfall events result in landslides, flash floods and damage to crops that have major impacts on the society, economy and environment. Although prediction of such extreme weather events is still fraught with uncertainties, a proper assessment of likely future trends would help in setting up infrastructure for disaster preparedness.

The non-hydrostatic mesoscale models are capable for simulation/prediction of high impact weather systems which lead to heavy rainfall episodes over India (Mohanty et al. 2012; Routray et al. 2005, 2010; Deb et al. 2008; Kumar et al. 2008). However, the forecast skill of these models is very limited, particularly for important variables like rainfall (Roy Bhowmik and Prasad 2001; Rama Rao et al. 2007; Sikka and Rao 2008; Das et al. 2008).

Gujarat state is located in the extreme western part of India. It is situated roughly between 20°15′N and 24°30′N. It has Sind (Pakistan) and Rajasthan to the north, Maharashtra to the south, Madhya Pradesh to the east and the Thar desert in the north-east. Based on meteorological conditions the state is divided into two subdivisions: (i) Saurashtra-Kutch and (ii) Gujarat region (Fig. 1).

S.I. Laskar (✉) • S.D. Kotal • S.K. Bhattacharya • S.K.R. Bhowmik
NWP Division, India Meteorological Department, Lodhi Road, New Delhi, India
e-mail: drsebul@gmail.com

Fig. 1 The geographical location of Gujarat region and Saurashtra & Kutch (region of study) in sub-division map of India

During 22–29th September 2013, there was widespread heavy rainfall activities which gripped Gujarat region and Saurashtra & Kutch which brought normal life to a standstill. The number of stations reporting heavy rainfall (7 cm or more) on 23, 24, 25, 26, 27, 28 and 29th September 2013 were 49, 74, 66, 89, 47, 18 and 16 respectively. The highest rainfall recorded during the period was 43 cm over Quant (Gujarat region) on 24th September and 45 cm over Khambhalia (Saurashtra & Kutch) on 27th September 2013.

The prime objective of the study is to evaluate the performance of IMDGFS T574 model by analyzing various model products viz., wind, vorticity, vertical velocity, moisture convergence flux at 850 hPa level that led to intense precipitation events over Gujarat region, Saurashtra & Kutch. The following section presents an overview of our study. The description of the Numerical Weather Prediction model has been given in Sect. 2. Section 3 gives data and methodology. The synoptic situations of the event are presented in Sect. 4. Section 5 provides the results and discussion while Sect. 6 briefly mentions the broad overall conclusions of this study.

2 Model Description

The Global Forecast System GFS T574L64 (~25 km in horizontal over the tropics), adopted from National Centre for Environmental Prediction (NCEP), was implemented at India Meteorological Department (IMD), New Delhi on IBM-based High Power Computing Systems (HPCS). The GFS T574L64 (GSI 3.0.0 and GSM 9.1.0) model is run twice in a day (0000 UTC and 1200 UTC). In horizontal, it resolves 574 waves ($\approx$23 km) in spectral triangular truncation representation (T574), for which the Gaussian grid of 1,760×880 dimensions are used. The model has 64 vertical levels (hybrid, sigma and pressure). There are 15 levels below 800 hPa (in lower troposphere), 14 layers in the middle troposphere (800–350 hPa), eight levels are in jet-stream region (350–150 hPa) and 27 levels above 150 hPa in the stratosphere. Sigma level is considered from surface to 100 hPa (40 levels) and 24 P-level above upto 0.27 hPa. It has time step of 2 min and Global Data Assimilation System (GDAS) runs four times per day (0000, 0600, 1200 and 1800 UTC). The GFS (T574L64) model runs two times per day (0000, 1200 UTC) and gives forecast upto 168 h.

3 Data and Methodology

Synoptic features associated with this heavy rainfall event and rainfall data have been collected from IMD, New Delhi. IMD GFS (T574L64) dataset has been used to plot and analyze wind at 850 hPa level, vorticity, vertical velocity, moisture flux convergence and rainfall forecast. The observed rainfall has been plotted by using IMD gridded rainfall data set.

4 Synoptic Features

A low pressure area formed over northwest Bay of Bengal off Odisha-West Bengal coasts on 19 Sept. morning. It lay over Odisha and adjoining areas of Gangetic West Bengal and Jharkhand on 19 Sept. evening and over Chhattisgarh and adjoining areas of east Madhya Pradesh and Vidarbha on 20 Sept. It continued to move in west-northwesterly direction and lay over south Madhya Pradesh and adjoining Vidarbha on 21 Sept; over central parts of south Madhya Pradesh and neighbourhood on 22 Sept. and weakened on 23 Sept. However, its associated upper air cyclonic circulation extending up to mid and upper tropospheric level lay over southwest Madhya Pradesh and adjoining Gujarat region on 23 Sept. and over southwest Madhya Pradesh and adjoining north Gujarat region and southeast Rajasthan on 24 Sept. It was seen over north Gujarat region and neighbourhood on 25 Sept. and over Saurashtra & Kutch and neighbourhood on 26 Sept. On 27 Sept. morning, with fresh moisture supply from Arabian Sea, it developed into a low pressure area over Kutch and neighbourhood with associated cyclonic circulation extending upto mid-tropospheric levels. Under the influence of an approaching mid-latitude trough in the westerlies, the low pressure area re-curved towards south Rajasthan and became less marked on 30 Sept. However, the associated cyclonic circulation still persisted over southeast Rajasthan and neighbourhood on 30 Sept. morning and over Saurashtra and neighbourhood till 2 September 2013. This synoptic feature is very well captured by GFST574 model. The analysis of wind by GFST574 model at 850 hPa level during the period 0000 UTC of 22 Sept. to 0000 UTC of 28 September 2013 is illustrated in Fig. 2a–g. The model also shows that the southerly/southwesterly wind speed was maximum (30–40 knot) mainly over

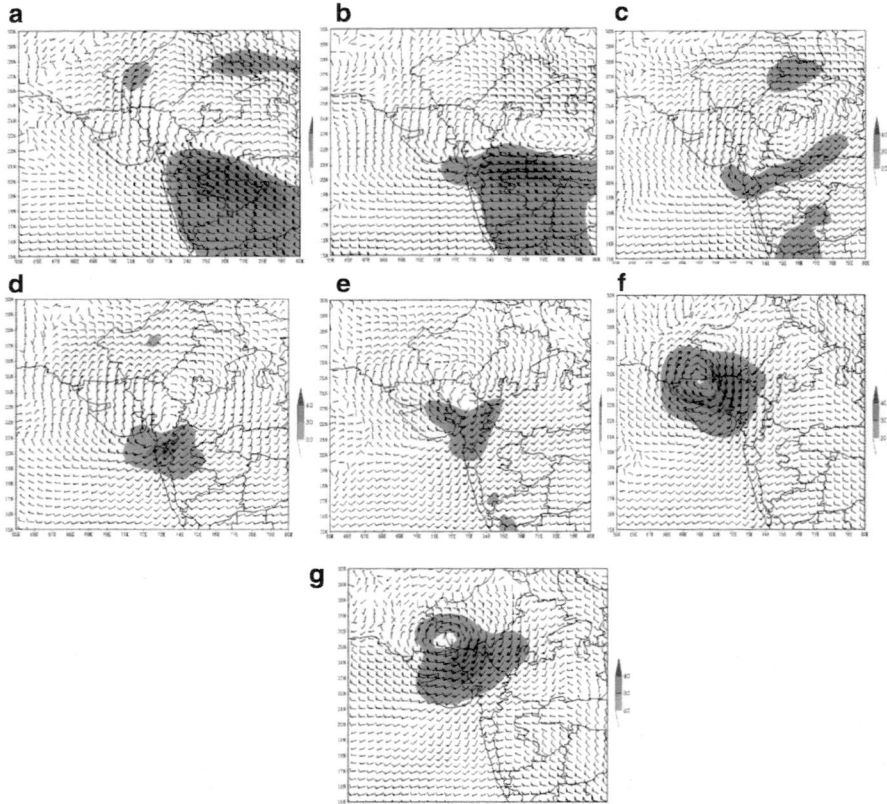

Fig. 2 850 hPa level wind analysis at 0000 UTC of (**a**) 22.09.2013, (**b**) 23.09.2013, (**c**) 24.09.2013, (**d**) 25.09.2013 (**e**) 26.09.2013, (**f**) 27.09.2013 and (**g**) 28.09.2013

southern parts of Gujarat during 22–25 Sept. and on 26 Sept. covered entire Gujarat region and Saurashtra & Kutch on 27 and 28 September.

5 Results and Discussion

5.1 Vorticity Analysis

The vorticity analysis by GFST574 model at 850 hPa level during the period 0000 UTC of 22 Sept. to 0000 UTC of 28 September 2013 is illustrated in Fig. 3a–g. The cyclonic vorticity of the order of 20×10^{-5} s^{-1} or more emerged over extreme south western parts of Gujarat at 0000 UTC of 23 September. The cyclonic vorticity moved slightly in northward direction with increase in intensity and was found over central parts of Gujarat at 0000 UTC of 26 September with intensity (30×10^{-5} s^{-1} or more). The cyclonic vorticity further intensified (40×10^{-5} s^{-1} or more) with

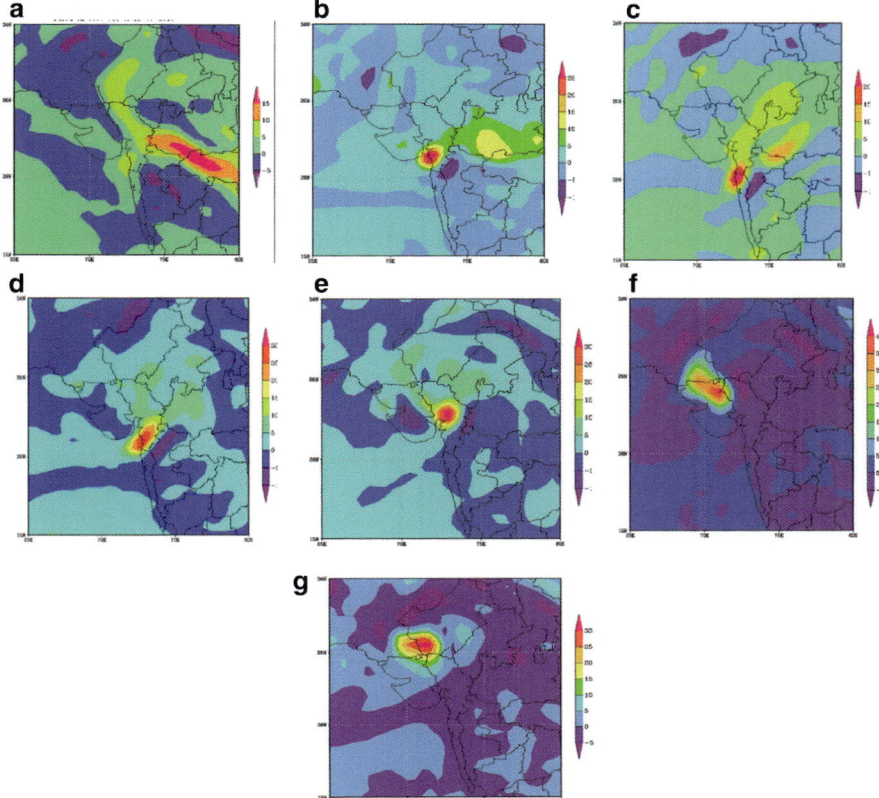

Fig. 3 850 hPa level vorticity analysis at 0000 UTC of (**a**) 22.09.2013, (**b**) 23.09.2013, (**c**) 24.09.2013, (**d**) 25.09.2013 (**e**) 26.09.2013, (**f**) 27.09.2013 and (**g**) 28.09.2013

movement in northwestward direction and found to be over Saurashtra & Kutch region at 0000 UTC of 27 Sept. which further moved in northward direction towards west Rajasthan.

5.2 Vertical Velocity

The vertical velocity analysis at 850 hPa level by GFST574 model during the period 0000 UTC of 22–0000 UTC of 28 September 2013 is shown in Fig. 4a–g. From the figure the region of rising motion is found to be over southwestern parts of Gujarat at 0000 UTC of 23 Sept. which moved slightly in a northwestward direction with increase in intensity till 0000 UTC of 25 September. The region of rising motion further moved in a northwestward direction from 0000 UTC of 26 Sept. onwards but with decrease in intensity. The highest value of the vertical velocity has been observed as 3×10^{-5} Pa/s over southwestern parts of Gujarat on 24 and 25 September, 2013.

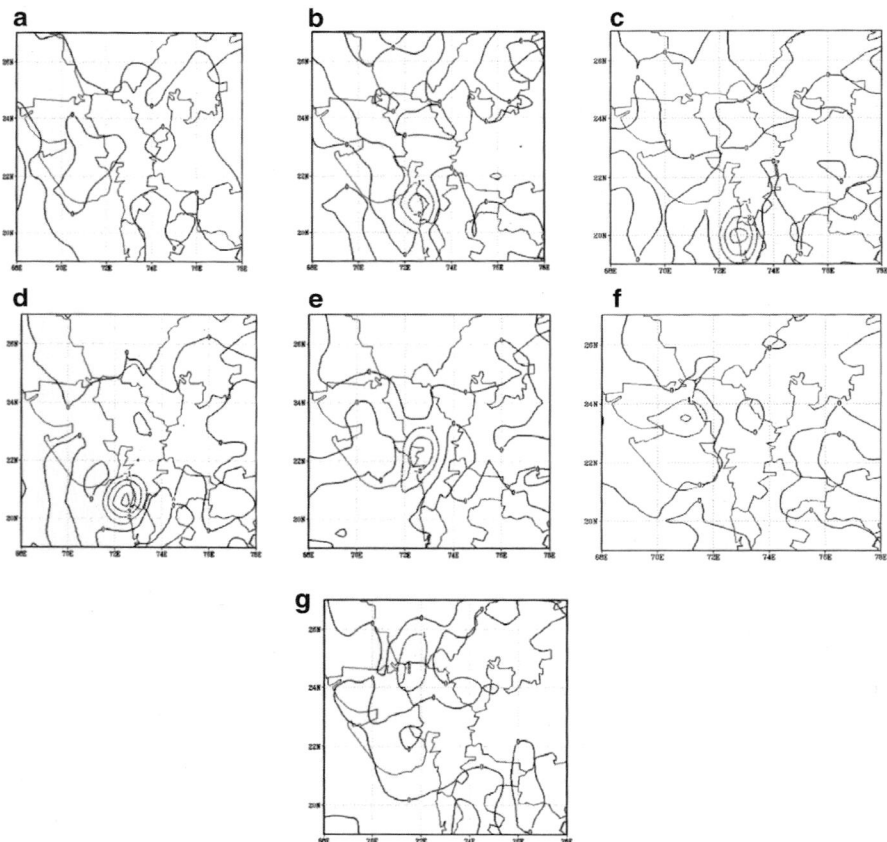

Fig. 4 850 hPa level vertical velocity analysis at 0000 UTC of (**a**) 22.09.2013, (**b**) 23.09.2013, (**c**) 24.09.2013, (**d**) 25.09.2013 (**e**) 26.09.2013, (**f**) 27.09.2013 and (**g**) 28.09.2013

5.3 Moisture Flux Convergence

Efficient moisture transport and accumulation are important factors for MCS development and the subsequent heavy rainfall (Chang et al. 2008). Figure 5a–g gives the moisture flux in the lower level (850 hPa) based on 0000 of 22–28 September 2013 respectively. The accumulation of moisture occurred due to constant currents approaching from Arabian Sea. Accumulation of moisture started over Gujarat from 0000 UTC of 23 September and the region of maximum moisture convergence restricted to only southernmost parts of Gujarat till 0000 UTC of 25 September. The region of maximum moisture convergence increased gradually and shifted towards northwestward direction and lay over Saurashtra & Kutch region and adjoining western parts of Gujarat on 26 and 27 September. It moved further in northward direction and lay over west Rajasthan on 28 September, 2013. Maximum convergence of the moisture (18×10^{-4} g/kg/s or more) took place over southwestern parts of Gujarat at 0000 UTC of 25 September.

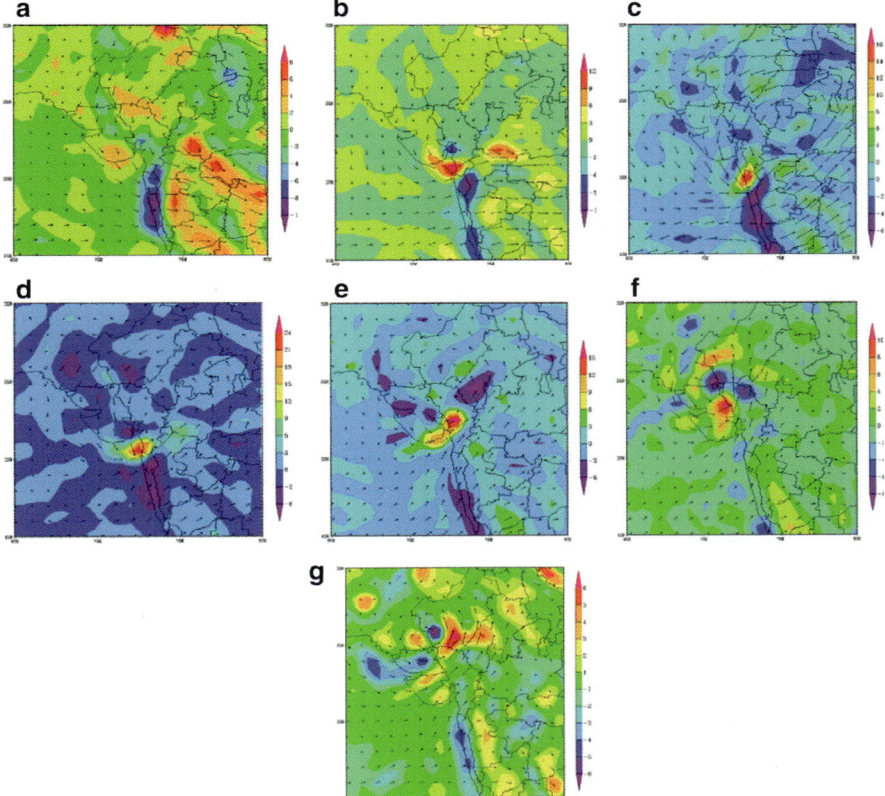

Fig. 5 850 hPa level moisture flux convergence at 0000 UTC of (**a**) 22.09.2013, (**b**) 23.09.2013, (**c**) 24.09.2013, (**d**) 25.09.2013 (**e**) 26.09.2013, (**f**) 27.09.2013 and (**g**) 28.09.2013

5.4 Rainfall Forecast

Figure 6a1–c7 represents the observed rainfall in the left side and their 24 and 48-h forecast in the right side respectively for the period 23–29 September, 2013. For the 23 and 24 September the heavy rainfall over south Gujarat region was well predicted in 24-h forecast whereas in 48-h forecast the model underestimated the heavy rainfall event for both the days. The GFST574 model overestimated the rainfall of

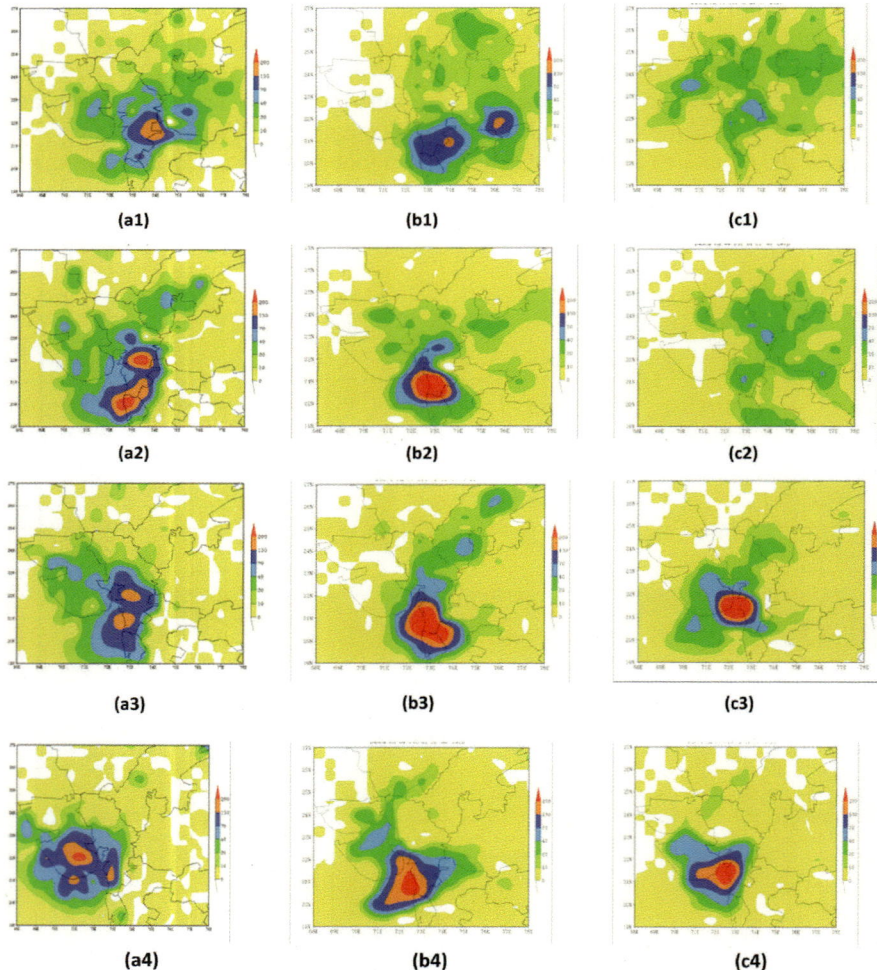

Fig. 6 Observed rainfall on (**a1**) 23.09.2013, (**a2**) 24.09.2013, (**a3**) 25.09.2013, (**a4**) 26.09.2013, (**a5**) 27.09.2013, (**a6**) 28.09.2013 and (**a7**) 29.09.2013; 24-h forecast valid for 0000 UTC of (**b1**) 23.09.2013, (**b2**) 24.09.2013, (**b3**) 25.09.2013, (**b4**) 26.09.2013, (**b5**) 27.09.2013, (**b6**) 28.09.2013 and (**b7**) 29.09.2013; and 48-h forecast valid for 0000 UTC of (**c1**) 23.09.2013, (**c2**) 24.09.2013, (**c3**) 25.09.2013, (**c4**) 26.09.2013, (**c5**) 27.09.2013, (**c6**) 28.09.2013 and (**c7**) 29.09.2013

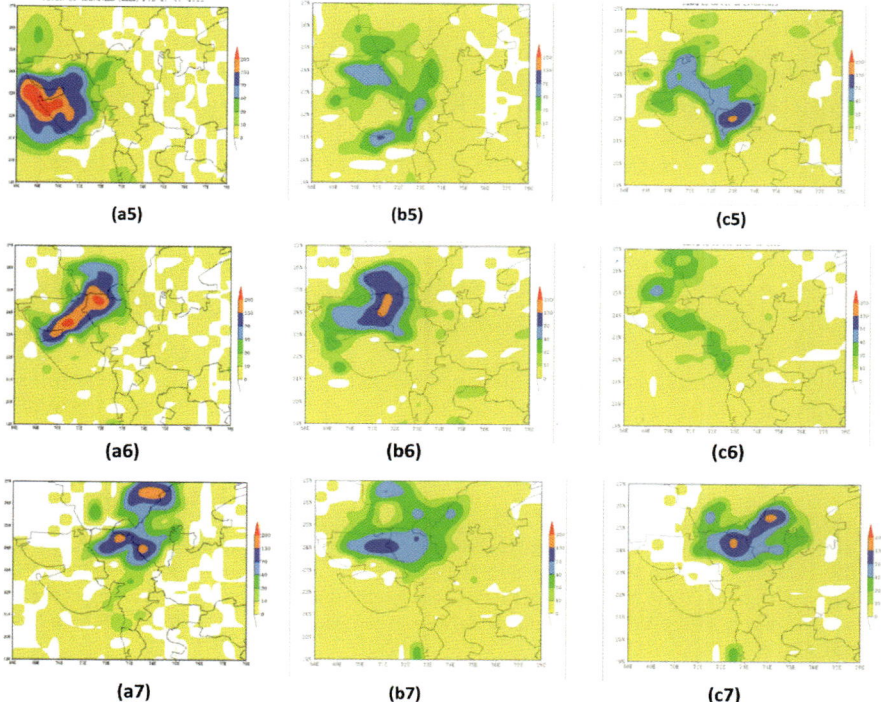

Fig. 6 (continued)

25 September in both the 24 and 48-h forecasts. The GFST574 model could capture the heavy rainfall event of 26 September in both the 24 and 48-h forecasts but underestimated the heavy rainfall of 27. For 28 September the 24-h forecast of rainfall could capture the heavy rainfall event but with less intensity whereas 48-h forecast could not capture the heavy rainfall event. The 24-h forecast overestimated the rainfall of 29 September over Saurashtra & Kutch whereas it underestimated the same over Gujarat. The 48-h forecast of GFST574 model could capture the heavy rainfall event of 29 September very well.

Conclusions
(i) Heavy rainfall event over Gujarat and Saurashtra & Kutch was under the influence of low pressure area. This low pressure area enhanced the moisture incursion from the Arabian Sea and supported to drift the moisture to higher latitudes causing heavy rainfall over the area.
(ii) The intensity of cyclonic vorticity increased rapidly from 0000 UTC of 26–0000 UTC of 27 Sep. and again decreased rapidly at 0000 UTC

(continued)

of 28 Sep. This showed that the rapid growth of convective system took place between 0000 UTC 26 and 0000 UTC of 27 September.
(iii) The vertical velocity at 850 hPa level helped to carry moisture content up to sufficient high level to form convective clouds.
(iv) Sufficient moisture convergence took place over the area during the period of heavy rainfall.
(v) On almost all the days the model could capture in 24–48 h in advance heavy rainfall event though in some cases intensity and location varied slightly from the actual observation. The performance of 24-h rainfall forecast for this particular event by GFST574 model is better than that of 48 h.

References

Benson CL, Rao GV (1987) Convective bands as structural components of an Arabian Sea convective cloud cluster. Mon Weather Rev 115:3013–3023
Chang LT-C, Chen GT-J, Cheung KKW (2008) Mesoscale simulation and moisture budget analyses of a heavy rain event over southern Taiwan in the Meiyu season. Meteorol Atmos Phys 101:43–63
Das S, Ashrit R, Iyengar GR, Mohandas S, Dasgupta M, Gupta AS, George JP, Dutta SK (2008) Skills of different mesoscale models over Indian region during monsoon season: forecast errors. J Earth Syst Sci 117:603–620
Deb SK, Srivastava TP, Kishtawal CM (2008) The WRF model performance for the simulation of heavy precipitating events over Ahmedabad during August 2006. J Earth Syst Sci 117:589–602
Kodama KR, Barnes GM (1997) Heavy rain events over the south-facing slopes of Hawaii: attendant conditions. Weather Forecast 12:347–367
Kumar A, Dudhia J, Rotunno R, Dev N, Mohanty UC (2008) Analysis of the 26 July 2005 heavy rain event over Mumbai, India using the Weather Research and Forecasting (WRF) model. Q J R Meteorol Soc 134:1897–1910
Mohanty UC, Routray A, Osuri KK, Kiran Prasad S (2012) A study on simulation of heavy rainfall events over Indian region with ARW-3DVAR modeling system. Appl Geophys 169:381–399
Rama Rao YV, Hatwar HR, Salah AK, Sudhakar Y (2007) An experiment using the high resolution eta and WRF models to forecast heavy precipitation over India. Pure Appl Geophys 164:1593–1615
Routray A, Mohanty UC, Das AK, Sam NV (2005) Study of heavy rainfall event over the west-coast of India using analysis nudging in MM5 during ARMEX-I. Mausam 56(1):107–120
Routray A, Mohanty UC, Niyogi D, Rizvi SRH, Osuri KK (2010) Simulation of heavy rainfall events over Indian monsoon region using WRF-3DVAR data assimilation system. Meteorol Atmos Phys 106:107–125
Roy Bhowmik SK, Prasad K (2001) Some characteristics of limited area model precipitation forecast of Indian monsoon and evaluation of associated flow features. Meteorol Atmos Phys 76:223–236
Sikka DR, Gadgil S (1980) On the maximum cloud zone and ITCZ over the Indian longitudes during the southwest monsoon. Mon Weather Rev 108:1122–1135
Sikka DR, Sanjeeva Rao P (2008) The use and performance of mesoscale models over the Indian region for two high-impact events. Nat Hazards 44(3):353–372

Simulation of Rainfall over Uttarakhand, India, in June Using WRF-ARW Model and Impact of AIRS Profiles

Sqn Ldr Prabodh Shukla, Wg Cdr Anil Kumar Devrani, and Sqn Ldr Himanshu Singh

1 Introduction

The south west monsoon normally sets in over Uttarakhand by last week of June. However, in 2013, the monsoon current reached the region by mid-June. June 16 and 17 of this year were considered as one of the saddest days in the history of Uttarakhand, when unprecedented rains swelled the rivers both upstream and downstream, leading to devastating landslides and floods. This led to massive loss of life and extensive damage to property. The reason for this has been attributed to the extreme rain, along-with rampant unauthorized and mindless construction work with total disregard to nature and environment.

The first principle of disaster management is prevention—by taking necessary precautionary measures. One step would be to initiate measures to stop the haphazard human intervention with nature. Such extreme weather events can neither be controlled nor ruled out, but timely prediction with some accuracy would be an important step in providing adequate notice to the administrative authorities to initiate appropriate action to mitigate the damage.

With the Chardham pilgrimage centre in the region, there is a mad rush of religious tourism during the summer season. Nowcasting in such a complex hilly terrain would be of no help in case of such a natural catastrophe. The best available option is to provide a short-range forecast through Numerical Weather Prediction (NWP).

NWP is basically an initial value problem, which means that the simulated atmosphere has "sensitive dependence on initial conditions," i.e., small differences in the initial state of the atmosphere ultimately result in large differences in the forecast. Hence accurate initial condition is the foremost pre-requisite for accurate numerical simulation and prediction. Data assimilation has been recognized as a

---

Sqn Ldr P. Shukla (✉) • Wg Cdr A.K. Devrani • Sqn Ldr H. Singh
Directorate of Meteorology, Air Headquarters, New Delhi, India
e-mail: prab.shuks@gmail.com

useful way to obtain better "consistent" initial conditions for NWP. One of the most attractive and effective methods of data assimilation is the variational technique (Gelb et al. 1974). In this study an attempt has been made to improve the NWP output using this technique.

2 Data and Methodology

Uttarakhand experienced flash floods between 14 June 2013 and 17 June 2013. During this period most of the Automatic Weather Stations (AWS) recorded heavy to exceptionally heavy rainfall in Uttarakhand. The list of actual recorded rainfall by different observatories in Uttarakhand is given in Table 1. In the current study, an attempt has been made to compare the rainfall prediction of WRF-ARW model with Kain-Fritsch (KF) and Grell-Devenyi (GD) convective parameterization schemes and also examine the impact of assimilation of conventional data and AIRS profiles on rainfall forecast over Uttarakhand.

Table 1 Rainfall (mm) data from AWS (Uttarakhand)

Sl. no.	Date	14 June 13	15 June 13	16 June 13	17 June 13
District: Almora					
1.	Almora	1.0	32.4	89.3	100.0
2.	Ranikhet (G)	0.0	16.0	38.0	120.0
District: Bageshwar					
3.	Bageshwar (Thmo)	3.0	61.0	161.0	63.0
4.	Kosani (U Prob)	20.2	105.0	205.0	83.2
District: Chamoli					
5.	Chamoli	37.0	58.0	76.0	100.0
6.	Joshimath	31.4	41.9	113.8	78.6
7.	Karnaprayag	7.0	88.0	89.6	82.3
8.	Tharali	15.0	58.0	173.0	80.0
District: Champawat					
9.	Bambasa	0.0	3.0	99.0	230.0
10.	Champawat	1.0	34.0	97.0	222.0
District: Dehra Dun					
11.	Dehra Dun	53.5	219.9	370.2	11.8
12.	Mussoorie	44.0	137.0	155.0	8.0
District: Garhwal Pauri					
13.	Kotdwara	9.0	73.0	23.0	52.2
14.	Landsdown	0.0	64.0	51.0	28.0
15.	Pauri	0.0	44.0	51.0	38.0

(continued)

Table 1 (continued)

Sl. no.	Date	14 June 13	15 June 13	16 June 13	17 June 13
District: Garhwal Tehri					
16.	Deoprayag	7.3	129.5	163.3	69.5
17.	Tehri	33.5	121.9	168.9	53.4
District: Hardwar					
18.	Hardwar	20.0	107.6	218.0	14.0
19.	Roorkee	5.0	51.0	147.0	15.0
District: Nainital					
20.	Haldwani	13.0	91.0	200.0	278.3
21.	Mukteshwar	0.4	78.4	236.8	183.0
22.	Nainital	18.6	43.6	175.6	170.2
District: Pithoragarh					
23.	Munsiyari	25.0	44.0	85.0	75.0
24.	Pithoragarh	0.0	11.2	85.5	117.2
District: Rudraprayag					
25.	Jakholi	71.0	121.0	108.0	65.0
26.	Rudraprayag	11.8	89.4	92.2	59.2
District: Udham Singh Nagar					
27.	Kashipur	65.0	2.0	31.0	35.0
28.	Pantnagar	0.0	5.6	62.1	113.0
District: Uttarkashi					
29.	Barkot	15.4	112.6	20.0	20.0
30.	Dunda	80.0	118.0	185.0	16.0
31.	Purola	36.0	165.0	60.0	104.0
32.	Uttar Kashi	35.0	129.0	162.0	19.0

The conventional data used in this study included surface station reports and upper air observation. This data included observations from Indian Air Force as well as civil observatories. Observations from civil observatories were obtained from OGI MET (http://ogimet.com/home/pntml.en). Period of data used was from 0130 UTC of 13 June 13 to 0130 UTC of 16 June 13 for assimilation in the model.

For this study, we have also used AIRS level 2, atmospheric temperature and moisture profiles. The data has a horizontal resolution of 45 km. Globally, the AIRS level 2 retrieved profiles compared to rawinsonde collocated in time and space exhibit root-mean-square error of 1 K in 1 km layers of temperature and 10–15 % in relative humidity in 2 km layer (Tobin et al. 2006; Divakarla et al. 2006; Fetzer 2006; Susskind et al. 2006). The lowest errors occur for clear sky cases over water with degradation in profile accuracy in cloudy regions and/or over land. AIRS data was obtained from (http://mirador.gsfc.nasa.gov/). Period of data used was from 0130 UTC of 13 June 13 to 0130 UTC of 16 June 13.

2.1 Model Configuration

The configuration of WRF-ARW model used in this study are given below.

Model	WRF-ARW Model Version 3.1.1
Microphysics	Thompson Scheme (Thompson et al. 2004)
Convective parameterization	Grell-Devenyi Ensemble Scheme (Grell and Devenyi 2002) and Kain-Fritsch (Kain 2004)
Planetary boundary layer	Mellor-Yamada-Janjic TKE Scheme (Janjic 1990,1996, 2002)
Long wave and short wave	The Rapid Radiative Transfer Model (RRTM; Mlawer et al. 1997)

Simulations were conducted with the two-way nested domains as shown in Fig. 1. Coarser resolution domain was consisting of 356×376 grid points with 18 km resolution and 739×649 grid points with 6 km resolution for inner domain. The model had 28 vertical levels with the top of the model atmosphere located at 10 hPa. Physics options were kept same in both the domains. Terrain data resolution in both the domains was 2 min. The model was initialized using NCEP GFS data of 0.5 degree resolution and boundary conditions (1.0 degree resolution) were updated at every 6 h.

Background covariance matrix was generated using NMC method (now known as NCEP method).In the WRF-3DVar, all observation errors are assumed to be uncorrelated and constant in space and time. The observation error estimates for AIRS temperature profiles varies in the vertical, with maxima in the lower and upper levels (2.0 K) and minima in the mid levels (1.0 K). The observation error estimates for the AIRS moisture profiles also varies in the vertical, with maxima in the middle and upper levels (20 %) and a minima in the mid levels (10 %).

2.2 Experiment Design

Total four cases were selected, from 14 June 13 to 17 June 13. Model was integrated for 51 h in all the cases. The study was divided in two phases:

(a) **Phase I:** The aim in this phase was to ascertain the better convective parameterization scheme over mountainous terrain. In the first phase, separate model runs were carried out for all the cases with KF and GD convective parameterization scheme, using NCEP GFS initial condition from 13 June 13 to 16 June 13. In each case, 27–51 h accumulated rainfall forecast produced by finer resolution domain (domain 2) with different schemes were compared with the IMD rainfall gridded data. IMD rainfall gridded data is a merged product of Automatic Weather Stations & TRMM TB42, available at 50 km resolution. The convective scheme which gave better results was selected for use in phase II.

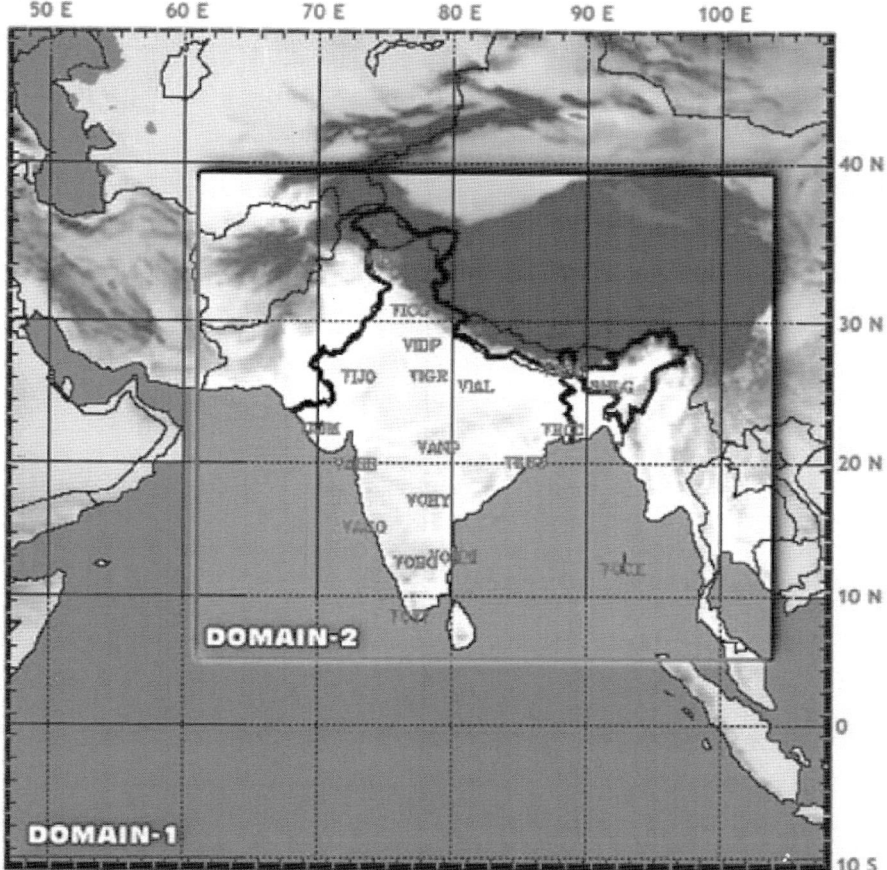

Fig. 1 Nested domain of the model used in study

(b) **Phase II:** In this phase model runs were carried out after assimilating conventional data and AIRS temperature and moisture profiles, using the better convective parameterization scheme. Instead of directly using NCEP GFS data as the first guess in the 3DVar experiments, a WRF-ARW simulation was first integrated at 0000 UTC of D-1 days for 3 h, using GFS data. Here D day means the day from where model run in forecasting mode was initiated for each case. Data valid for 90 min prior and 90 min after 0300 UTC was assimilated with this 03-h forecast. Using this new initial condition valid at 0300 UTC, again model was integrated for 3 h and forecast valid at 0600 UTC was generated. Assimilation was carried out at this 0600 UTC forecast and same cycle continued at every 3 h, till 0000 UTC of D day as shown in Fig. 2.

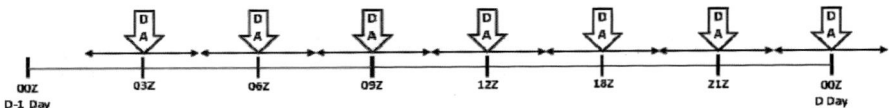

Fig. 2 Assimilation cycle of WRF-ARW model used for study

At 0000 UTC of D days, using this new initial condition, model run was initiated for 51 h and boundary conditions were updated using NCEP GFS data valid for D day. For each case, 27–51 h forecast accumulated rainfall was compared with IMD's gridded rainfall and also location specific rainfall was compared with the actual observations from 32 AWS observations in Uttarakhand. For comparison of location specific rainfall, the maximum rainfall predicted by the model within approximately 12 km radius of the grid point was compared with actual AWS observation. This was done to cater to the effect of poor terrain resolution in the model environment. For all the cases, assimilation was carried out in the coarser, 18 km resolution domain. The model initial condition for finer resolution domain was interpolated from domain 1. Prior to data assimilation, the conventional data and AIRS data underwent quality checking processes in order to reduce the possibility of assimilating bad observations. A set of quality assurance (QA) flags provided by AIRS science team members (Susskind et al. 2006) were used to filter the cloud contaminated temperature and moisture data. Each profile of AIRS contains level-specific quality indicators allowing us to select the highest quality data (products quality flag = 0).

3 Results and Discussion

3.1 Comparison of Convective Parameterization Schemes

Figure 3a, b show the 27–51 h forecast accumulated rainfall based on initial condition of 13 June 13, with KF and GD convective parameterization scheme, respectively. Figure 3c shows the IMD's gridded actual rainfall data reported on 0300 UTC of 15 June 13. Predictions with both the schemes were different from the actual; however KF scheme produced better result, especially over northern parts of Uttarakhand. Rainfall over southern part of Uttarakhand was missed by both the schemes.

Figure 4a, b show the 27–51 h forecast accumulated rainfall based on initial condition of 14 June 13, with KF and GD convective parameterization scheme, respectively. Figure 4c shows the IMD's gridded actual rainfall data reported on 0300 UTC of 16 June 13. Rainfall forecast produced with both the schemes were different from the actual; however KF scheme produced better result, especially over western and eastern parts of Uttarakhand. Heavy rainfall over western Uttarakhand was missed by both the schemes.

Figure 5a, b show the 27–51 h forecast accumulated rainfall based on initial condition of 15 Jun 13, with KF and GD convective parameterization scheme, respectively. Figure 5c shows the IMD's gridded actual rainfall data reported on

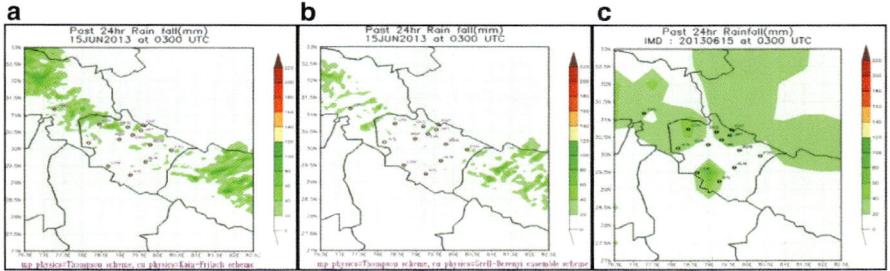

Fig. 3 27–51 h forecast accumulated rainfall (mm) valid at 0300 UTC of 15 June 13: (**a**) with KF; (**b**) with GD; and (**c**) IMD's gridded actual rainfall (mm) reported on 0300 UTC of 15 June 13

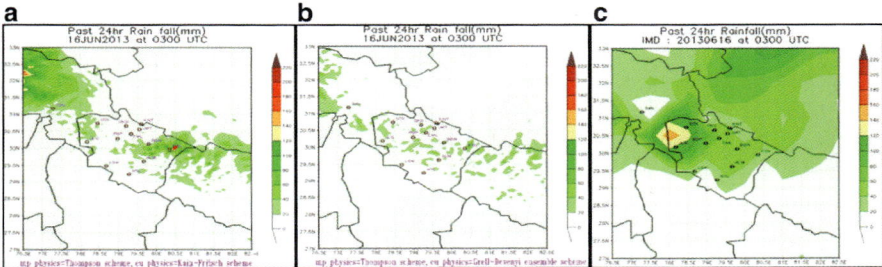

Fig. 4 27–51 h forecast accumulated rainfall (mm) valid at 0300 UTC of 16 June 13: (**a**) with KF; (**b**) with GD; and (**c**) IMD's gridded actual rainfall (mm) reported on 0300 UTC of 16 June 13

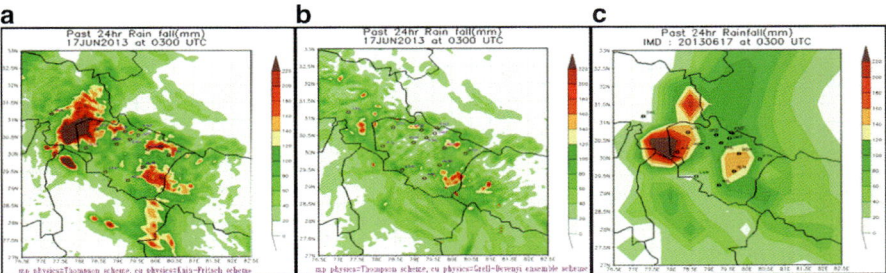

Fig. 5 27–51 h forecast accumulated rainfall (mm) valid at 0300 UTC of 17 June 13: (**a**) with KF; (**b**) with GD; and (**c**) IMD's gridded actual rainfall (mm) reported on 0300 UTC of 17 June 13

0300 UTC of 17 June 13. On this day also KF rainfall predictions were better than GD rainfall forecast. KF could simulate the heavy rainfall incidences over western and eastern parts of Uttarakhand. Although there is slight difference in location and amount but that could be because of difference in resolution of model (06 km) and IMD's rainfall data (50 km). GD could pick up heavy rainfall cases, but in isolated pockets. It could not forecast the widespread heavy rainfall activity.

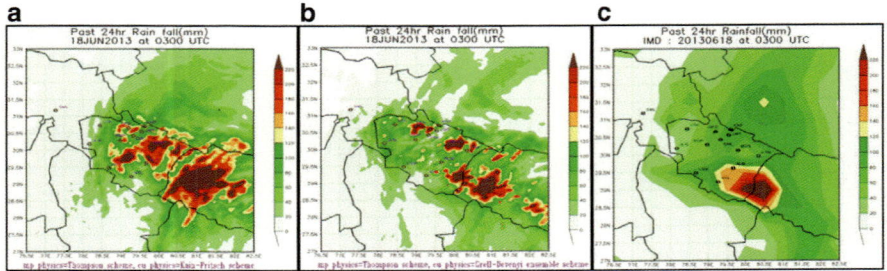

Fig. 6 27–51 h forecast accumulated rainfall (mm) valid at 0300 UTC of 18 June 13: (**a**) with KF; (**b**) with GD; and (**c**) IMD's gridded actual rainfall (mm) reported on 0300 UTC of 18 June 13

Figure 6a, b show the 27–51 h forecast accumulated rainfall based on initial condition of 16 June 13, with KF and GD convective parameterization scheme, respectively. Figure 6c shows the IMD's gridded actual rainfall data reported on 0300 UTC of 18 June 13. On this day, IMD's data show widespread activity over eastern parts of Uttarakhand and adjoining Nepal. Rainfall simulated by GD was closer to actual. KF predictions also match with the heavy rainfall incidences, but it overestimated the rainfall in terms of area and intensity.

From the above comparisons, it is evident that KF performed better on 3 out of 4 days. Though there is slight mismatch in terms of location and intensity, but overall KF performs better than GD in predicting rainfall over mountainous terrain.

3.2 Evaluation of Rainfall Forecast After Assimilation

3.2.1 Areal Rainfall Forecast Comparison

Figure 7a, b show the 27–51 h forecast accumulated rainfall based on initial condition of 13 June 13, without assimilation and with assimilation using KF convective parameterization scheme, respectively. Figure 7c shows the IMD's gridded actual rainfall data reported on 0300 UTC of 15 June 13. After assimilation, the rainfall forecast improved, both in terms of location and intensity. Heavy rainfall over Garhwal district was picked up after assimilation, though the intensity was over estimated. After assimilation another rainfall maxima appeared over western part of Uttarakhand and over Pithoragarh district.

Figure 8a, b show the 27–51 h forecast accumulated rainfall based on initial condition of 14 June 13, without assimilation and with assimilation using KF convective parameterization scheme, respectively. Figure 8c shows the IMD's gridded actual rainfall data reported on 0300 UTC of 16 June 13. On this day also, rainfall forecast improved after assimilation. Without assimilation, widespread rainfall activity over Uttarakhand was predicted with lesser intensity than actual.

Simulation of Rainfall over Uttarakhand, India

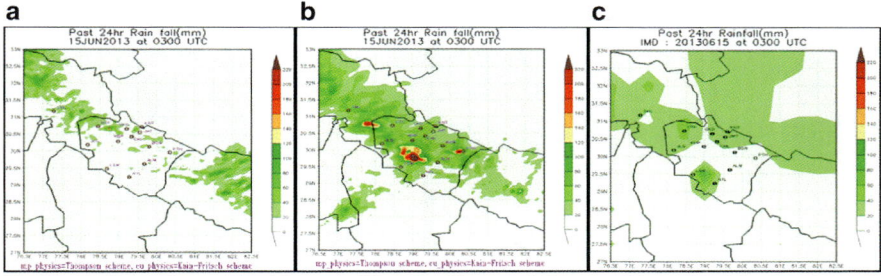

Fig. 7 27–51 h forecast accumulated rainfall (mm) valid at 0300 UTC of 15 June 13: (**a**) without DA; (**b**) with DA; and (**c**) IMD's gridded actual rainfall (mm) reported on 0300 UTC of 15 June 13

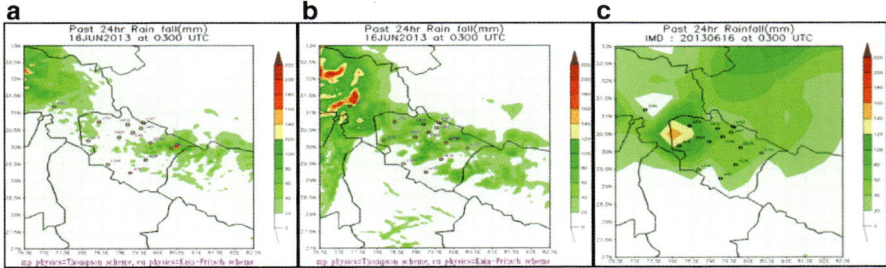

Fig. 8 27–51 h forecast accumulated rainfall (mm) valid at 0300 UTC of 16 June 13: (**a**) without DA; (**b**) with DA; and (**c**) IMD's gridded actual rainfall (mm) reported on 0300 UTC of 16 June 13

After assimilation, rainfall intensity increased all over the Uttarakhand. Although even after assimilation, model could not simulate the rainfall maxima over western part of Uttarakhand.

Figure 9a, b show the 27–51 h forecast accumulated rainfall based on initial condition of 15 June 13, without assimilation and with assimilation using KF convective parameterization scheme, respectively. Figure 9c shows the IMD's gridded actual rainfall data reported on 0300 UTC of 17 June 13. On this day, forecast produced before assimilation was close to the actual with almost correct location of rainfall maxima. After assimilation, model overestimated the rainfall, though it did not change the locations of rainfall maxima with respect to the actual location of rainfall maxima, but predicted few more locations of rainfall maxima.

Figure 10a, b show the 27–51 h forecast accumulated rainfall based on initial condition of 16 June 13, without assimilation and with assimilation using KF convective parameterization scheme, respectively. Figure 10c shows the IMD's gridded actual rainfall data reported on 0300 UTC of 18 June 13. On this day, rainfall predictions without assimilation were close to the actual heavy rainfall incidences

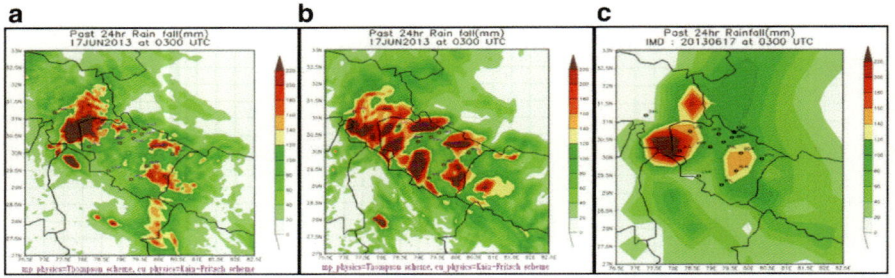

Fig. 9 27–51 h forecast accumulated rainfall (mm) valid at 0300 UTC of 17 June 13: (**a**) without DA; (**b**) with DA; and (**c**) IMD's gridded actual rainfall (mm) reported on 0300 UTC of 17 June 13

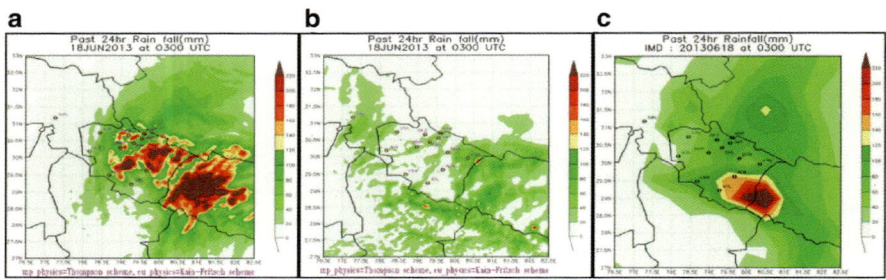

Fig. 10 27–51 h forecast accumulated rainfall (mm) valid at 0300 UTC of 18 June 13: (**a**) without DA; (**b**) with DA; and (**c**) IMD's gridded actual rainfall (mm) reported on 0300 UTC of 18 June 13

with overestimation in terms of area as well as intensity. After assimilation, the intensity of rainfall reduced significantly over entire Uttarakhand. Though occurrence of rainfall was picked up but the intensity did not match with the actual.

3.2.2 Location Specific Rainfall Forecast Comparison

Table 2 shows the daywise location specific actual rainfall and rainfall forecast before and after assimilation for 32 AWSs in Uttarakhand. On 14 June 13, after assimilation over most of the AWSs location specific rainfall forecast predicted overestimation of rainfall. Before assimilation the location specific rainfall forecast of 26 stations was closer to actual. In case of only six stations, the forecast improved after assimilation by approx. 40–50 %. On 15 June 13, there is a significant improvement in the location specific rainfall forecast after assimilation. Rainfall forecast of 22 stations improved by approx. 20–50 %. On this day there was no false alarm. On 16 June 13, location specific forecast of 25 stations was closer to actual before assimilation. On this day also overestimation in location specific rainfall forecast was observed after assimilation. On 17 June 13, almost 50 % location specific

Table 2 Actual rainfall (mm) and forecast rainfall (mm) without and with assimilation From 14 June 13 to 17 June 13

Sl. no.	Stations	14 June 13			15 June 13			16 June 13			17 June 13		
		Actual	Without DA	With DA	Actual	Without DA	With DA	Actual	Without DA	With DA	Actual	Without DA	With DA
District: Almora														
1.	Almora	1.0	23.1	54.9	32.4	25.8	50.3	89.3	131.0	118.0	100.0	24.7	18.3
2.	Ranikhet (G)	0.0	7.7	149.0	16.0	30.6	84.0	38.0	114.0	125.0	120.0	140.0	34.5
District: Bageshwar														
3.	Bageshwar (Thmo)	3.0	10.6	57.5	61.0	88.2	87.3	161.0	118.0	71.6	63.0	283.0	17.0
4.	Kosani (U Prob)	20.2	13.7	75.5	105.0	88.2	87.8	205.0	139.0	71.6	83.2	283.0	38.6
District: Chamoli														
5.	Chamoli	37.0	15.6	58.5	58.0	19.4	60.3	76.0	61.1	101.0	100.0	148.0	57.4
6.	Joshimath	31.4	48.1	53.5	41.9	23.5	34.2	113.8	71.1	157.0	78.6	134.0	40.6
7.	Karnaprayag	7.0	17.4	72.0	88.0	27.0	92.1	89.6	61.3	109.0	82.3	161.0	35.4
8.	Tharali	15.0	46.5	65.8	58.0	44.9	126.0	173.0	105.0	86.3	80.0	195.0	34.0
District: Champawat														
9.	Bambasa	0.0	0.6	32.7	3.0	14.4	94.0	99.0	149.0	251.0	230.0	93.9	77.9
10.	Champawat	1.0	28.4	37.4	34.0	27.2	80.7	97.0	226.0	280.0	222.0	150.0	58.8
District: Dehradun														
11.	Dehradun	53.5	23.0	99.2	219.9	30.3	74.1	370.2	195.0	253.0	11.8	91.0	28.0
12.	Mussoorie	44.0	25.0	76.0	137.0	62.4	83.1	155.0	254.0	252.0	8.0	9.9	32.0
District: Garhwal Pauri														
13.	Kotdwara	9.0	0.7	13.1	73.0	4.9	4.4	23.0	43.0	112.0	52.2	65.6	21.7
14.	Landsdown	0.0	0.0	44.1	64.0	8.5	2.4	51.0	81.8	227.0	28.0	114.0	8.2
15.	Pauri	0.0	0.8	157.0	44.0	20.4	62.6	51.0	110.0	113.0	38.0	160.0	50.3

(continued)

Table 2 (continued)

Sl. no.	Stations	14 June 13 Actual	14 June 13 Without DA	14 June 13 With DA	15 June 13 Actual	15 June 13 Without DA	15 June 13 With DA	16 June 13 Actual	16 June 13 Without DA	16 June 13 With DA	17 June 13 Actual	17 June 13 Without DA	17 June 13 With DA
District: Garhwal Tehri													
16.	Deoprayag	7.3	28.4	133.0	129.5	16.7	59.6	163.3	96.4	240.0	69.5	121.0	32.3
17.	Tehri	33.5	14.9	95.8	121.9	23.7	68.0	168.9	195.0	252.0	53.4	96.6	20.2
District: Hardwar													
18.	Hardwar	20.0	19.6	47.5	107.6	47.8	53.1	218.0	88.8	271.0	14.0	43.4	13.5
19.	Roorkee	5.0	0.8	22.4	51.0	12.2	2.7	147.0	94.4	55.8	15.0	6.0	33.1
District: Nainital													
20.	Haldwani	13.0	13.7	39.6	91.0	28.5	47.1	200.0	125.0	135.0	278.3	70.1	13.8
21.	Mukteshwar	0.4	20.4	27.6	78.4	19.1	49.3	236.8	182.0	149.0	183.0	62.2	20.3
22.	Nainital	18.6	0.1	26.5	43.6	3.8	1.4	175.6	52.1	160.0	170.2	62.7	31.9
District: Pithoragarh													
23.	Munsiyari	25.0	29.6	24.4	44.0	93.9	83.4	85.0	181.0	261.0	75.0	258.0	61.2
24.	Pithoragarh	0.0	26.6	96.7	11.2	24.0	18.2	85.5	141.0	215.0	117.2	154.0	51.2
District: Rudraprayag													
25.	Jakholi	71.0	28.4	96.4	121.0	14.4	70.5	108.0	137.0	229.0	65.0	98.5	21.0
26.	Rudraprayag	11.8	23.3	84.0	89.4	23.7	58.0	92.2	115.0	118.0	59.2	175.0	73.5
District: Udham Singh Nagar													
27.	Kashipur	65.0	0.3	56.8	2.0	4.9	2.6	31.0	51.2	273.0	35.0	52.4	53.0
28.	Pantnagar	0.0	0.2	5.7	5.6	2.5	3.6	62.1	118.0	214.0	113.0	29.9	20.5
District: Uttarkashi													
29.	Barkot	15.4	62.7	124.0	112.6	33.0	93.5	20.0	356.0	271.0	20.0	20.9	56.3
30.	Dunda	80.0	33.4	59.8	118.0	27.0	58.1	185.0	233.0	241.0	16.0	30.7	30.6
31.	Purola	36.0	52.2	115.0	165.0	30.7	93.5	60.0	340.0	188.0	104.0	11.1	56.3
32.	Uttarkashi	35.0	50.6	51.4	129.0	23.2	51.6	162.0	190.0	162.0	19.0	76.7	27.0

rainfall forecast improved after assimilation. In the IMD's gridded rainfall data the only maxima was observed over SE Uttarakhand and location specific rainfall data over this region was not available. Hence comparison of location specific rainfall forecast couldn't be carried out for SE Uttarakhand. Actually after assimilation underestimation in rainfall forecast was observed on this day.

Although there are incidences of overestimation of rainfall, comparison of districtwise rainfall reveals that areas of heavy rainfall incidences are captured well after assimilation, except on 17 June 13. Comparison of location specific rainfall forecast was carried out using the maximum rainfall observed within approximately 12 km radius of grid point, but still results didn't show any significant improvement. Over complex mountainous terrain, the resolution of terrain data used, play a significant role. Probably poor terrain resolution caused these deviations in the location specific forecast.

3.2.3 Impact of Resolution

In addition to above mentioned comparisons, the rainfall forecast produced by 18 km resolution domain was also studied for all the cases (results are not shown here). The model produced rainfall with westward geographical shift. Quantum of rainfall forecast was of the same order but the location was different. This clearly indicates that model may have capability of producing unusual heavy rainfall occurrences, but correct location forecast requires high resolution terrain data and high resolution model domain, especially over such complex mountainous region.

Conclusions
In this study an attempt has been made to carry out sensitivity experiment for convective parameterization schemes and also to examine the impact of data assimilation in rainfall forecast over Uttarakhand region. Sensitivity experiment revealed that Kain-Fritsch produces better rainfall forecast than Grell-Devenyi Ensemble Scheme. Uttarakhand is the region of hilly terrain, where most of the weather occurrences are attributable to orographic lifting, which is probably better simulated by Kain-Fritsch.

Assimilation of conventional data and AIRS temperature and moisture profiles improved the rainfall forecast on three cases, in terms of geographical spread and intensity, though there are incidences of overestimation also. Actual conventional surface data was available over Uttarakhand, but upper air conventional as well as satellite data was not available over Uttarakhand. In the AIRS profiles lowest errors occur for clear sky cases over water with degradation in profile accuracy in cloudy regions and/or over land. Hence during

(continued)

quality check most of the AIRS data over land has been rejected. Hence whatever impact the actual observation would have made was because of surface data only. Location specific rainfall forecast could not produce any remarkable improvement. Uttarakhand is a region of complex topography, where elevation changes abruptly with slight horizontal displacement. Perfect representation of terrain in the model environment has been a challenge, especially over Indian region. Poor representation of topography has led to deviation in the location specific rainfall forecast. In this study, grid point rainfall, along with average and maximum rainfall within approximately 12 km radius of grid point have been analyzed, but the results did not improve significantly.

One of the important findings was that models have capability to forecast such unusual rainfall incidences over mountainous terrain with some geographical error. Improvement in terrain data and model resolution can provide better location specific forecast. NWP can improve severe weather forecasting and thus improve the lead time of warnings, it is equally important to interact with the media and disaster management authorities to convey the warnings and identify areas for further improvement, including skills of NWP products i.e. strengthen the forecast engine and warning services to save precious lives and reduce the damage to property.

Acknowledgement The authors of this study show their gratitude to India Meteorological Department for providing the actual daywise location specific rainfall data over Uttarakhand region and all-India gridded rainfall data.

References

Divakarla MG, Barnet CD, Goldberg MD, Mcmillin LM, Maddy E, Wolf W, Zhou L, Liu X (2006) Validation of Atmospheric Infrared Sounder temperature and water vapor retrievals with matched radiosondes measurements and forecasts. J Geophys Res 111:D09S15. doi:10.1029/2005JD006116

Fetzer EJ (2006) Preface to special section: validation of atmospheric infrared sounder observations. J Geophys Res 111:D09S01. doi:10.1029/2005JD007020

Gelb A, Kasper JF, Nash RA, Price CF, Sutherland AA (eds) (1974) Applied optimal estimation. The MIT Press, Cambridge, MA

Grell GA, Devenyi D (2002) A generalized approach to parameterizing convection combining ensemble and data assimilation techniques. Geophys Res Lett 29(14):Article 1693

Janjic ZI (1990) The step-mountain coordinate: physical package. Mon Weather Rev 118:1429–1443

Janjic ZI (1996) The surface layer in the NCEP Eta Model. In: Eleventh conference on Numerical Weather Prediction, Norfolk, VA, 19–23 August. American Meteorological Society, Boston, MA

Janjic ZI (2002) Nonsingular Implementation of the Mellor-Yamada Level 2.5 Scheme in the NCEP Meso model, NCEP Office Note, No. 437

Kain JS (2004) The Kain-Fritsch convective parameterization: an update. J Appl Meteorol 43:170–181

Mlawer EJ, Taubman SJ, Brown PD, Iacono MJ, Clough SA (1997) Radiative transfer for inhomogeneous atmosphere: RRTM, a validated correlated-k model for the longwave. J Geophys Res 102(D14):16663–16682

Susskind CB, Blaisdell J (2003) Retrieval of atmospheric and surface parameters from AIRS/AMSU/HSB under cloudy conditions. IEEE Trans Geosci Remote Sens 41:390–409

Susskind J, Barnet C, Blaisdell J, Iredell L, Keita F, Kouvaris L, Molnar G, Chahine M (2006) Accuracy of geophysical parameters derived from Atmospheric Infrared Sounder/Advanced Microwave Sounding Unit as a functional cloud cover. J Geophys Res 111:D09S17. doi:10.1029/2005JD006272

Thompson G, Rasmussen RM, Manning K, Manning K (2004) Explicit forecasts of winter precipitation using an improved bulk microphysics scheme. Part I: description and sensitivity analysis. Mon Weather Rev 132:519–542

Tobin DC et al (2006) Atmospheric Radiation Measurement site atmospheric state best estimates for Atmospheric Infrared Sounder temperature and water vapor retrieval validation. J Geophys Res 111:D09S14. doi:10.1029/2005JD00610

Incessant Rainfall Event of June 2013 in Uttarakhand, India: Observational Perspectives

M.R. Ranalkar, H.S. Chaudhari, G.K. Sawaisarje, A. Hazra, and S. Pokhrel

1 Introduction

The Indian summer monsoon is a spectacular seasonal phenomenon on the globe. Large part of Indian summer monsoon rainfall is an outcome of various synoptic scale disturbances, many of which are extremely intense. Monthly and seasonal rainfall amounts are considered for describing general behaviour of summer monsoon. These amounts may give a misleading impression that monsoon is a robust and slowly evolving system. The mean rainfall, however, is the rainfall averaged over many sporadic weather events having spatial scales from 100 to 1,000 km (Stephenson et al. 1999). In past decade, there have been many instances of very heavy and extreme rainfall events over India (Rajeevan et al. 2008). Recently, in the month of June 2013, the state of Uttarakhand experienced heavy rainfall event owing to passage of well-formed synoptic systems. Incessant rainfall from 14 to 17 June 2013 accompanied with melting snow and Himalayan terrain resulted in flash floods in this region especially over Kedarnath valley and adjoining areas.

IMD reported departure of about 847 % in the rainfall volume over Uttarakhand region for the week 13–19 June 2013. This incessant rainfall caused floods and ultimately resulted in tremendous loss of life and infrastructure. Indian Daily Weather Report states that monsoon trough passed through Bikaner, Gwalior, Gaya and Imphal and across the Gangetic West Bengal. Low pressure area which originated over northwest Bay of Bengal moved eastwards and was seen over Odisha region on 13 June 2013. The low pressure intensified into a well marked low pressure area and sustained its northwestward movement till 18 June 2013 and thereafter it weakened

---

M.R. Ranalkar (✉) • G.K. Sawaisarje
India Meteorological Department, Pune, India
e-mail: mr.ranalkar@imd.gov.in

H.S. Chaudhari • A. Hazra • S. Pokhrel
Indian Institute of Tropical Meteorology, Pune, India

into a cyclonic circulation over Haryana and adjoining west Uttar Pradesh. Western disturbances (WD) in the form of a trough in mid-tropospheric level was observed around west Rajasthan on 16 June. This WD moved eastwards (towards east Rajasthan) and it was observed near northern regions of India (Punjab, Haryana, Uttarakhand and adjoining areas) on 17 and 18 June. This system finally moved away eastwards on 19 June 2013. Detailed analysis of this Uttarakhand event is performed in this study.

This study has for the first time utilized the records of recently established automated surface observational network in the state of Uttarakhand along with validation of observational network record with the widely accepted TRMM rainfall (Level I and Level II) dataset. The interaction of trough of mid-latitude westerlies penetrating southward over India and monsoon lows had a major role in triggering such extreme rainfall events.

2 Data and Methodology

A fairly dense network of AWS and ARG stations is now available in India for operational utilization (Ranalkar et al. 2013). The hourly and daily rainfall data recorded at AWS and ARG stations in Uttarakhand have been used in this study. Romatschke and Houze (2011) have demonstrated that convective systems of various types contribute to precipitation of south Asian monsoon. To explore this aspect in Uttarakhand rainfall event of 14–17 June 2013, we have used minimum polarization corrected temperature (PCT) at 85.5 GHz microwave channel as a proxy for convection. The PCT (formula for PCT = 1.818 T_{BV} – 0.818 T_{BH}) was computed using TRMM 1B11 data set (Kummerow et al. 2000) following Spencer et al. (1989). Here, T_{BV} is brightness temperature for vertically polarized channel at 85.5 GHz and T_{BH} is brightness temperature for horizontally polarized channel at 85.5 GHz. A PCT range of 250–260 K is generally taken as threshold below which precipitating systems are found. PCT value of 250 K corresponds to moderate rain rate of 3 mm h^{-1}. In this study, rain rates as revealed by TRMM PR were analyzed using 2A25 dataset.

Trenberth (1998, 1999) have argued that increase in moisture content of the atmosphere should increase rate of precipitation locally by invigorating the storms and depressions through latent heat release and further by supplying more moisture. The possible link between moisture flux convergence and occurrence of heavy rainfall over Uttarakhand were examined using a measure of tropospheric forced lifting called Vertically Integrated Moisture Flux Convergence (VIMFC) which is defined as

$$\mathrm{VIMFC} = \frac{-1}{g} \int_{Surface}^{300hPa} \left(\frac{\partial uq}{\partial x} + \frac{\partial vq}{\partial y} \right) dp$$

where symbols have their usual meaning. The VIMFC was computed using MERRA reanalysis field. The VIMFC also serves as a tool to track movement of a low/well marked low along the monsoon trough. We have also used NCEP reanalysis dataset to evaluate the monsoon trough and mid-latitude interactions.

3 Discussion

3.1 Observed Rainfall (AWS and ARG)

The hourly and daily rainfall events are well recorded by AWS and ARG stations during the period 14–18 June 2013, although some stations were non-functional. The daily rainfall recorded at different AWS and ARG stations are presented in Table 1. The hourly rainfall recorded at Dehradun AWS and Kalsi ARG station from 0300 UTC of 16 June 2013 to 0300 UTC of 17 June 2013 are shown in Fig. 1a–b. It can be concluded from daily and hourly variation of rainfall recorded at AWS and ARG stations that the region experienced incessant rainfall during the period but the rain rate was well below 100 mm/h and hence the event cannot be termed as cloudburst.

3.2 Role of Convection (Polarization Corrected Temperature (PCT) as Convective Proxies)

Study by Nesbitt et al. (2000) suggests that PCT can be used as convective proxy. Hence, the polarization corrected temperature (PCT) derived from 85 GHz channel using TRMM 1B11 data set is utilized in the present analysis. PCT for the period 15–17 June 2013 is presented in Fig. 2a–f. As local convection and orographic upliftment is expected to cause rainfall from the weather system, the PCT was monitored in an area bounded by 29°N–32°N and 77°E–81°E. The channel is sensitive to precipitation sized ice particles which scatter the upwelling radiation and reduce the brightness temperature. Thus, low brightness temperature implies increased updraft strength. The low brightness temperature also indicates increased latent heat release and precipitation. In Fig. 2, the pixels with PCT >250 K are shaded in gray colour and those with PCT <250 K are shaded in colour (as discussed above it is an indicative of large convection). As the low pressure area moved over central India on 15 June 2013, the pixels centered on Rudra Prayag, Chamoli, Uttar Kashi and Karnaprayag revealed that PCT was less than 250 K and hence increased rainfall could be inferred over these stations. The first pass of TRMM on 16 June 2013 shown in Fig. 2c reveals that during 0810 UTC to 0811 UTC many pixel values were dominated by PCT well below 250 K. The second pass of TRMM over the region on 16

Table 1 Daily rainfall (mm) recorded at selected AWS and ARG stations in Uttarakhand

No	Station name	Latitude	Longitude	14/06/2013	15/06/2013	16/06/2013	17/06/2013	18/06/2013	Total
1	Bharsar	30.05	79.00	43	0	145	122	80	390
2	Champawat	29.34	80.09	0	1	39	194	222	456
3	Dehradun	30.32	78.05	89	54	210	338	9	700
4	Dhanauri	29.93	77.97	0	17	203	151	11	382
5	Jollygrant	30.19	78.18	43	28	203	224	29	527
6	Matela	29.62	79.62	9	1	33	97	100	240
7	Nainital	29.36	79.46	9	18	62	216	188	493
8	Pantnagar	29.02	79.48	0	0	2	58	115	175
9	Pithoragarh	29.57	80.23	0	0	12	69	123	204
10	Rani_Chawri	30.20	78.52	4	2	163	205	47	421
11	Roorkee	29.84	77.92	0	5	51	148	15	219
12	Ramnagar	29.39	79.11	0	1	16	56	58	131
13	Ranikhet	29.64	79.42	1	0	17	43	50	111
14	Haridwar	29.92	78.12	10	20	86	218	14	348
15	Rishikesh	30.11	78.28	16	2	178	145	40	381
16	Kalsi	30.53	77.84	20	95	175	391	33	714
17	Pati	29.40	79.93	3	0	82	206	128	419
18	Lohaghat	29.40	80.09	2	0	21	139	181	343
19	Devprayag	30.14	78.60	0	7	129	116	50	302
20	Srinagar	30.22	78.77	0	0	129	133	58	320
21	Bageshwar	29.83	79.77	15	3	61	161	63	303
22	Jakholi	30.39	78.89	25	72	121	108	65	391
23	Sitarganj	28.93	79.70	0	0	8	75	174	257
24	Gangolihat	29.65	80.04	0	0	24	103	125	252

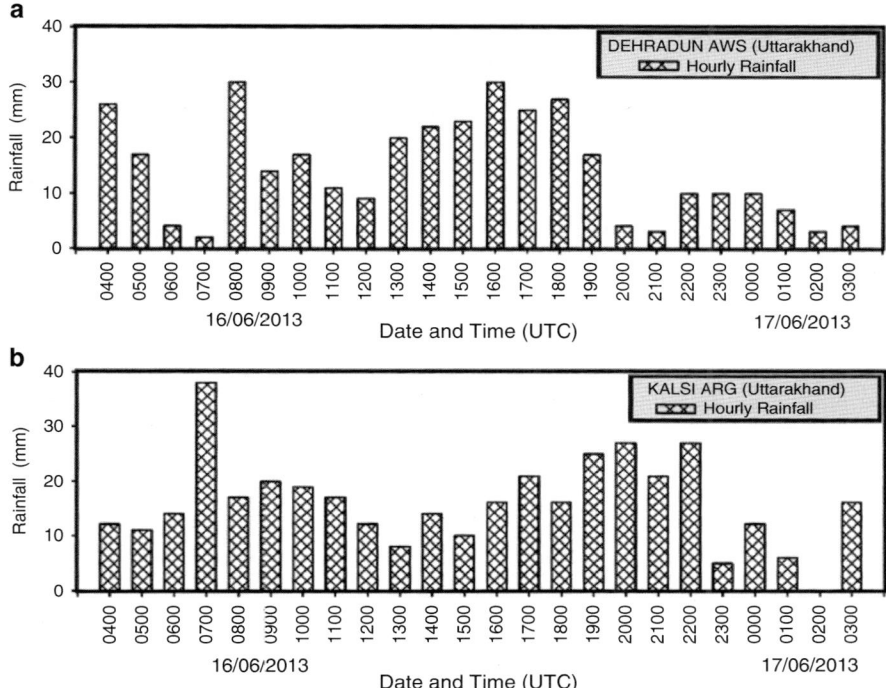

Fig. 1 Hourly rainfall recorded at Dehradun AWS and Kalsi ARG station on 16th June 2013

June 2013 (Fig. 2d; 1126 UTC to 1127 UTC) also reveals low PCT values over the entire region. In fact, incessant rainfall was experienced by many stations in Uttarakhand due to the passage of the system. On 17 June 2013, both TRMM passes have depicted many pixels with values of PCT <250 K (see Figs 2e–f). A snapshot taken by Precipitation Radar aboard TRMM on 17 June 2013 is shown in Fig. 3. The pixel values indicate that wide area had experienced a rainrate of the order of 6–20 mm/h. Few pixels indicated a rainrate exceeding 40 mm/h, however, none had value in excess of 100 mm/h.

3.3 Vertically Integrated Moisture Flux Convergence (VIMFC) and Intrusion of Mid-tropospheric Westerlies over Indian Latitudes

On 14 June 2013, a low was developed over north coastal Andhra Pradesh and adjoining Odisha which resulted in moisture convergence over North Coastal Andhra Pradesh and adjoining Odisha. On 15 June 2013, two anticyclonic circulation

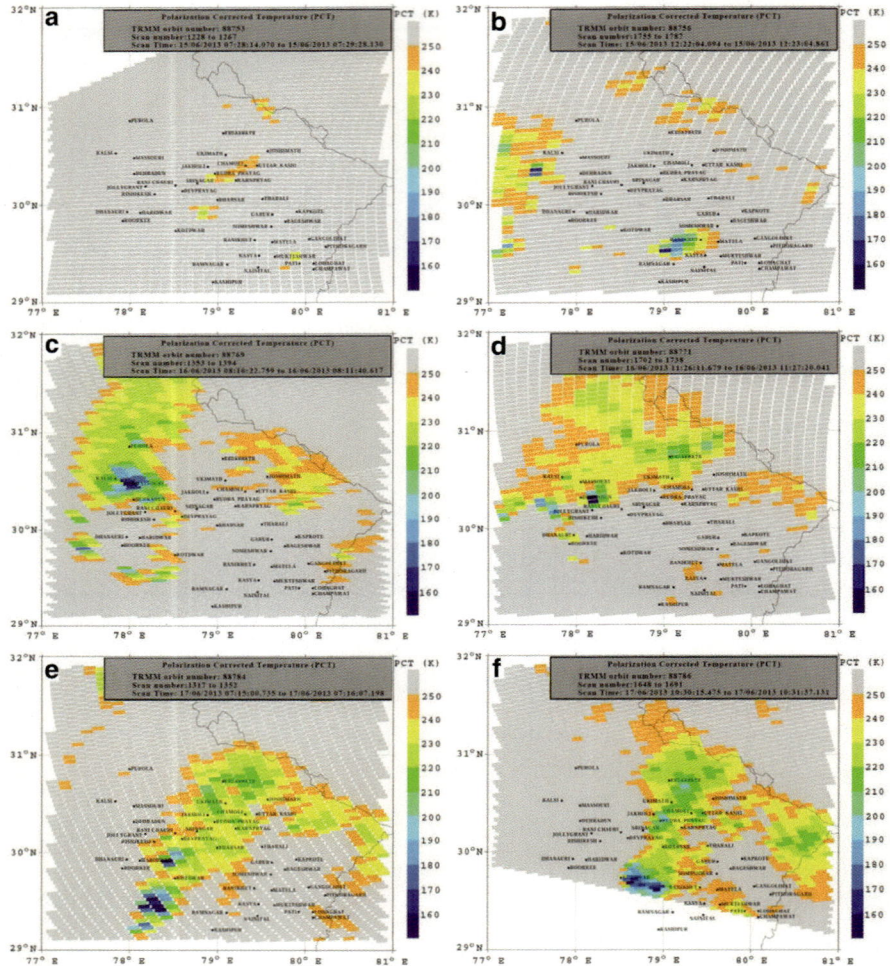

Fig. 2 Polarization Corrected Temperature (PCT) over Uttarakhand from 15th June 2013 to 17th June 2013 for daily two TRMM overpass (PCT <250 are *colour shaded* to indicate conditions favourable for convection)

centres at 500 hPa level one along 50°E between 25°N–30°N and other along 98°E as shown in Fig. 4 indicates shifting of these centres towards east and meridional shift towards south.

On 16 and 17 June 2013, the cyclonic circulation at 500 hPa level over central India moved in north northeasterly direction under the influence of upper level winds. These upper level winds at 200 hPa level might have provided necessary steering current to mid-tropospheric cyclonic circulation observed in 500 hPa level.

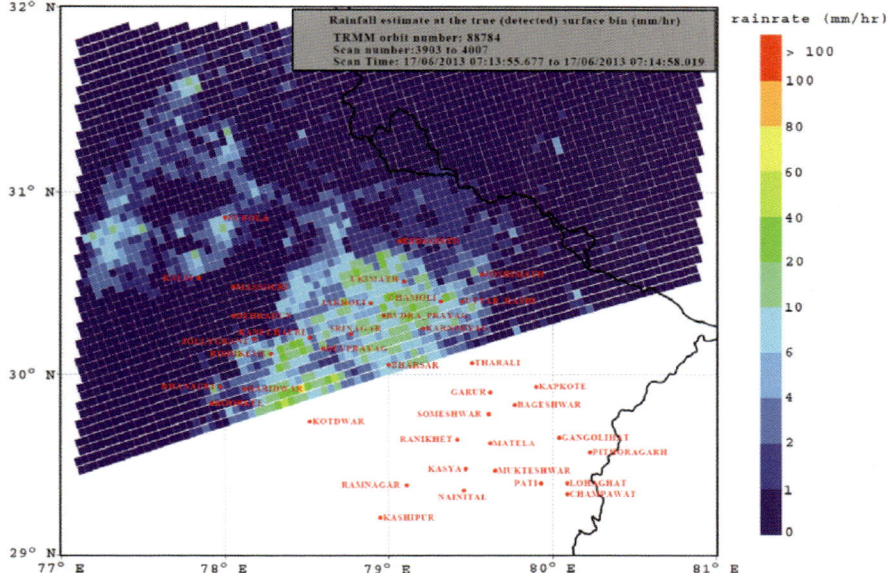

Fig. 3 Rain rate as estimated by TRMM Precipitation Radar on 17th June 2013 as it passed over Uttarakhand

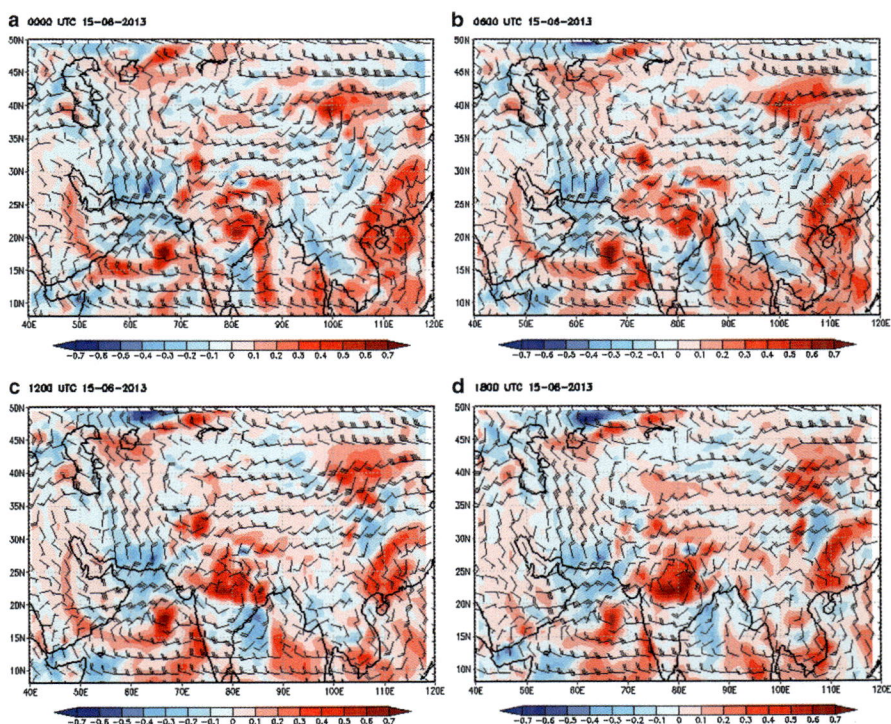

Fig. 4 Vertically Integrated Moisture Flux Convergence ($\times 10^{-3}$ kg m^{-2} s^{-1}) and wind field at 500 hPa on 15th June 2013

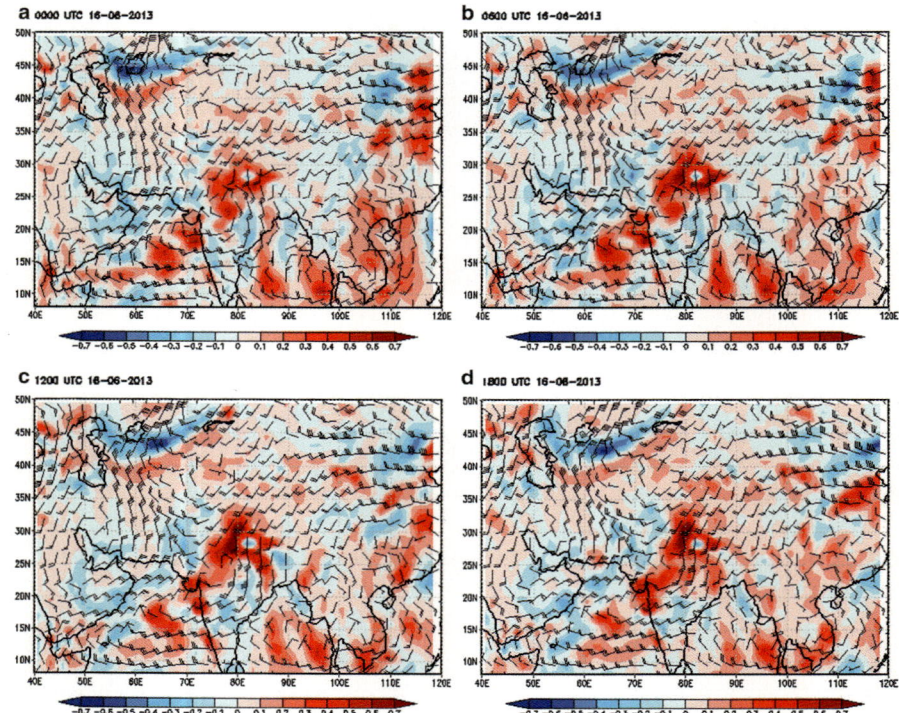

Fig. 5 Vertically Integrated Moisture Flux Convergence (×10^{-3} kg m^{-2} s^{-1}) and wind field at 500 hPa on 16th June 2013

A southward shift of upper air cyclonic circulation at 500 hPa in mid-latitude on 16 June 2013 was seen (Fig. 5) with deep penetration of winds over Indian region up to 20°N. It can also be inferred that there could be a constant feeding of moisture from both Arabian Sea and Bay of Bengal to upper air cyclonic circulation at 500 hPa level along east of 80°E between 25°N–30°N, which resulted in increased VIMFC over the region east of 80°E between 25°N–30°N. However, on 17 June 2013, upper level anticyclones are clearly seen at 500 hPa level (Fig. 6) in mid-latitudes with a cyclonic circulation along 75°E between 40°N–45°N which resulted in a deep trough from this cyclonic circulation with its extension up to 20°N. From synoptic point of view, it is inferred that a low index cycle prevailed during the period in which westerlies were weak and maximum belt of westerlies is displaced towards subtropics with wind field oriented more in meridional direction.

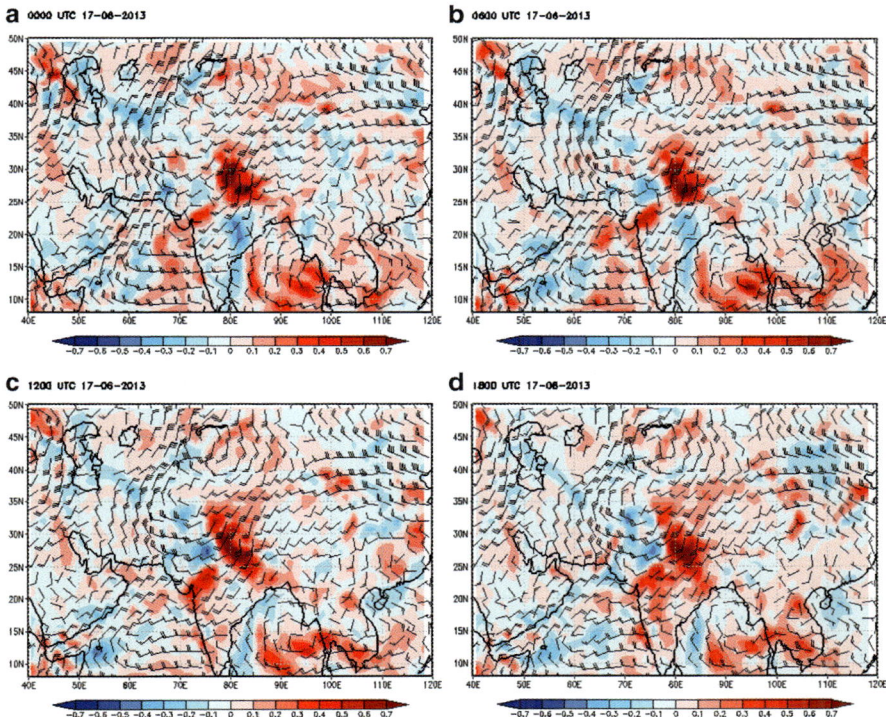

Fig. 6 Vertically Integrated Moisture Flux Convergence ($\times 10^{-3}$ kg m^{-2} s^{-1}) and wind field at 500 hPa on 17th June 2013

Conclusion

The interaction of trough of mid-latitude westerlies penetrating southward over India, feeding of moisture from Arabian Sea and Bay of Bengal to the monsoon system, resulting in convergence of moisture over Uttarakhand and the orography of the state of Uttarakhand played a significant role during 14 and 17 June 2013 in causing incessant rainfall and landslides over the region.

Acknowledgements The authors wish to thank Dr. L.S. Rathore, Director General of Meteorology, India Meteorological Department, New Delhi for encouragement in research and being an enduring source of inspiration. Authors are thankful to Dr. S. Dutta, Scientist-E, IMD, Pune for helpful discussion.

References

Kummerow C et al (2000) The status of the Tropical Rainfall Measuring Mission (TRMM) after two years in orbit. J Appl Meteorol 39:1965–1982

Nesbitt SW, Zipser EJ, Cecil DJ (2000) A census of precipitation features in the tropics using TRMM: radar, ice scattering, and lightning observations. J Clim 13:4087–4106

Rajeevan M, Bhate J, Jaswal AK (2008) Analysis of variability and trends of extreme rainfall events over India using 104 years of gridded daily rainfall data. Geophys Res Lett 35:1–6

Ranalkar MR, Anajan A, Mishra RP, Mali RR, Krishnaiah S (2013) Network of automatic weather stations: time division multiple access type. *Mausam* (Accepted)

Romatschke U, Houze RA Jr (2011) Characteristics of precipitating convective systems in the South Asian monsoon. J Hydrometeorol 12:1–26

Spencer RW, Goodman HG, Hood RE (1989) Precipitation retrieval over land and ocean with the SSM/I: identification and characteristics of the scattering signal. J Atmos Ocean Technol 6:254–273

Stephenson DB, Kumar KP, Doblas-Reyes FJ, Royer JF, Chauvin F (1999) Extreme daily rainfall events and their impact on ensemble forecasts of the Indian Monsoon. Mon Weather Rev 127:1954–1966

Trenberth KE (1998) Atmospheric moisture residence times and cycling: implications for rainfall rates with climate change. Clim Chang 39:667–694

Trenberth KE (1999) Conceptual framework for changes of extremes of hydrological cycle with climate change. Clim Chang 42:327–339

Convergence of Synoptic and Dynamical Conditions Responsible for Exceptionally Heavy Rainfall over Uttarakhand, India

Charan Singh

1 Introduction

Fairly widespread to widespread rainfall with scattered heavy falls is rare over Uttarakhand during onset phase of southwest monsoon. This year southwest monsoon onset over Kerala was on 1 June, which is its normal onset date. During initial days of onset of monsoon over peninsular and adjoining central and east India, a low pressure area formed over Bay of Bengal and it moved west-northwestwards. It was one of the causes of early onset of southwest monsoon over the NW India, i.e. on 16 June, 2013 (Fig. 1). The advance southwest monsoon onset over the country during 2013 is the earliest as per IMD record. SW monsoon reached Uttarakhand almost 2 weeks in advance from its normal onset date. Hence the state received early monsoon rainfall. It also caused heavy rain over a larger area over Uttarakhand during 14–17 June, 2013, creating a catastrophic situation at Kedarnath, Badrinath, Joshimath and nearby areas. Sometimes, short wave perturbations moving in the broad mid-latitude westerlies amplify the long wave troughs creating new baroclinic zones in relatively southern latitudes. These baroclinic zones interact with the low latitude circulations and thus leading to development of new circulations in which low level easterlies extend northward over central and northwest India.

Kalsi (1980) linked interaction between tropics and mid-latitude during monsoon to the large amplitude troughs in mid-latitude westerlies at 500 hPa level. Wang et al. (2011) studied the summer precipitation in northern Pakistan during 2010. According to them, it comprises two distinct phases: a pre-monsoon trough phase with more intense rainfall occurring without the monsoon trough and another monsoon trough phase, when large area gets rainfall but of low intensity. In pre-monsoon trough phase, convective nature of the pre-monsoon trough phase enhanced

---

C. Singh (✉)
India Meteorological Department, Lodhi Road, New Delhi 110003, India
e-mail: csingh1964@gmail.com

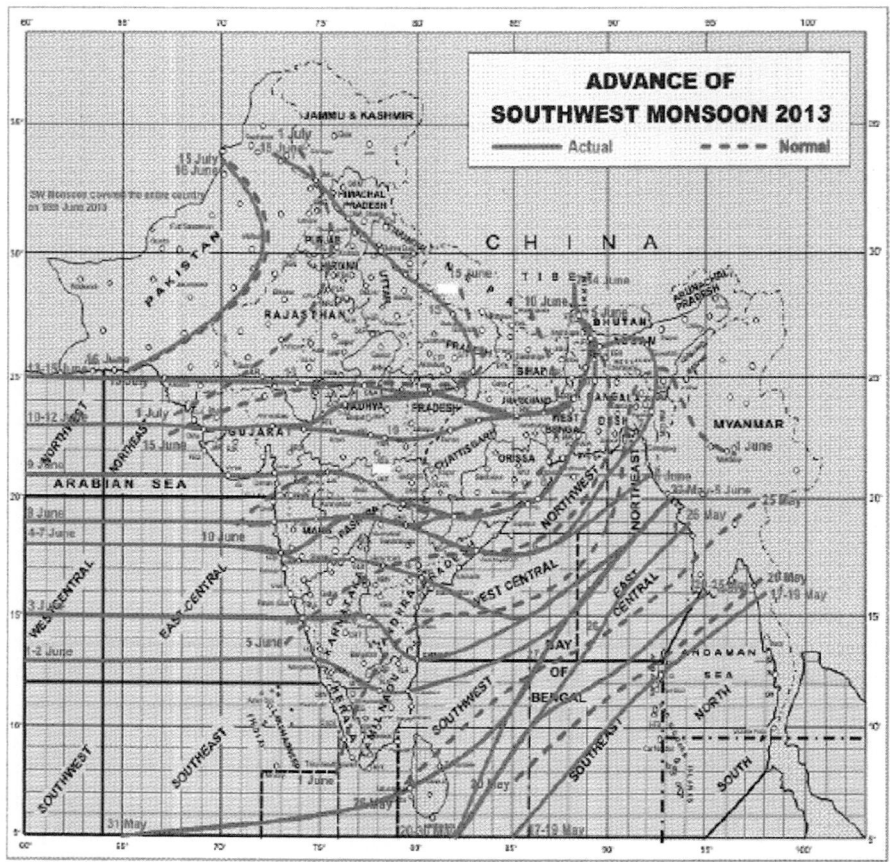

Fig. 1 Map showing onset and advance of southwest monsoon during 2013

unstable environment. Rao (1976) described the effect of mid-latitude systems on monsoon as of four kinds: (i) intensifying or developing lower tropospheric lows or troughs, (ii) enhancing rainfall in pre-existing systems, (iii) causing re-curvature of depressions and lows and (iv) leading to onset of break conditions.

It also leads to intensification of western disturbances, outbreak of convection, oscillation of monsoon trough and deformation of circulations including cyclonic storms. Kalsi and Haldar (1992) suggested that the interaction between tropics and mid-latitudes plays an important role in many of the weather producing processes in and around India.

Dimri (2011) described in his study that whenever convergence reaches peak, a precipitation maximum occurs too. Moreover, temporal distribution shows a periodic 2–3 days association with convergence peak and precipitation maxima. Nandargi and Dhar (2012) carried out a detailed analysis of rainstorms that affected the northwest region of the Himalayas and made an attempt to assess the orographic effects of the Himalayas on precipitation by using 135 years data from 1875 to 2010. Barros and Lang (2003) study revealed that the large scale monsoon flow is presumed to be roughly constant during the day and night, though it is of course subject to the variability in the monsoon trough over northern India.

Floods in Uttarakhand are mainly caused by heavy rainfall in the catchments during the summer monsoon season (July-September), which is augmented by snowmelt flows (Das et al. 2006). The Himalayas is characterized by topographic heterogeneity and land use variability (Dimri 2009). De et al. (2005) had listed major rainstorms and compiled rain producing system in their study, which were associated with extreme damage and huge depth of flood water submerging vast areas. A study done by Ugnar (1999) indicates that the losses due to extreme weather events are increasing steeply, especially in the last decade of the twentieth century.

2 Precipitation over Uttarakhand During 15–17 June 2013

The rainfall activity over western Himalaya increased gradually from 10 June onwards, rainfall pattern showed a sudden rise and heavy rainfall over the western Himalayan region particularly over Uttarakhand during 14–17 June, 2013. The major amounts of rainfall (in cm) of preceding 24 h recorded on date at 0,830 h IST were: (15 June) Dunda – 8, Jakholi and Kashipu – 7 each and Dehradun – 5; (16 June) Dehradun – 22, Purola – 17, Devprayag – 13, Uttarkashi – 13 and Tehri – 12; (17 June) Dehradun – 37, Mukteshwar – 24, Hardwar – 22 and Uttarkashi – 21 and (18 June) Haldwani – 28, Champawat – 22, Nainital – 17 and Ranikhet – 12 cm. Isohyetal analysis of cumulative rainfall during 15–17 June, 2013 over Uttarakhand shows two peaks of rainfall one over west-northwest of Kedarnath with cumulative rainfall of the order of 600 mm and another over south Uttarakhand with more than 500 mm (Fig. 2).

Tropical rain measuring mission (TRMM) images show that on 14 June rainfall was less, the rainfall areas as well intensity increased on 15 June and continued over a large area with peak around Kedarnath ($30.7°N/79.1°E$). It decreased around Kedarnath and area of higher rainfall shifted towards southeast on 16 June, 2013 (Figs. 3 and 4).

3 Data Source and Methodology

Rainfall data network enhanced during southwest monsoon season, as part-time observatories have also supplied the rainfall data to India Meteorological Department (IMD) in addition to IMD's own network. These data are used for day-to-day monitoring of monsoon activities. Late received data are also taken into consideration for final assessment of monsoon status. The daily area weighted rainfall as well as daily station rainfall data for Uttarakhand were taken from Hydrology and National Weather Forecasting Centre (NWFC) divisions of IMD. The pressure field and wind

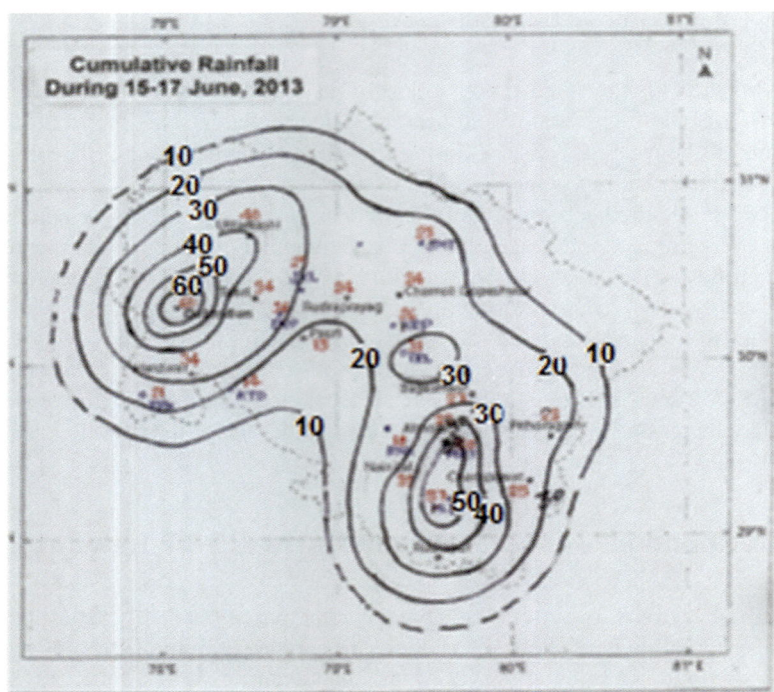

Fig. 2 Isohyetal analysis of cumulative rainfall during 15–17 June, 2013

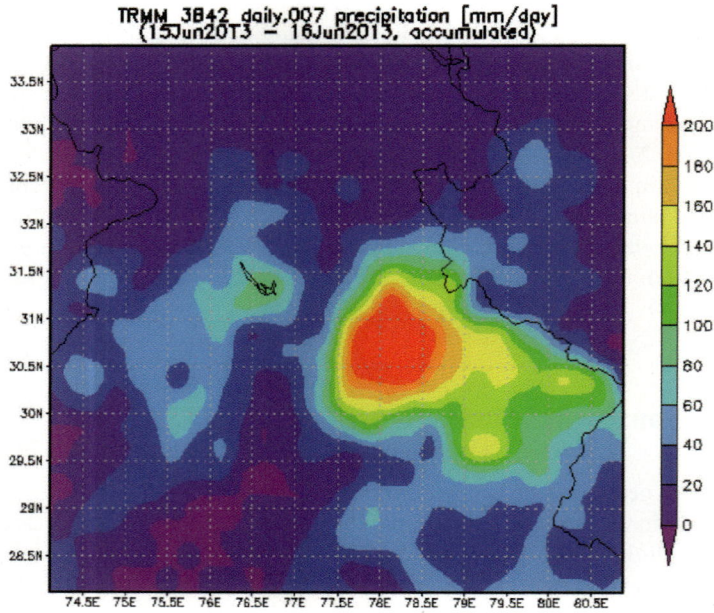

Fig. 3 Tropical Rain Measuring Mission (TRMM) images on 15–16 June, 2013

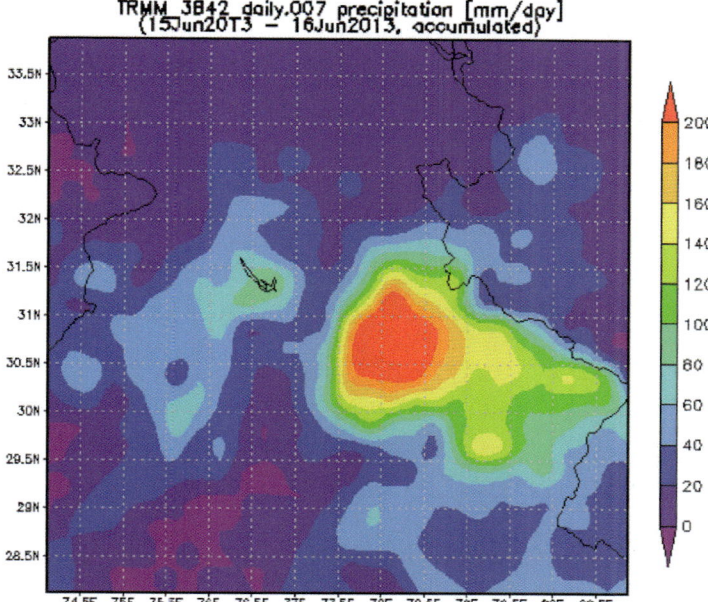

Fig. 4 Tropical Rain Measuring Mission (TRMM) images on 16–17 June, 2013

flow pattern were analyzed by using synoptic and upper air data of IMD available in synergy. Satellite images {Infra Red (IR) Visible (VIS)} and derived products namely water vapour wind, outgoing long-wave radiation (OLR) and quantitative precipitation estimation (QPE) etc. were also used to monitor the clouds convection and their movements. These products and images were obtained from synergie system of NWFC and from satellite division of IMD. Patiala Doppler Weather Radar (DWR) reflectivity images were used. The dynamic parameters i.e. upper level divergence, moisture flux, OLR, and precipitable water contents were taken from NOAA website (www.esrl.noaa.gov/psd/data/grided/data.ncep.reanalysis). To show the relations between some dynamical parameters and area weighted cumulative rainfall during 10–20 June, 2013 over Uttarakhand, different graphs have been prepared. TRMM rainfall images have also been taken into account to assess the daily rainfall http://pmm.nasa.gov/trmm/realtime). These data are in image form and the values used in the paper are estimated.

4 Synoptic Conditions Responsible for Heavy Rainfall over Uttarakhand

The southwest monsoon has set in over Kerala on 1 June, 2013 and covered whole country by 16 June, 2013. A low pressure area was formed over northwest Bay of Bengal on 12 June and it moved west-northwestwards and lay over west

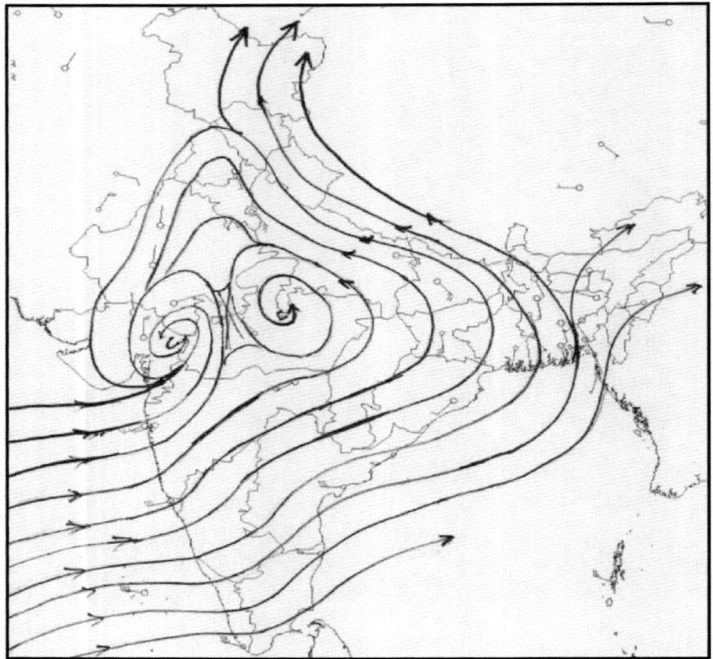

Fig. 5 850 hPa level wind analyzed chart at 00 UTC of 16 June, 2013

Madhya Pradesh and adjoining east Rajasthan on 15th and 16th June, 2013. Easterly winds were prevailing over northern plains in lower levels. The wind flow pattern at 850 hPa level is shown in Fig. 5. A western disturbance as a trough in mid-level westerlies lay over north Pakistan and neighbourhood on 15 June and as a cyclonic circulation over north Pakistan and adjoining Jammu & Kashmir on 16 June, 2013 (Fig. 6).

In present case the southern end of trough in mid-levels westerlies dipped into Arabian Sea and a low pressure system lay over northeast Rajasthan and neighbourhood which interacted with westerly trough over the region. Tibetan High was located west of its normal position, which provides upper level divergence over the region. Due to these synoptic conditions moisture incursion was taking place over Western Himalayan Region (WHR) particularly over Uttarakhand from Arabian Sea as well as from Bay of Bengal. The movement of westerly trough was very slow and easterly winds in lower levels continued to prevail over the region upto 18 June, 2013. This situation caused heavy to exceptionally heavy rainfall over WHR with some spells of very heavy rainfall.

The cumulative rainfall during 15–17 June, 2013 over Uttarakhand is shown in isohyetal analysis (Fig. 2).

Satellite (IR) and VIS imagery suggests intense to very intense cloud on 16 and 17 June, 2013 over Uttarakhand and neighbourhood. Intense convective clouds over northeast Arabian Sea and Madhya Pradesh on 16 (Fig. 7), cloud cluster over

Fig. 6 500 hPa level wind analyzed chart of 00 UTC of 16 June, 2013

Madhya Pradesh moved to west Uttar Pradesh and adjoining Uttarakhand on 17 while intense convective clouds over northeast Arabian Sea remained over the same area. The clouds over Uttarakhand in association with westerly trough have merged with the clouds associated with a low pressure area. Visible image of 0600 UTC of 16 June showed intense to very intense convection over Uttarakhand and neighbourhood. This type of dense clouds are rarely seen in onset phase of monsoon over WHR. Doppler weather radar of Patiala images show dBz value more than 38 over Uttarakhand particularly over the region to the north of Kedarnath at 06:00 UTC of 16 June 2013 (Fig. 8).

5 Dynamical and Thermodynamical Conditions Responsible for Heavy Rainfall over Uttarakhand

Dynamical and thermodynamical conditions including moisture flux, precipitable water contents, lower level convergence, relative humidity, upper air divergence, OLR etc. are discussed in detail. Most of the dynamical parameter values have started to increase gradually from 12th onwards and it became favourable for widespread rainfall activities over the regions, which are responsible for heavy precipitation. Divergence between 150 and 300 hPa rose from 0 on 10 and 11 June to about

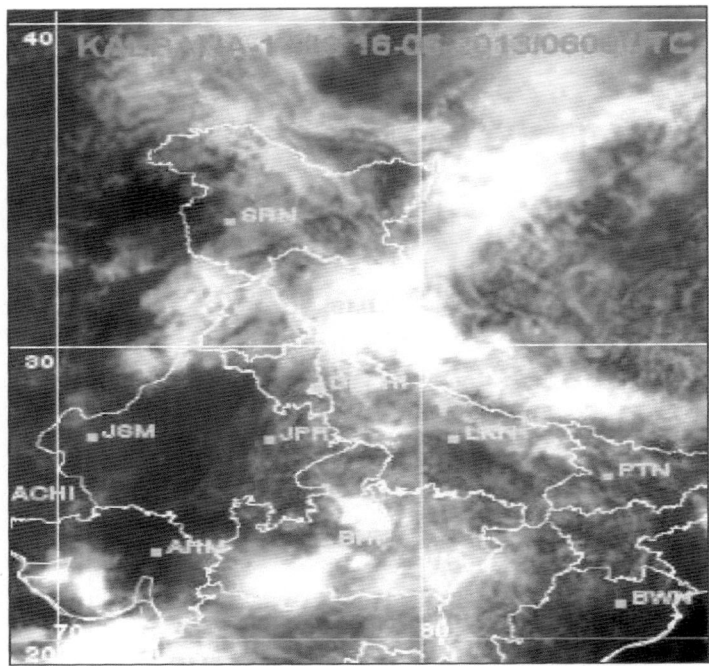

Fig. 7 Satellite (Kalpana-1) visible imagery at 06 UTC of 16 June, 2013

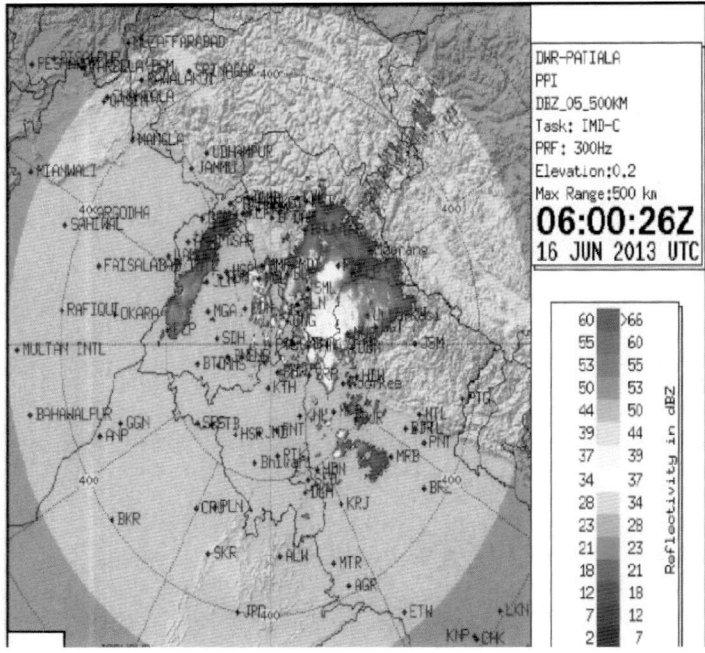

Fig. 8 Patiala dopler weather radar (DWR) reflectivity images at 06:00 UTC of 16 June, 2013

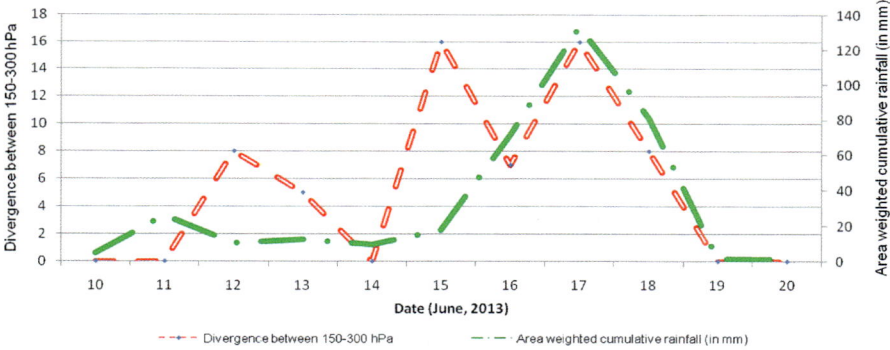

Fig. 9 Area weighted cumulative rainfall (AWCR) (in mm) of Uttarakhand verses divergence between 150 and 300 hPa level during 10–20 June, 2013

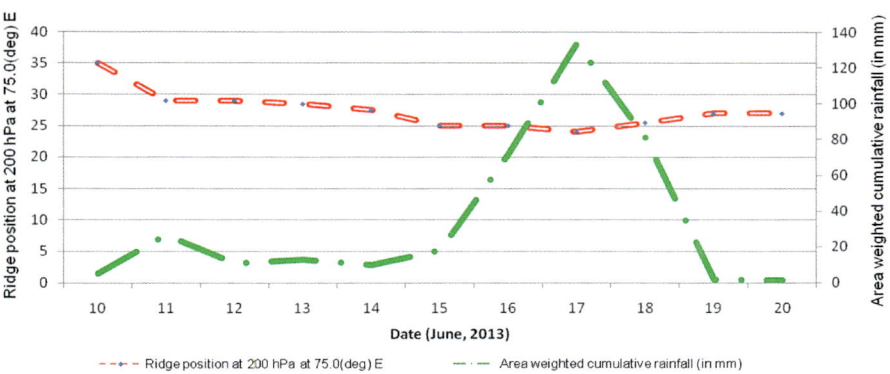

Fig. 10 AWCR (in mm) of Uttarkhand verses ridge position at 200 hPa level along 75.0°E longitude during 10–20 June, 2013

8×10^{-5} s^{-1} on 12 and again on 14 June onward. It was more than 16 10^{-5} s^{-1} on 15 June and about 17 10^{-5} s^{-1} on 17 and 18 June, 2013. The atmosphere remained very much unstable due to upper air divergence coupled with lower level convergence (Fig. 9) in the region. The ridge line at 200 hPa level ran along latitude 25.0°N during 15–18 June and it again shifted northwards to north of 27.0°N and persisted between 27.0° and 28.0°N during 18–20 June, 2013 (Fig. 10).

The OLR is inversely correlated to the albedo and cloudiness on a broad scale. The OLR is the measure of cloud mass over a particular area. It is presumed that the OLR value less than 200 w/m^2 indicates overcast sky conditions. Figure 11 clearly indicates that the OLR value was between 190 and 250 w/m^2 during 10–20 June 2013, and it was around 200 w/m^2 during 15–17 June with slight increase on 18 June.

Moisture flux at 700 hPa level and precipitable water contents are shown in Figs. 12 and 13. These figures clearly indicate that the values of these parameters were significantly high. The moisture flux at 700 hPa level was 700 g/kg and precipitable water

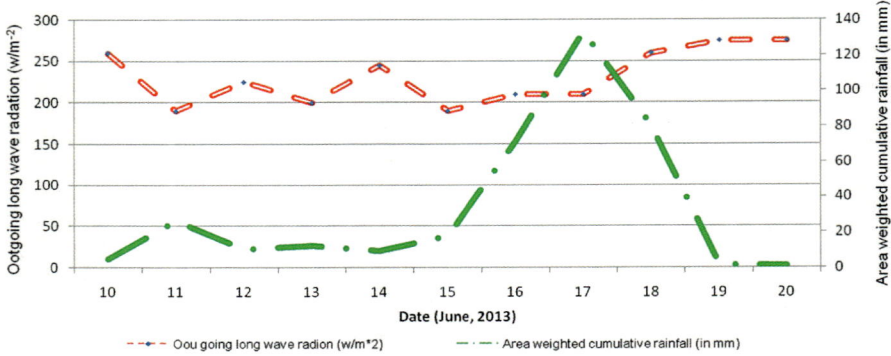

Fig. 11 Area weighted cumulative rainfall (in mm) of Uttarakhand verses outgoing long-wave radiation (w/m^2) during 10–20 June, 2013

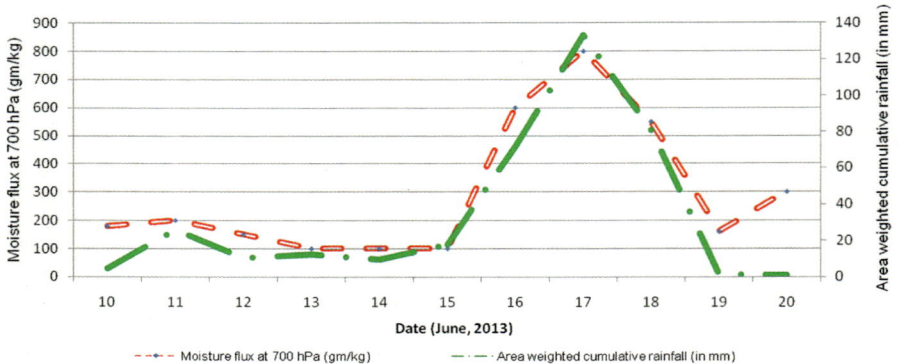

Fig. 12 AWCR (in mm) of Uttarakhand verses moisture flux at 700 hPa level showing positive correlation during 10–20 June, 2013

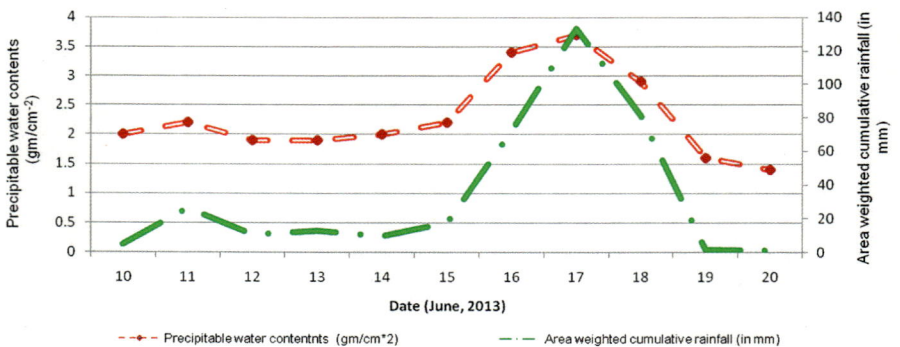

Fig. 13 AWCR (in mm) of Uttarakhand verses precipitable water contents during 10–20 June, 2013

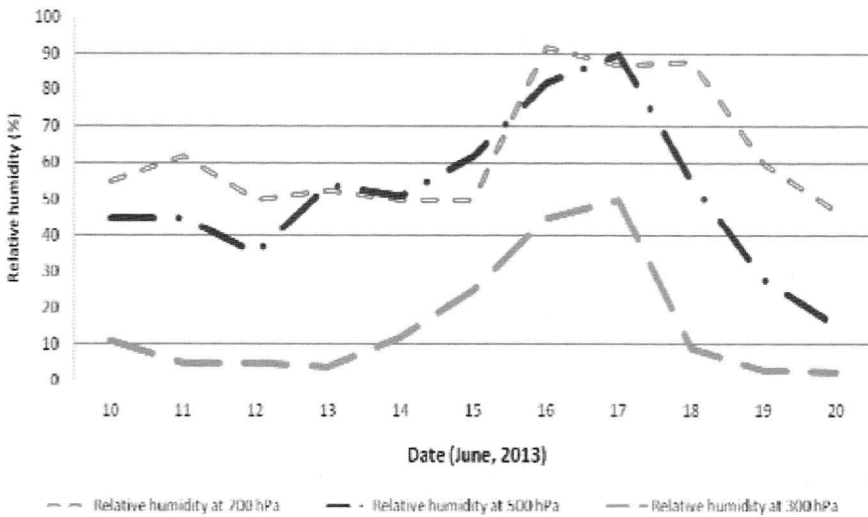

Fig. 14 Relative humidity over Kedarnath, Uttarakhand (lat. 30.7°N/long. 79.1°E) at 700, 500 and 300 hPa levels during 10–20 June, 2013

contents 3.5 g/cm^2 on 17 June, whereas on 16 and 18, it was little less. Singh et al. (2007) found that total precipitable water (TPW) of the order of 70 mm accompanied by favourable wind speed over the eastern Arabian Sea is one of the important factors contributing to heavy rainfall during next 24 h over the west coast of India and neighbourhood. Both, moisture flux and precipitable water contents over Uttarakhand area started to rise from 12 June onwards gradually. As per Asnani (2005) the maximum value of total water vapour contents of vertical column with unit area of horizontal cross section is around 4.5 g/cm^2 and minimum value 2.5 g/cm^2 in tropical latitude. Relative humidity over Kedarnath, Uttarakhand (lat. 30.7°N/long.79.1°E) at 700, 500 and 300 hPa levels during 10–20 June, 2013 are shown in Fig. 14.

Conclusions
I. Western Himalayan region experienced fairly widespread rainfall with onset of southwest monsoon due to interaction of westerlies with monsoonal winds i.e. easterlies.
II. Interaction between tropics and mid latitude led to intensification of western disturbance, outbreak of convection and enhancement of rainfall activity over Uttarakhand.
III. The favourable synoptic conditions were: the Tibetan High much west of its normal position, approaching deep westerly trough between 65.0° and 70.0°E, a low pressure area over east Rajasthan and neighbourhood and moisture incursion from Bay of Bengal as well from Arabian Sea over western Himalayan region particularly over Uttarakhand.

(continued)

IV. The OLR was about 190–250 w/m^2 during 10–20 June and it was around 200 w/m^2 during 15–17 June with slight increase on 18 June, 2013.
V. Upper level divergence was 15×10^{-5} s^{-1} and lower level convergence about 20×10^{-5} s^{-1} which were favourable for heavy rainfall over the region. Due to moisture incursion over the region relative humidity increased from 10 June onwards and it was 90 % during 15–18 June.
VI. TRMM images during 14–17 June indicate that the rainfall was of the order of 40–80 mm/day on 14–15 and 120–160 mm/day on 16 June over vast area of Uttarakhand.
VII. Due to above favourable synoptic, dynamical and thermodynamical conditions western Himalayan region experienced widespread rainfall, particularly Uttarakhand experienced exceptionally heavy rainfall in some area.

References

Asnani GC (2005) Tropical meteorology, vol 1, 2nd edn. Indian Institute of Tropical Meteorology, Pune

Barros AP, Lang TJ (2003) Monitoring the monsoon in the Himalaya: observations in central Nepal, June 2001. Am Meteorol Soc 131(7):1408–1427

Das S, Ashrit R, Moncrieff MW (2006) Simulation of a Himalayan cloudburst event. J Earth Syst Sci 115(3):299–313

De US, Dube RK, Rao GSP (2005) Extreme weather events over India in the last 100 years. J Indian Geophys Union 9(3):173–187. http://pmm.nasa.gov/trmm/realtime

Dimri AP (2009) Impact of subgrid scale scheme on topography and land use for better regional scale simulation of meteorological variables over the western Himalayas. Clim Dyn 32:565–574

Dimri AP (2011) Atmospheric water budget over the western Himalaya in a regional climate model. J Earth Syst Sci 121(2):963–973

Kalsi SR (1980) On some aspects of interaction between middle latitude westerlies and monsoon circulation. Mausam 31:305–308

Kalsi SR, Haldar SR (1992) Satellite observations of interaction between tropics and mid latitude. Mausam 43(1):59–64

Nandargi S, Dhar ON (2012) Extreme rainstorm events over the Northwest Himalayas during 1875-2010. J Hydrometeorol Am Meteorol Soc 13(4):1383–1388

Rao YP (1976) Meteorological Monogram Synop. Met. No. 1/1976 on South-West Monsoon, pp. 224–247

Singh D, Singh V, Malik DK (2007) Retrieval of TPW over ocean from locally received AMSU measurement. Mausam 58(3):375–380

Ugnar S (1999) Is strange weather in the air? A study of U.S. national network news coverage of extreme weather events. Clim Change 41(2):133–150

Wang SY, Davies R, Gillies R, Jin J (2011).Changing monsoon extremes and dynamics: example in Pakistan. In: 36th NOAA annual climate diagnostics and prediction workshop, Fort Worth, TX, 3–6 October, 2011. www.esrl.noaa.gov/psd/data/grided/data.ncep.reanalysis

Changes in Rainfall Concentration over India During 1871–2011

Naresh Kumar and A.K. Jaswal

1 Introduction

Change in rainfall pattern and its impact on the water resources is one of the important climatic problem. There are many studies in the literature that indicate about the change in rainfall pattern globally (De Luís et al. 2000; Hulme et al. 1998; Rodríguez-Puebla et al. 1998). Many studies carried out in India also showed variability in the rainfall on different temporal and spatial scales (Kripalani et al. 2003; Sahai et al. 2003). In India, rainfall is highly variable from place to place. Some regions like Western Ghats as well as Sub-Himalayan areas in North East and Meghalaya receive annual rainfall more than 200 cm while northern part of Kashmir, West Rajasthan and Punjab receive annual rainfall less than 50 cm.

Goswami et al. (2006) found increasing trend in the frequency of extreme precipitation events over central parts of India. Krishnamurthy et al. (2009) also observed statistically significant increasing trends in rainfall extremes over many parts of India. They also observed decreasing trends in rainfall extremes over some parts of India. The changes in rainfall extremes may be due to change in rainfall concentration over the different regions of the India. Higher rainfall concentration shows higher percentage of annual and seasonal total rainfall in a few rainy days. Changes in rainfall concentration may cause floods and droughts and may put wide-ranging pressure on water resources (Zhang et al. 2009). Therefore, studies related to rainfall concentrations are very important. To study the rainfall concentration, Precipitation Concentration Index (PCI) (Oliver 1980) is suggested by many researchers (De Luis et al. 2010, 2011; Apaydin et al. 2006; Michiels et al. 1992) as

---

N. Kumar (✉)
India Meteorological Department, Lodhi Road, New Delhi 110003, India
e-mail: naresh.nhac@gmail.com

A.K. Jaswal
India Meteorological Department, Shivajinagar, Pune 411005, India

this index provides information on seasonal and annual variability on long-term in total amount of rainfall. There are hardly any study related to rainfall concentration over India and its regions. Therefore a study is carried out to analyse the annual and seasonal rainfall concentration over India by using PCI for the period 1871–2011. To see the long-term changes in seasonal and annual rainfall, trends of PCI series have been studied by using linear regression method and Mann-Kendall test. The trend is said to be significant if the confidence level is 95 % or more.

2 Data and Methodology

To carry out the study, monthly rainfall data series for India as a whole and its five regions (namely northwest, west-central, central northeast, northeast and peninsular India) for the period 1871–2011 have been collected from IITM Pune website (http://www.tropmet.res.in/) and shown in Fig. 1. A detail about the data and regions is given in Parthasarathy et al. (1995). From monthly rainfall data, PCI has been calculated on annual and seasonal scale. The year has been divided into four seasons viz. winter (January to February), pre-monsoon (March to May), monsoon (June to September) and post-monsoon (October to December). The PCI as proposed by Oliver (1980) is calculated as:

$$PCI_{Annual} = \frac{\sum_{i=1}^{12} pi^2}{\left(\sum_{i=1}^{12} pi\right)^2} \times 100 \quad PCI_{Winter} = \frac{\sum_{i=1}^{12} pi^2}{\left(\sum_{i=1}^{12} pi\right)^2} \times 16.67$$

$$PCI_{Pre-monsoon} = \frac{\sum_{i=1}^{3} pi^2}{\left(\sum_{i=1}^{3} pi\right)^2} \times 25 \quad PCI_{Monsoon} = \frac{\sum_{i=1}^{4} pi^2}{\left(\sum_{i=1}^{4} pi\right)^2} \times 33.33$$

$$PCI_{Post-monsoon} = \frac{\sum_{i=1}^{3} pi^2}{\left(\sum_{i=1}^{3} pi\right)^2} \times 25$$

According to Oliver (1980),

(a) $PCI \leq 10$: Uniform distributed precipitation
(b) $11 \leq PCI \leq 15$: Moderate distributed precipitation
(c) $16 \leq PCI \leq 20$: Irregular distributed precipitation
(d) $PCI \geq 20$: Strongly irregular distributed precipitation

Finally, trend analysis of PCI series has been carried out using linear regression method and significance of trend is tested by using non-parametric Mann Kendall test (Kendall 1976; Kumar and Jain 2010; Subash et al. 2011).

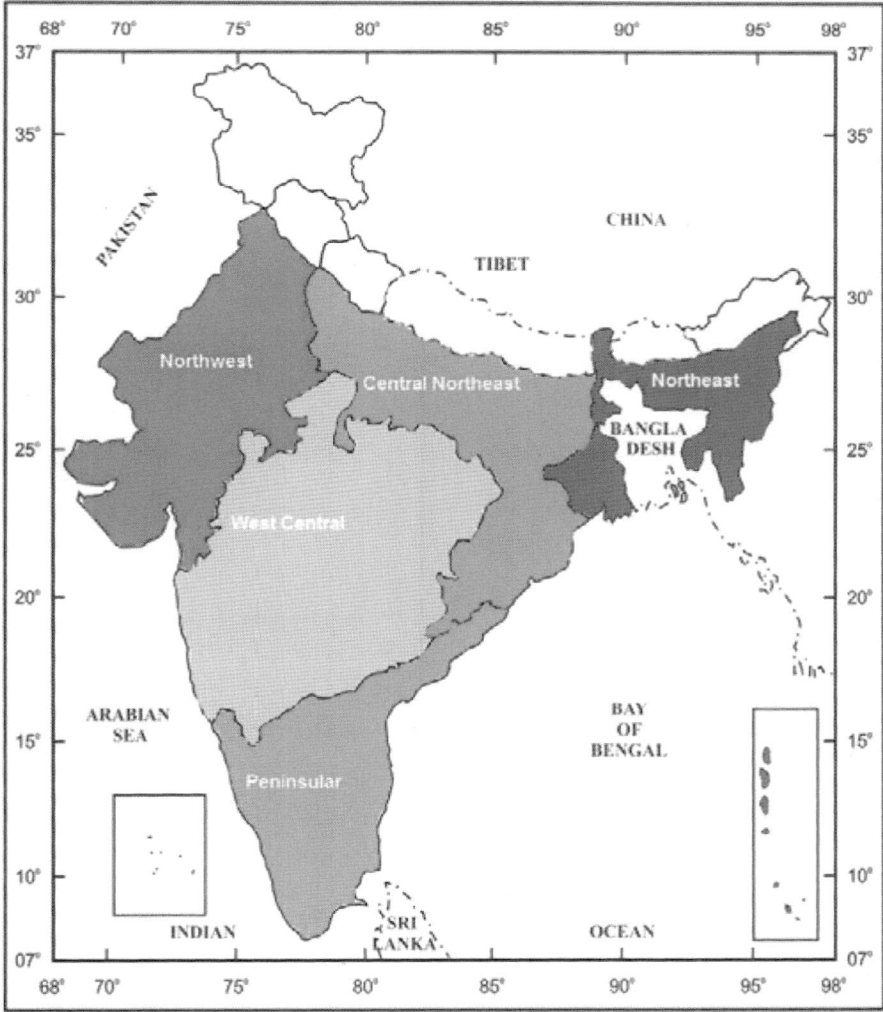

Fig. 1 Five homogeneous regions of India considered for study

3 Results and Discussions

3.1 PCI for Annual Rainfall

Analyses of data indicate that annually in India as a whole rainfall is irregularly distributed, among all the five regions: rainfall over northwest, westcentral and central northeast India are strongly irregular and rainfall over northeast and peninsular India is irregularly distributed. Among all the regions, rainfall over peninsular India is more uniform.

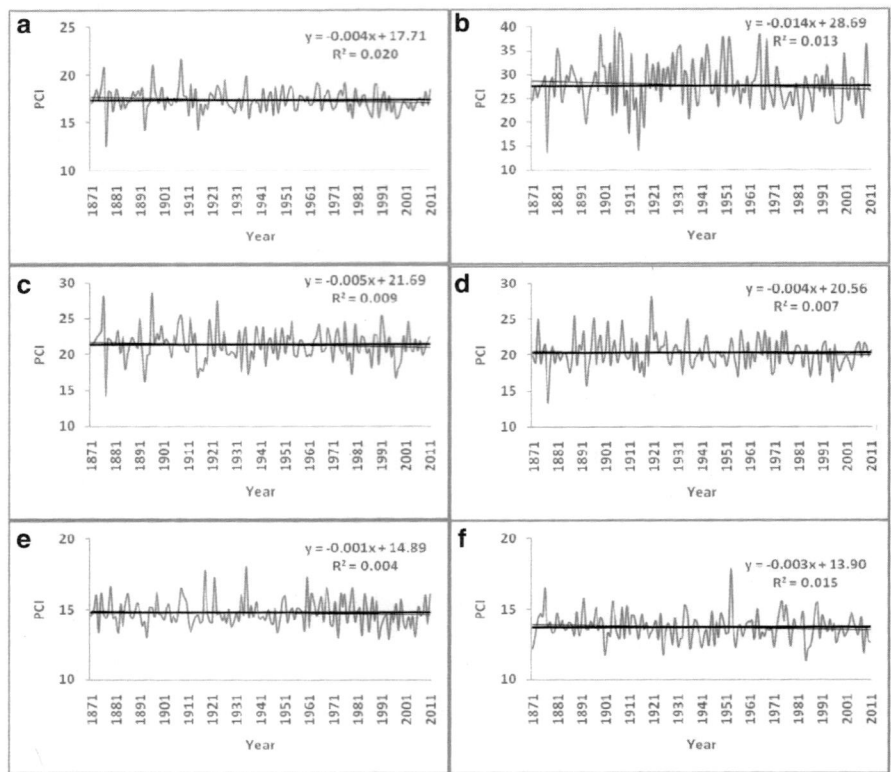

Fig. 2 Annual PCI trend for (**a**) India as a whole, (**b**) northwest, (**c**) west-central, (**d**) central northeast, (**e**) northeast and (**f**) peninsular India

Annual PCI trend for India as a whole, northwest, west-central, central northeast, northeast and peninsular India is given in Fig. 2. In trend analysis, significant decreasing trend in PCI of annual rainfall for India as a whole is observed. This indicates that rainfall is becoming more uniform. Decreasing trends in annual rainfall PCI also is observed in all the five regions. However, they are not statistically significant.

3.2 PCI for Winter Rainfall

Analysis of winter PCI indicate the uniform distribution of rainfall during winter season over India as a whole. In regional analysis, it indicates that rainfall is moderately distributed in all the five regions of India.

Winter PCI trend for India as a whole, northwest, west-central, central northeast, northeast and peninsular India is given in Fig. 3. In trend analysis, decreasing trend

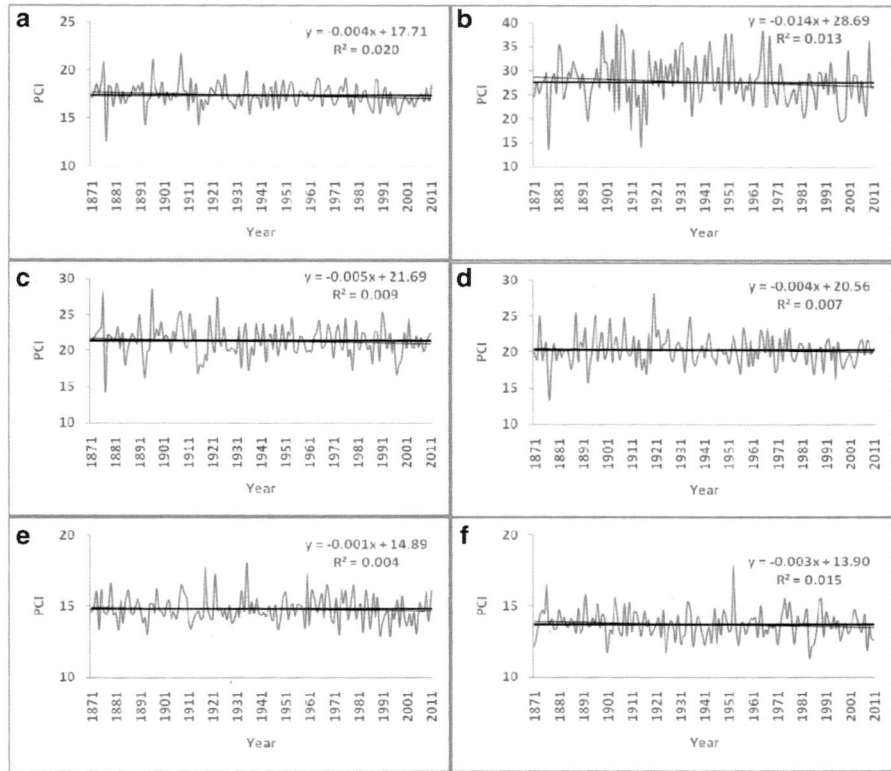

Fig. 3 Winter PCI trend for (**a**) India as a whole, (**b**) northwest, (**c**) west-central, (**d**) central northeast, (**e**) northeast and (**f**) peninsular India

in India as a whole winter rainfall PCI is observed. Decreasing trends in winter rainfall PCI is also observed over northwest, central northeast and peninsular India. Increasing trends are observed over west-central and northeast India.

3.3 PCI for Pre-monsoon Rainfall

Analysis of pre-monsoon PCI indicates the moderate distribution of rainfall during the season over India as a whole as well as over all the five regions of India.

Trend of pre-monsoon season PCI for India as a whole, northwest, west-central, central northeast, northeast and peninsular India is given in Fig. 4. Non-significant decreasing trend is observed for India as a whole pre-monsoon PCI. Decreasing trends in pre-monsoon PCI is also observed over all the regions except northeast India, which shows increasing trend. This indicates that pre-monsoon rainfall is becoming more irregular.

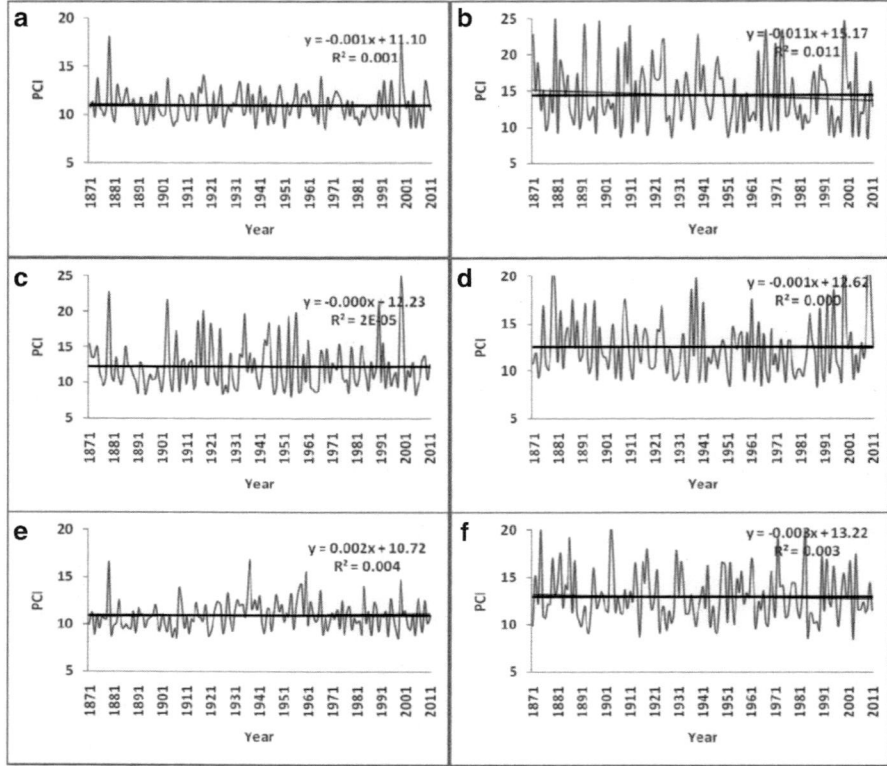

Fig. 4 Pre-monsoon PCI trend for (**a**) India as a whole, (**b**) northwest, (**c**) west-central, (**d**) central northeast, (**e**) northeast and (**f**) peninsular India

3.4 PCI for Monsoon Rainfall

PCI analyses of data show that monsoon season rainfall over India as a whole is uniformly distributed, among all the five regions except northwest India, in which rainfall is moderately distributed.

Monsoon season PCI trend for India as a whole, northwest, west-central, central northeast, northeast and peninsular India is given in Fig. 5. In trend analysis, nearly no trend is observed in India as a whole monsoon rainfall. This indicates that no change in rainfall distribution is observed. Decreasing trends in annual rainfall PCI is also observed in all the five regions except west-central and northeast India, in which increasing trends in PCI are seen.

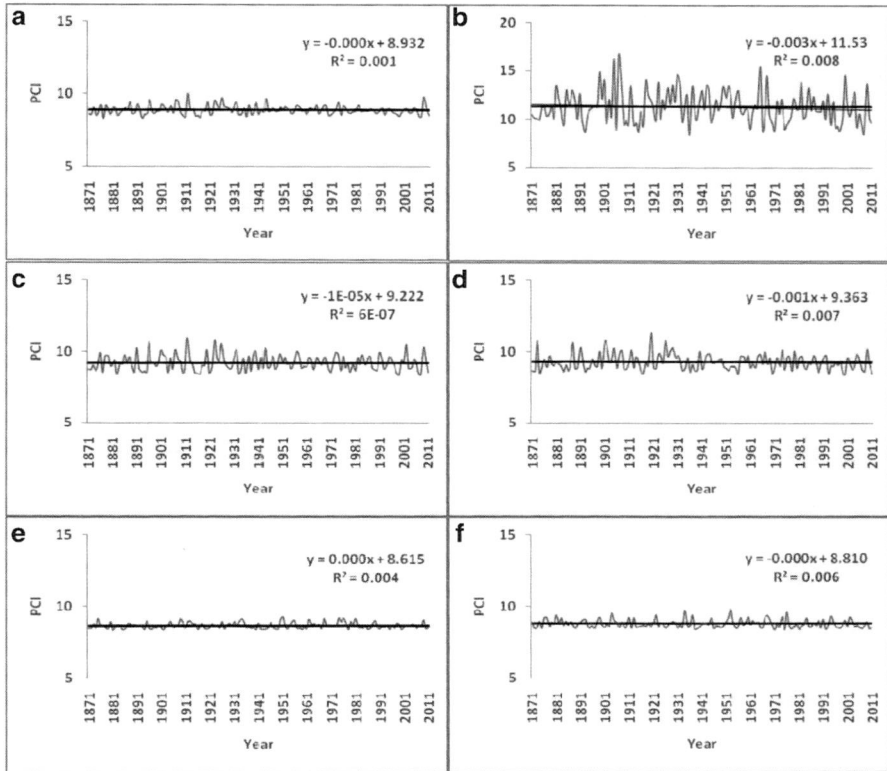

Fig. 5 Monsoon PCI trend for (**a**) India as a whole, (**b**) northwest, (**c**) west-central, (**d**) central northeast, (**e**) northeast and (**f**) peninsular India

3.5 PCI for Post-monsoon Rainfall

PCI analysis for rainfall over India as a whole shows that rainfall during post-monsoon season is moderately distributed. Considering region-wise analysis, it is observed that rainfall is moderately distributed in northeast India and irregularly distributed in the remaining four regions.

Trends of post-monsoon season PCI for India as a whole, northwest, west-central, central northeast, northeast and peninsular India are shown in Fig. 6. No trend is observed in India as a whole rainfall PCI and northeast India. Decreasing trend is observed in northwest and peninsular India and increasing trend in west-central and central northeast India.

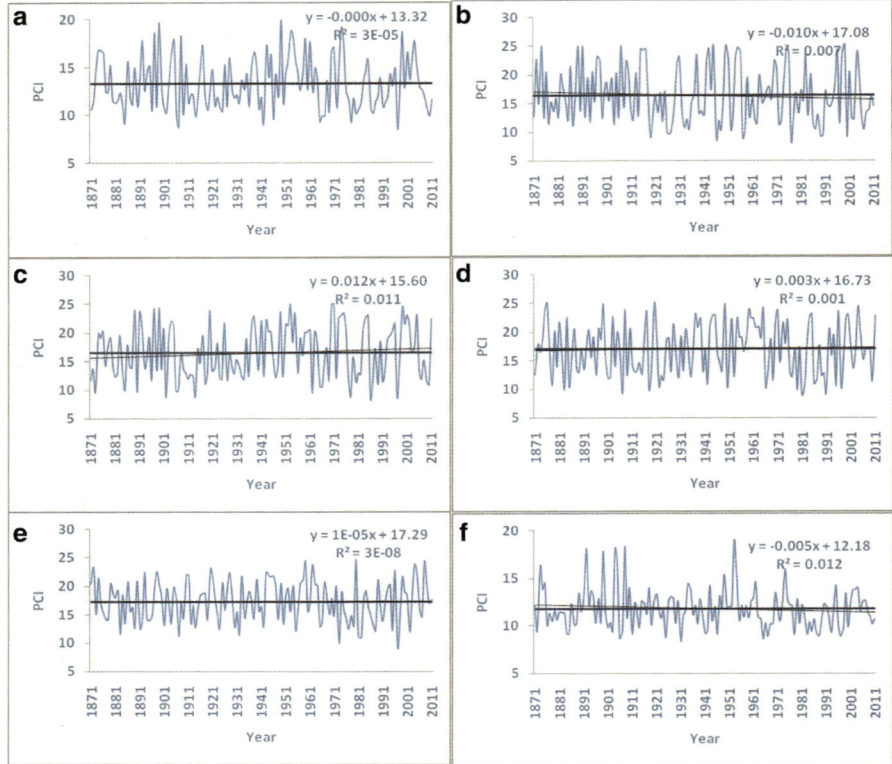

Fig. 6 Post-monsoon PCI trend for (**a**) India as a whole, (**b**) northwest, (**c**) west-central, (**d**) central northeast, (**e**) northeast and (**f**) peninsular India

Conclusions

Broad conclusions of the study are as follows:

- The annual rainfall of peninsular India is more uniformly distributed as compared to other regions. The annual rainfall over northwest India is strongly irregular. In peninsular India, May to November months are wet months, whereas in northwest India, wet months are from June to September. Due to this, annual rainfall over peninsular India is more uniform.
- In seasonal analysis, it is observed that in India as a whole as well as in all the five regions, monsoon rainfall is uniformly distributed except northwest India, where rainfall is moderately distributed. It may be due to fact that in northwest India, contribution of June rainfall is only 12 %, whereas it is not true for other four regions and India as a whole.

(continued)

- The trend analyses of annual PCI for India as a whole and all the five regions show decreasing trends with significant decrease for India as a whole. This shows that annual rainfall for India as a whole is becoming more uniform. In seasonal analysis, no significant trend is observed in India as a whole or any of its regions during study period.

References

Apaydin H, Erpul G, Bayramin I, Gabriels D (2006) Evaluation of indices for characterizing the distribution and concentration of precipitation: a case for the region of Southeastern AnatoliaProject, Turkey. J Hydrol 328:726–732

De Luís M, Raventós J, González-Hidalgo JC, Sánchez JR, Cortina J (2000) Spatial analysis of precipitation trends in the region of Valencia (East Spain). Int J Climatol 20:1451–1469

De Luis M, Brunetti M, González-Hidalgo JC, Longares LA, Martín-Vide J (2010) Changes in seasonal precipitation in the Iberian Peninsula during 1946–2005. Glob Planet Change 74:27–33

De Luis M, González-Hidalgo JC, Brunetti M, Longares LA (2011) Precipitation concentration changes in Spain 1946-2005. Nat Hazards Earth Syst Sci 11:1259–1265

Goswami BN, Venugopal V, Sengupta D, Madhusoodanan MS, Xavier PK (2006) Increasing trend of extreme rain events over India in a warming environment. Science 314:1442–1445

Hulme M, Osborn TJ, Johns TC (1998) Precipitation sensitivity to global warming: comparison of observations with HADCM2 simulations. Geophys Res Lett 25:3379–3382

Kendall M (1976) Time series. Griffin, London

Kripalani RH, Kulkarni A, Sabade SS, Khandekar ML (2003) Indian monsoon variability in a global warming scenario. Nat Hazards 29:189–206

Krishnamurthy C, Kiran B, Lall U, Kwon H-H (2009) Changing frequency and intensity of rainfall extremes over India from 1951 to 2003. J Clim 22:4737–4746

Kumar V, Jain SK (2010) Trends in seasonal and annual rainfall and rainy days in Kashmir Valley in the last century. Quat Int 212:64–69

Michiels P, Gabriels D, Hartmann R (1992) Using the seasonal and temporal precipitation concentration index for characterizing monthly rainfall distribution in Spain. Catena 19:43–58

Oliver JE (1980) Monthly precipitation distribution: a comparative index. Prof Geogr 32:300–309

Parthasarathy B, Munot AA, Kothawale DR (1995) Monthly and seasonal rainfall series for all-India homogeneous regions and meteorological subdivisions 1871-1994. Research Report No. RR-065, Indian Institute of Tropical Meteorology, Pune

Rodríguez-Puebla C, Encinas AH, Nieto S, Garmendia J (1998) Spatial and temporal patterns of annual precipitation variability over the Iberian Peninsula. Int J Climatol 18:299–316

Sahai AK, Pattanaik DR, Satyan V, Grimm Alice M (2003) Teleconnections in recent time and prediction of Indian summer monsoon rainfall. Meteorol Atmos Phys 84:217–227

Subash N, Sikka AK, Ram Mohan HS (2011) An investigation into observational characteristics of rainfall and temperature in Central Northeast India—a historical perspective 1889–2008. Theor Appl Climatol 103:305–319

Zhang Q, Xu C-Y, Gemmer M, Chen YD, Liu C-L (2009) Changing properties of precipitation concentration in the Pearl River basin, China. Stoch Env Res Risk A (SERRA) 23:377–385

Identifying the Changes in Rainfall Pattern and Heavy Rainfall Events During 1871–2010 over Cherrapunji

Pulak Guhathakurta, Preetha Menon, and N.B. Nipane

1 Introduction

Cherrapunji holds world record of highest recorded point rainfall for different durations i.e. 1 month, 2 months, 12 months, 2 years (WMO 1994). It is situated in the southern slopes of Khasi and Jayantia Hills of 'Meghalaya' province of India overlooking the plains of the Sylhet district in Bangladesh. The meaning of Meghalaya is the abode of cloud. The altitude of Cherrapunji (Fig. 1) is 1,313 m above mean sea level. Rain occurs almost regularly throughout the year over this region due to its peculiar orographic features. Synoptically, this region is the convergence zone of westerly, easterly and moist southerly winds causing lots of rain and therefore the average monthly rainfall is also very high over this region. The monsoon clouds fly unhindered over the plains of Bangladesh for about 400 km. Thereafter, they hit Khasi hills which abruptly erupt out of the plains to reach a height of about 1,370 m above m.s.l. within a short distance of 2–5 km. The orography of the hills, with many deep valleys, channels the low flying (150–300 m) moisture laden clouds from a wide area to converge over Cherrapunji which falls in the middle of the path of this stream. The winds push the rain clouds through these gorges and up the steep slopes. The rapid ascendance of the clouds into the upper atmosphere hastens the cooling and helps vapour to condense. Most of Cherrapunji's rain is the consequence of air being lifted as a large body of water vapour. Extremely large amount of rainfall at Cherrapunji is perhaps the most well-known feature of orographic rain in northeast India.

According to Blanford (1886) the dynamic cooling of the saturated southwest winds which are forced to rise vertically due to the steep slopes of the region is the main cause of the enormous rainfall at Cherrapunji. Das (1951), however, feels that

---

P. Guhathakurta (✉) • P. Menon • N.B. Nipane
India Meteorological Department, Shivajinagar, Pune, India
e-mail: pguhathakurta@rediffmail.com

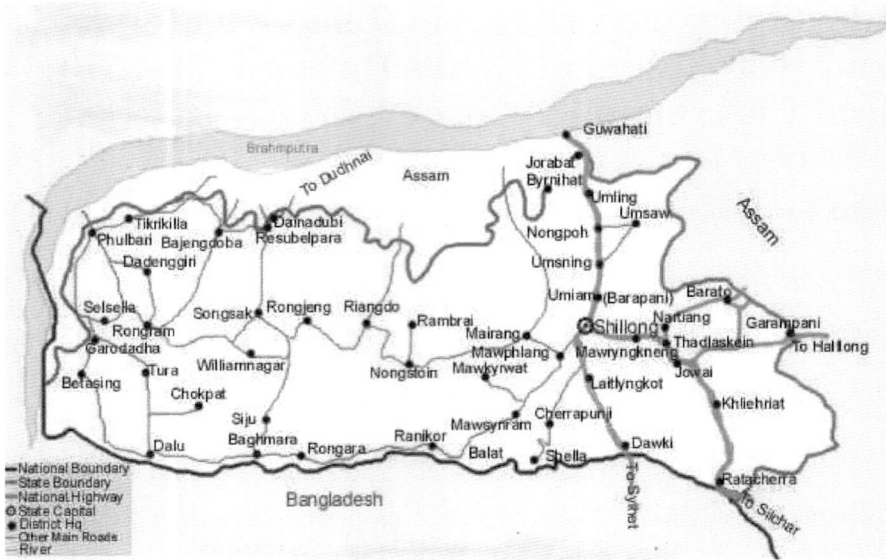

Fig. 1 Location map of Cherrapunji

besides orography, the high rainfall at this station is due to its proximity to the line of discontinuity between the comparatively dry easterlies and north easterlies and the moist southerlies and south westerlies. Dhar and Farooqui (1973) studied the maximum rainfall at Cherrapunji with 57 years of rainfall data from 1903 to 1959. Guhathakurta (2007) has shown that highest 1 day and highest 2-days rainfall in India occurred in Cherrapunji on 16 June, 1995 and 15–16 June, 1995.

There are two observational long records over Cherrapunji viz. Cherrapunji Police Station (25°17′ N, 91°43′ E) which is under the control of state authorities of Meghalaya and the observatory (25°15′ N, 91°44′ E) of India Meteorological Department. The present paper involves for the first time a detailed analysis of long series of rainfall data of Cherrapunji. The study identified some noticeable patterns in the frequency of rainfall events of different categories. Trend analysis was done on monthly and seasonal rainfall series as well as on frequency of rainfall events of different categories in order to find out the existence of changes in mean rainfall, its intensity and frequency.

2 Data Used and Methodology

Daily and monthly rainfall data of the following two stations and periods are considered:

1. Cherrapunji Met Observatory (25°15′ N, 91°44′ E) 1902–2010
2. Cherrapunji Police Station (25°17′ N, 91°43′ E) 1871–1957

From the monthly rainfall data series, rainfall for the seasons winter (January and February), pre-monsoon (March, April and May), southwest monsoon (June to September) and post-monsoon (October-December) are computed.

For the trend analysis of monthly and seasonal rainfall most widely used linear regression and 't' test for significance of the trend are used. To analyze the rainfall data of different intensities and frequencies we have used the following seven classifications of daily rainfall into different intensities:

 (i) Dry day if rainfall is zero,
 (ii) very light rain if rainfall is between 0.1 and 2.4 mm,
 (iii) light rain if rainfall is between 2.5 and 7.5 mm,
 (iv) moderate rain if rainfall is between 7.6 and 64.4 mm,
 (v) heavy rain if rain is between 64.5 and 124.4 mm,
 (vi) very heavy if rain is between 124.5 and 244.4 mm, and
 (vii) extremely heavy if rain is 244.5 mm or more.

The well known non-parametric Mann Kendal test is used for trend analysis of frequencies of daily rainfall of different categories.

3 Trends in Monthly and Seasonal Rainfall

Mean rainfall for Cherrapunji observatory station for all the 12 months for the period 1902–2010 are shown in Fig. 2. It is interesting to see that compared to most of the stations in India where southwest monsoon is predominant and July being the most rain-bearing month, Cherrapunji gets maximum rainfall during June followed by July and August. Mean rainfall of May which gets rainfall mostly due to

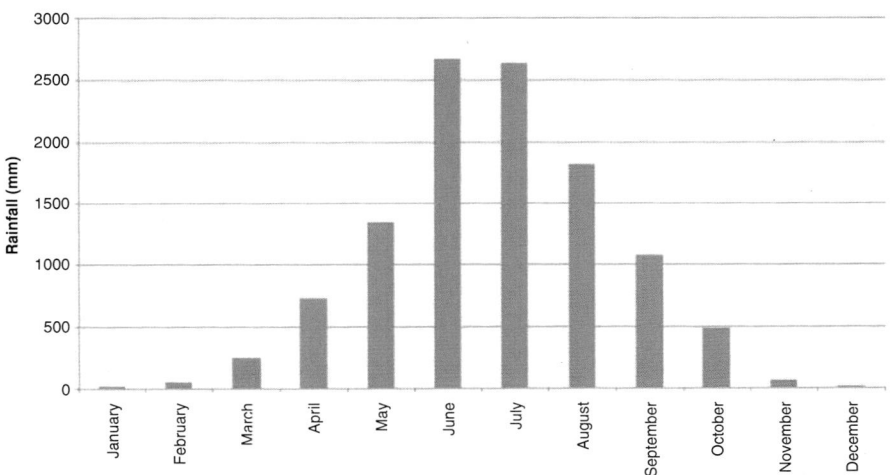

Fig. 2 Mean monthly rainfall of Cherrapunji observatory during the period 1902–2010

Table 1 Mean rainfall (mm) of Cherrapunji observatory and Cherrapunji P.S. for the period 1902–2010, 1902–1957 and 1871–1957 for all the seasons and annual

Station	Period	Jan–Feb	Mar–May	June–Sep	Oct–Dec	Annual
Cherrapunji Obsy	1902–2010	71.52	2322.71	8283.57	568.72	11246.53
	1902–1957	68.81	2298.35	8162.29	559.09	11088.54
Cherrapunji P.S.	1871–1957	69.46	2198.05	7981.37	490.94	10739.81

convective activity is higher than September. Lowest rainfall is received during December followed by January. Table 1 shows the mean rainfall for both the stations for all the four seasons and year as whole for the period 1871–1957 for Cherrapunji P.S. and 1902–2010 and 1902–1957 for Cherrapunji observatory. It can be seen that the mean rainfall values are higher for all the seasons and the years in recent period 1902–2010 compared to 1871–1957. This indicates increase of rainfall over Cherrapunji in recent years. To see the increasing trends of rainfall over Cherrapunji we have done the trend analysis of monthly rainfall (Table 2). March and July rainfall for Cherrapunji observatory has shown significant increasing trends at the rate of 2.12 and 7.34 mm/year respectively during the period 1902–2010. Rainfall for the months of January, February, April, June, October and December has shown increasing trends while rainfall for other months has shown decreasing trends but none of these trends are significant. The trend analysis of rainfall for the period 1871–1957 identified significant decreasing trends in March and April rainfall while increasing trends in May and June rainfall. However no significant trend is noticed in seasonal or annual rainfall though rainfall in all the seasons and annual has shown increasing trends during the period 1902–2010.

4 Variation of Daily Rainfall of Different Intensities

Mean daily rainfall of each day is computed for Cherrapunji for the period 1902–2010. Figure 3 depicts the behaviour of mean daily rainfall for each of the 12 months. Daily rainfall is highest on 30th June (121.3 mm). The second highest is also in June (13th June with value of 118.4 mm). The third highest is on 1st July (114.9 mm). From 3rd June to 31st July daily normal are more than 60 mm on all the days. Variability of daily mean rainfall is lowest (17 %) in August as the daily mean rainfall varies from 39 mm to 69.9 mm. Highest variability in December (126 %) as mean values vary from 0 mm to 2.3 mm. Both in November and December months daily rainfall are very less.

Mean monthly frequency of days of rainfall of different intensities i.e. dry days, very light rain, light rain, moderate rain, heavy rain, very heavy rain and extremely heavy rain are shown in Fig. 4. Frequency of dry day is highest in December followed by January and lowest in July followed by June. Frequency of moderate rain is more compared to all types of rain events. This implies that in Cherrapunji

Table 2 Increase/decrease in monthly, seasonal and annual rainfall in mm/year along with significant level

Station	Period	Jan	Feb	Mar	Apr	May	Jun	Jul	Aug	Sep	Oct	Nov	Dec
Cherrapunji Obsy	1902–2010	0.005	0.112	2.119	1.966	−0.087	0.170	7.340	−0.282	−0.043	0.532	−0.041	0.082
Cherrapunji P.S.	1871–1957	0.063	0.062	−2.259	−3.676	5.511	9.861	−0.858	−1.109	−2.796	2.232	0.512	0.067

Station	Period	Jan–Feb	Mar–May	June–Sep	Oct–Dec	Annual
Cherrapunji Obsy	1902–2010	0.116	3.610	7.875	0.639	8.710
Cherrapunji P.S.	1871–1957	0.148	−0.362	2.226	2.772	2.547

🟩	Significant increase at 95 % confidence level
🟧	Significant decrease at 95 % confidence level

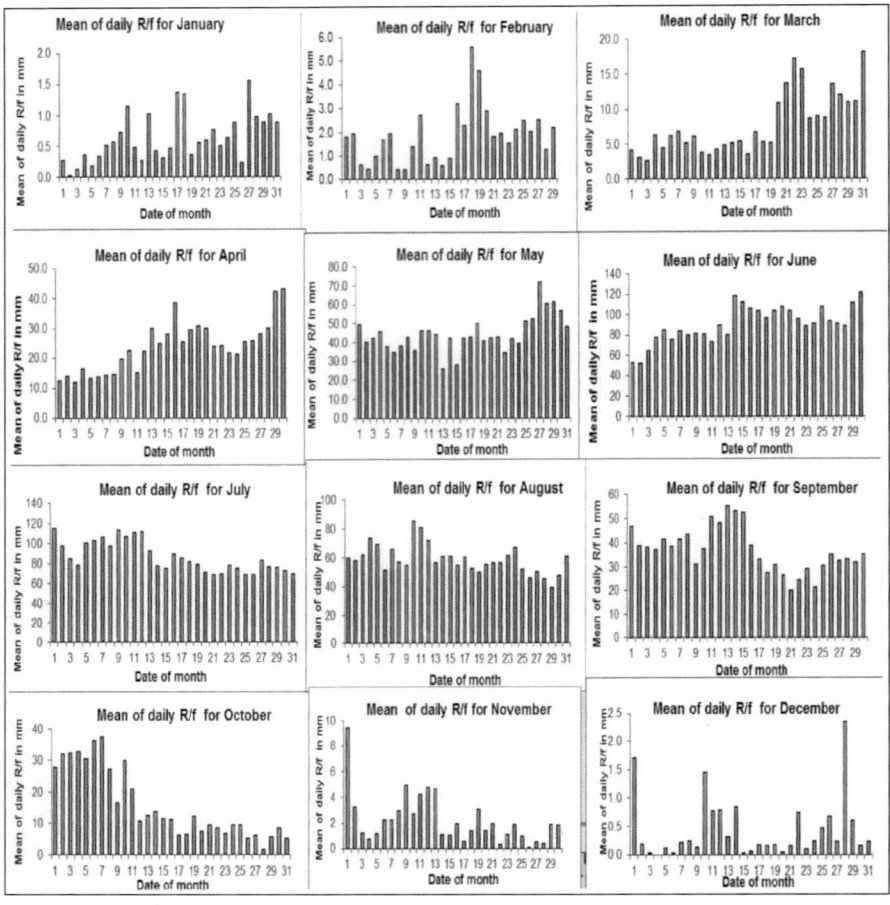

Fig. 3 Distribution of mean daily rainfall over the year in months

most of the rain days are of moderate intensity. Frequency of extremely heavy rainfall is highest in June (nearly 3). Frequencies of very heavy rain and heavy rain are highest in July. Frequencies of moderate rain is highest in August. Frequencies of all other categories i.e. light and very light rain are highest in September. Thus as the rainfall intensities decrease, maximum frequencies shift from June to September.

Table 3 shows the result of trend analysis of frequency of days of rainfall of different intensities including the dry days. Frequency of dry days has shown increasing trends in all the months except January, March and July being significant only in August. As a result annual frequency of dry days has shown significant increasing trends.

Frequencies of very light rain days have shown decreasing trends in all the months except May and August (significant increasing trend), being significant in

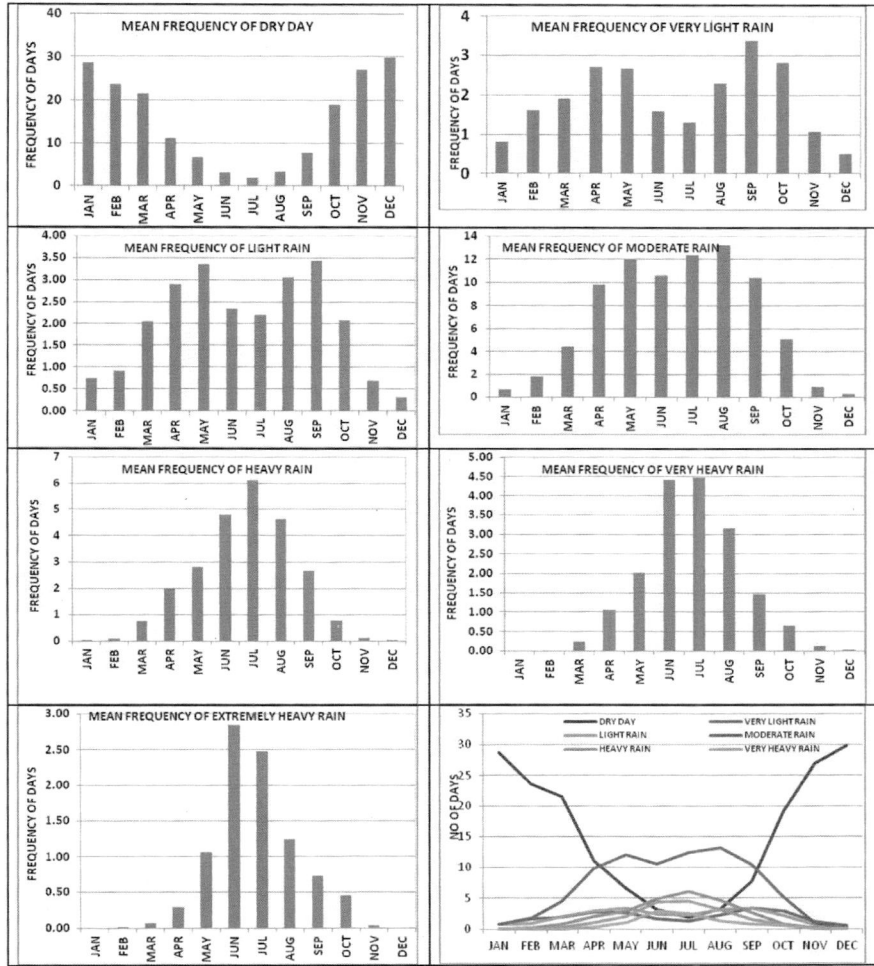

Fig. 4 Mean monthly frequency of rainfall of different intensities

February, March and July. As a result, annual frequency of very light rain days has shown significant decreasing trends.

Frequencies of light rain days have shown decreasing trends in all the months except July and November, being significant in February and April. As a result, annual frequency of light rain days has shown significant decreasing trends.

No significant trends are noticed in the frequency of moderate and very heavy rain days.

Frequencies of heavy rain days have shown significant decreasing trends for the months of April to September and November. As a result annual and SW monsoon frequencies also show significant decreasing trends.

Frequency of extremely heavy rain days has increased significantly in July.

Table 3 Increase/decrease in frequency of days per century

	Dry day	Very light rain	Light rain	Moderate rain	Heavy rain	Very heavy rain	Extremely heavy rain
January	−0.06	−0.03	−0.12	0.41	−0.04	0.00	0.00
February	1.65	−0.88	−0.81	−0.43	0.15	0.00	0.04
March	−0.32	−1.30	−0.34	0.81	0.77	0.48	0.05
April	1.62	−0.82	−1.04	−0.60	−0.01	0.68	0.19
May	1.55	0.65	−0.09	−2.05	−0.53	0.11	0.01
June	0.56	−0.84	−0.07	1.20	−0.82	−0.56	0.31
July	−0.30	−0.83	0.18	−1.22	−0.87	1.41	1.70
August	1.53	1.26	−0.13	−1.65	−1.65	0.82	−0.05
September	1.18	−0.39	−0.53	0.46	−0.52	−0.04	−0.12
October	1.89	−0.98	−0.30	−0.64	−0.23	0.31	0.09
November	0.29	−0.38	0.22	0.02	−0.07	0.11	−0.05
December	0.17	−0.28	−0.26	0.43	0.05	−0.03	0.00
SW monsoon	2.42	−0.61	−0.86	−0.30	−3.89	1.54	1.63
Annual	10.12	−4.70	−3.67	−3.04	−4.25	2.66	2.67

Significant levels	Increase	Decrease
90 %		
95 %		
99 %		

5 Trends in Annual 1-Day, 2-Day and 3-Day Maximum Rainfall Series

Country's ever recorded highest 1-day rainfall recorded at Cherrapunji on 16th June 1995 amounted to 156.3 cm. Though this record is not highest in the world record of 1-day point rainfall but the 2 days rainfall of 249.3 cm occurred on 15–16th June 1995 at Cherrapunji (WMO 1994; Guhathakurta 2007) is the highest ever recorded 2-day point rainfall. This motivates us to see whether any long-term trend exists in annual 1- , 2- and 3-day maximum rainfall series. Figure 5 shows the annual time series of 1- , 2- and 3-day maximum rainfall of Cherrapunji observatory. Trend analysis on these three series does not show any significant change in extreme rainfall. It may be mentioned that similar to the occurrence of 249.3 cm rainfall on 2 days i.e. 15th and 16th June 1995 and successive 3 days rainfall of 279.83 cm from 14th to 16th June 1995, there was earlier occurrence of 3 day rainfall of 276.05 cm from 12th–14th September 1974. Cherrapunji experiences extremely heavy rainfall on all the 3 days (i.e. 97.07 cm on day one, 98.55 cm on day two and 80.43 cm on day three).

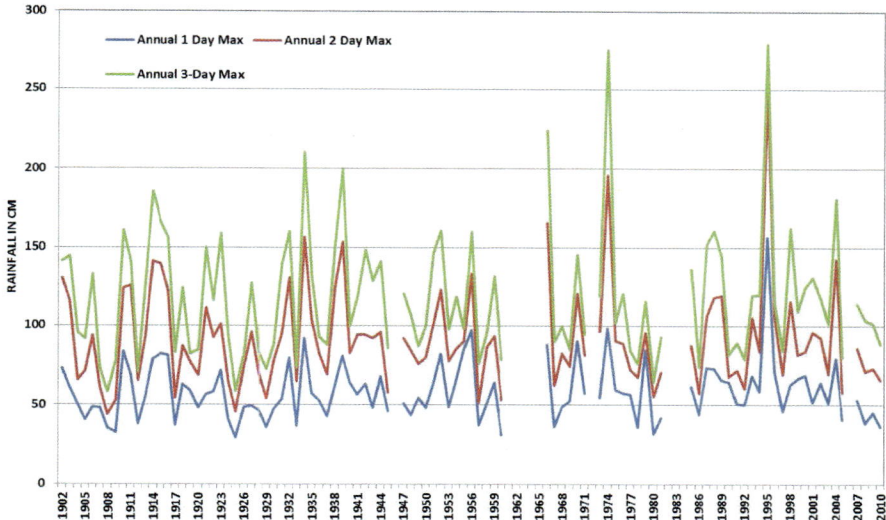

Fig. 5 Annual 1- , 2- and 3-day maximum rainfall of Cherrapunji observatory

Conclusions

A detailed analysis (including trends and variability) of rainfall data of two raingauge stations of Cherrapunji with longest available record are done. Study of mean monthly rainfall pattern reveals that Cherrapunji gets maximum rainfall in the month of June followed by July. Also the station gets good rain during pre-monsoon season. Rainfall in the month of May is much higher than in September.

Trend analysis of annual and seasonal rainfall shows increasing trend of rainfall for all the seasons during the period 1902–2010 but neither of these trends are significant. However in monthly scale March and July rainfall has shown significant increasing trend. During the period 1871–1957, March and April rainfall has shown significant decreasing trend whereas May and June rainfall has shown significant increasing trend.

Frequency analysis of days with rainfall of different intensities along with dry days reveals that the dry days are highest in December followed by January and lowest in July followed by June. Frequency of moderate rain is more compared to all types of rain events. Frequency of extremely heavy rainfall is highest in June (nearly 3). Frequencies of very heavy rain and heavy rain are highest in July while frequency of moderate rain is highest in August. Frequencies of light and very light rain are highest in September.

There is increasing trends in frequency of dry days in all the months except January, March and July being significant in August. There is also significant increasing trend in annual frequency of dry days

There is decreasing trend in frequency of very light rain days in all the months except May and August (significant increasing trend) being significant in February, March and July and significant decreasing trend in annual frequency. Frequency of light rain days has shown decreasing trends in all the months except July and November, being significant in February and April and significant decreasing trends in annual frequency. No significant trends are noticed in the frequency of moderate and very heavy rain days.

Significant decreasing trends in frequency of heavy rain days for the months April to September and November and significant decreasing trends in annual and SW monsoon frequencies are observed. Frequency of extremely heavy rain days has increased significantly in July.

In spite of occurrence of country's highest 1-day rainfall and world's highest 2-day rainfall in June 1995 there were no significant changes in annual 1-day, 2-day or 3-day maximum rainfall.

Acknowledgement The authors thank the Additional Director General of Meteorology (Research), IMD, Pune for the encouragement, data and all other facilities to bring out this result.

References

Blanford H (1886–1888) The rainfall of India, Indian meteorological memoirs, vol III. National Government Publication, Calcutta

Das JC (1951) On certain aspects of rainfall at Cherrapunji. Indian J Meteorol Geophys 2(3)

Dhar ON, Farooqui SMT (1973) A study of rainfalls recorded at the Cherrapunji observatory. Hydrol Sci Bull – des Sciences Hydrologiques XVIII:412

Guhathakurta P (2007) Highest recorded point rainfall over India. Weather 62(12):349

World Meteorological Organization (1994) Guide to hydrological practice. WMO no. 168. WMO, Geneva

Part IV
Drought and Temperature

Agricultural Drought Assessment: Operational Approaches in India with Special Emphasis on 2012

S.S. Ray, M.V.R. Sesha Sai, and N. Chattopadhyay

1 Introduction

Drought is a condition of moisture deficit sufficient to have an adverse effect on vegetation, animals and human being over a sizable area. In the past, India has experienced 24 large scale droughts in 1891, 1896, 1899, 1905, 1911, 1915, 1918, 1920, 1941, 1951, 1965, 1966, 1972, 1974, 1979, 1982, 1986, 1987, 1988, 1999, 2000, 2002, 2009 and 2012 with increasing frequencies during the periods 1891–1920, 1965–1990 and 1999–2012. During 1900–2012, 14 droughts have occurred, with 1,061 million people affected and 2,441 million dollar economic loss (Source: EMDAT). Drought can have economic, environmental and social impacts. Droughts are very different from the more attention grabbing hazards such as tsunamis and earthquakes in a variety of significant ways. It differs from other disasters, in following aspects (DAC 2009):

- No universal definition
- Very slow onset
- Duration ranges from months to years
- No single indicator or index can characterize drought
- Large spatial extent
- Impacts are generally non-structural and difficult to quantify
- Impacts are cumulative

S.S. Ray (✉)
Mahalanobis National Crop Forecast Centre, DAC, MoA, Pusa Campus, New Delhi, India
e-mail: shibendu.ncfc@nic.in

M.V.R. Sesha Sai
National Remote Sensing Centre (ISRO), Balanagar, Hyderabad, India

N. Chattopadhyay
Agricultural Meteorology Division, India Meteorological Department, Shivajinagar, Pune, India

Prolonged dry conditions are a natural part of life in the Southwest. Past records from tree-rings indicate that the Southwest has experienced protracted "mega" droughts that lasted about 50 years, a duration unlike the 1950s and 2000s droughts that caused economic losses to agriculture, ranching, industry and many other sectors. Because of recurring drought, many efforts have been made to minimize its impacts on human activity. The Colorado River, for example, stores about 4 years of water supply to buffer the effect of several consecutive years of below-average streamflows. However, some sectors are more vulnerable to drought than others, and drought can appear rapidly after just one season of below-average precipitation. It is critical to monitor drought to help local governments, resource managers, and many other groups make effective decisions. Several outputs produced at the national and regional level help to inform people living in the region about current and future conditions.

In India, National Commission on Agriculture (1976) has categorized drought into three types, viz., Meteorological Drought, Hydrological Drought and Agricultural Drought. In meteorological terms, a drought is "a sustained, regionally extensive, deficiency in precipitation". In quantitative terms, definitions could vary among countries and regions. In India, the definition for meteorological drought adopted by the India Meteorological Department (IMD) is a situation when the deficiency of rainfall at a meteorological sub-division level is 25 % or more of the long-term average (LTA) of that sub-division for a given period. The drought is considered "moderate", if the deficiency is between 26 and 50 %, and "severe" if it is more than 50 %. Hydrological drought is a prolonged meteorological drought situation resulting in depletion of surface water from reservoirs, lakes, streams, rivers and fall in groundwater levels causing severe shortage of water for livestock and human needs. Agricultural drought is a situation when rainfall and soil moisture are inadequate during the crop growing season to support healthy crop growth till maturity, causing crop stress and wilting. It is defined as a period of four consecutive weeks of severe meteorological drought with a rainfall deficiency of more than 50 % of the LTA or with a weekly rainfall of 5 mm or less during the period from mid-May to mid-October (the Kharif season) when 80 % of the country's total crop is planted, or six such consecutive weeks during the rest of the year.

1.1 Probability of Occurrence of Drought in Different Regions of India

- In the northwest region of India, the probability of moderate drought varies from 12 to 30 % and that of severe drought varies from 1 to 20 % in most of the parts and about 20–30 % in the extreme north-western parts.
- In west central India, the probability of moderate drought varies from 5 to 26 % and that of severe drought varies from 1 to 8 %.
- In the peninsular region, the probability of moderate drought varies from 3 to 27 %, and that of severe drought varies from 1 to 9 % in major parts.

Table 1 Frequency of deficient rainfall over different regions of the country

Regions	Frequency of deficient rainfall (75 % of normal or less)
Assam	Very rare, once in 15 years
West Bengal, Madhya Pradesh, Konkan, Bihar and Orissa	Once in 5 years
South interior, Karnataka, Eastern Uttar Pradesh and Vidarbha	Once in 4 years
Gujarat, East Rajasthan, Western Uttar Pradesh	Once in 3 years
Tamil Nadu, Jammu & Kashmir and Telangana	Once in 2.5 years
West Rajasthan	Once in 2 years

- In the central northeast region, the probability of moderate drought varies from 6 to 37 % and that of severe drought varies from 1 to 10 %.
- In the northeast region, the probability of moderate drought varies from 1 to 26 % and that of severe drought varies from 1 to 3 %.
- In the hilly region, the probability of moderate drought varies from 9 to 31 % and that of severe drought varies from 1 to 12 % except in Leh and Lahul & Spiti.

The frequency of deficient rainfall over different regions of the country is presented in Table 1.

In this article, the focus will be on agricultural drought and its monitoring, with special emphasis on the rainfall situation of 2012.

2 Agricultural Drought Assessment

Traditionally, as there is no definite and precise way to pinpoint the beginning and end of a drought event, the identification of certain critical parameters and indicators for drought and tracking these historical indicators to analyze for the frequency, duration and spatial extent of arid periods provide crucial means of monitoring drought events.

2.1 Meteorological Indicators and Indices for Drought Monitoring

Rainfall is the most important factor influencing the incidence of drought and practically all definitions use this variable either singly or in combination with other meteorological elements. Typically most of the indicators are based on meteorological and hydrological variables, such as precipitation, evapotranspiration, stream flows, soil moisture, reservoir storage, and groundwater levels. Various indices were developed to meet the needs for more accurate indication of the conditions leading to the different types of drought risks. Few of them are given below:

- *Aridity Anomaly Index:* Aridity Index (Thornthwaite and Mather 1955) is the percentage of annual water deficit to annual water need or annual potential evapotranspiration. Aridity Anomaly Index is the departure of aridity index value from normal, expressed in percentage.
- *Standardized Precipitation Index (SPI):* The SPI is an index based on the probability of recording a given amount of precipitation, and the probabilities are standardized so that an index of zero indicates the median precipitation amount. Half of the historical precipitation amounts are below the median, and half are above the median. The index is negative for drought, and positive for wet conditions.
- *Palmer Drought Severity Index (PDSI):* The PDSI, known operationally as the Palmer Drought Index (PDI), was devised by Palmer (1965). It uses temperature and precipitation data to calculate water supply and demand, incorporates soil moisture, and is considered most effective for unirrigated cropland.
- *Crop Moisture Index (CMI):* It measures the degree to which crop water requirements are met.
- *Moisture Adequacy Index (MAI):* MAI is based on calculation of weekly water balance and is equal to the ratio of Actual evapotranspiration and Potential evapotranspiration.

There is a limitation of using rainfall and its indices as agricultural drought indicator. Many studies have proved that spatial and temporal distribution of rainfall is more important than total rainfall in a season or month. Rainfall as an agricultural drought indicator is limited by the sparse ground observations as well as lack of spatially and temporally unique relationship between incident rainfall and vegetation development.

2.2 Drought Monitoring by India Meteorological Department

India Meteorological Department (IMD) monitors the incidence, spread, intensification and cessation of drought (near realtime basis) on a weekly time scale over the country based on Aridity Anomaly Index. It also issues weekly drought outlook, based on this index, which indicates the impending drought scenario in the country in the subsequent week. Based on aridity anomaly index, weekly aridity anomaly reports and maps for the southwest monsoon season for the whole country and for the northeast monsoon season for the five meteorological sub-divisions, viz. coastal Andhra Pradesh, Rayalseema, south interior Karnataka, Tamil Nadu & Pondicherry and Kerala, are prepared and sent to various agricultural authorities of State and Central governments and research institutes on operational basis for their use in agricultural planning purposes. The maps are also uploaded in the departmental website. These aridity anomaly maps/reports help to assess the moisture stress experienced by growing plants and to monitor agricultural drought situation in the country.

2.3 Space Technology for Agricultural Drought Monitoring

Unlike point observation of ground data, satellite-based sensors provide direct spatial information on vegetation stress caused by drought conditions and the information is useful to assess the spatial extent of drought situation. Satellite remote sensing is widely used for monitoring crops and agricultural drought assessment. Satellite observations can also have a high re-visit frequency over the same geographical area, allowing intra and inter seasonal comparisons of data collected over time. Space technology has become a global potential tool for detection and mitigation of natural disasters like drought and the availability of long-term satellite data enhance the accuracy of hazard detection, monitoring and impact assessment (Kogan 2001).

The drought monitor of USA using NOAA-AVHRR data (www.cpc.ncep.noaa.gov), Global Information and Early Warning System (GIEWS) and Advanced Real Time Environmental Monitoring Information System (ARTEMIS) of FAO using Meteosat and SPOT-VGT data (Minamiguchi 2005), International Water Management Institute (IWMI)'s drought assessment in southwest Asia using MODIS data (Thenkabail et al. 2004) and NADAMS drought monitoring in India with IRS-WiFS/AWiFS and NOAAAVHRR data (Murthy et al. 2007) are the proven examples for successful application of satellite remote sensing for operational drought assessment.

2.3.1 Various Indices from Satellite Data for Drought Monitoring

In optical remote sensing (RS), the typical reflectance pattern for a healthy vegetation shows high absorption due to chlorophyll, at red region and high reflection due to leaf internal structure, at near-infrared (NIR) region. Additionally, there are strong water absorption bands in short-wave infrared region (SWIR), i.e. at 1,450 and 1,950 nm. Using these contrast characteristics of near-infrared, red and short-wave infrared bands which indicate both the health and condition of the crops/vegetation, different types of vegetation indices (arithmetic combinations of reflectance in different bands) have been developed.

Some of the important vegetation indices which are typically used for crop stress detection and drought assessment are given below.

- *Normalized Difference Vegetation Index:* NDVI is derived as (NIR − Red)/(NIR + Red), where NIR and Red are the reflected radiation in visible and near-infrared channels. The NDVI values for vegetation generally range from 0.1 to 0.6, the higher index values being associated with greater green leaf area and biomass.
- *Crop Water Stress Index:* Jackson et al. (1977) used canopy temperature (Tc) minus air temperature (Ta) as an index to study the water status of the crops. The Crop Water Stress Index (CWSI), derived from $Tc - Ta$ versus the air vapour pressure deficit, was found to be a promising tool for quantifying crop water stress.

- *Vegetation Condition Index & Temperature Condition Index* (Kogan 1995): While VCI is the percentage of NDVI with respect to its maximum amplitude, TCI is the percentage in brightness temperature (derived from channel 4 of NOAA AVHRR) with respect to its maximum amplitude.
- *Vegetation Index/Temperature Trapezoid:* Moran et al. (1994) used NDVI to take into account vegetation cover and computed the Water Deficit Index (WDI) using the scatter plot of $Tc - Ta$ against NDVI.

Water absorption bands in short-wave infrared (SWIR) region have also been used to assess vegetation water status. Such indices are:

- *Normalized Difference Infrared Index* [NDII = $(R_{850} - R_{1650})/(R_{850} + R_{1650})$] by Hardisky et al. (1983)
- *Normalized Difference Water Index* [NDWI = $(R_{850} - R_{1240})/(R_{850} + R_{1240})$] by Gao (1996). Higher values of NDWI signify more surface wetness.
- *Shortwave Angle Slope Index* (SASI) parameterizes the general shape of the NIR–SWIR part of the spectrum. SASI, originally called as SANI, is based on a combination of NIR, SWIR1 and SWIR2 bands of MODIS (Palacios-Orueta et al. 2006).
- *ANIR (Angle at NIR)* is a combination of reflectance values in the Red, NIR and SWIR1 (1,240 nm) bands (Khanna et al. 2007).

2.4 National Agricultural Drought Assessment and Monitoring System (NADAMS)

'National Agricultural Drought Assessment and Monitoring System (NADAMS)' project was conceptualized and developed by the National Remote Sensing Centre (NRSC), ISRO, Department of Space. It provides near real-time information on prevalence, severity level and persistence of agricultural drought at state/district/sub-district level. Currently, it covers 13 states of India, which are predominantly agriculture based and prone to drought situation. Under this agricultural conditions are monitored at state/district levels using daily NOAA AVHRR data. AWiFS (Advanced Wide Field Sensor) of Resourcesat (56 m resolution) is used for detailed assessment of agricultural drought at sub-district level in four states. Fortnightly/monthly report of drought condition is provided to the Government agencies under NADAMS. From the year 2012, the NADAMS project is being implemented by the Mahalanobis National Crop Forecast Centre (MNCFC), Ministry of Agriculture, after the technology was transferred to MNCFC by NRSC.

In NADAMS, maximum value compositing of NDVI/NDWI for the period of 15 days and 1 month is carried out. Drought assessment is done on fortnightly and monthly basis. The assessment of agricultural drought in each district/taluk/block/mandal takes into consideration the following factors.

1. Seasonal NDVI/NDWI progression
2. Comparison of NDVI/NDWI with previous normal years by computing relative deviation and Vegetation Condition Index (VCI)

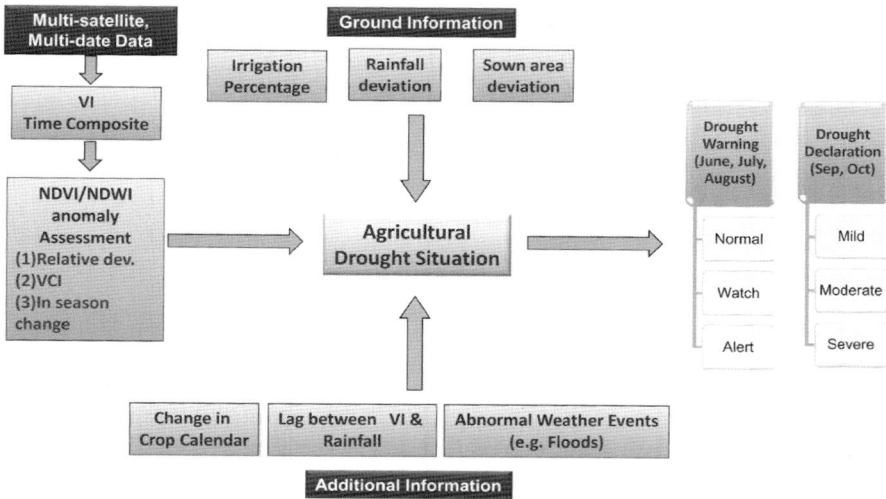

Fig. 1 Approaches for drought assessment under NADAMS programme

3. Weekly rainfall status compared to normal
4. Progression of sown area

The relative NDVI/NDWI deviation from that of normal and rate of progression of NDVI/NDWI from month to month gives the indication about the agricultural situation in the district or sub-district which is then complemented by ground situation as evident from rainfall and sown area. Concept of drought assessment in NADAMS is depicted in Fig. 1.

3 Agricultural Drought Assessment for 2012

3.1 Rainfall Situation in 2012

The monsoon onset over Kerala in 2012 was 4 days later than the normal date. The sluggish advance during the month of June resulted in below normal rainfall (Fig. 2) during the crucial sowing phase of agriculture in general. For the country as a whole seasonal rainfall at the end of south-west monsoon season (June–September) was 820 mm against 887 mm of normal rainfall i.e. deficit of 8 % at the country level (Pai and Bhan 2012). During the season, 13 out of 36 meteorological sub-divisions received deficient rainfall while 41 % of meteorological districts received deficient or scanty rainfall (Fig. 3). The rainfall activity over the country as a whole was below normal in June, slightly below normal in July, normal in August and above normal in September.

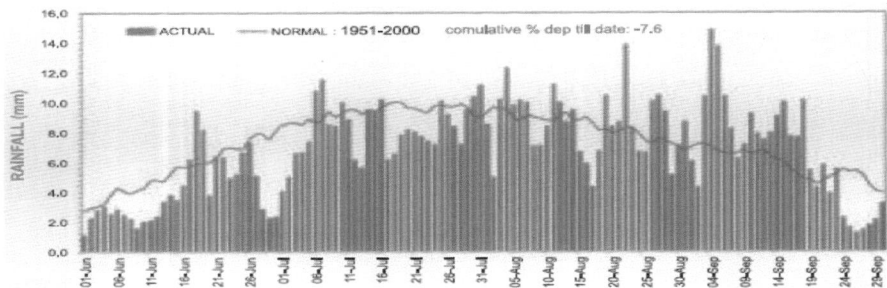

Fig. 2 Daily mean rainfall (mm) over the country as a whole (2012)

During 2012 monsoon season drought in the early part and flood in the latter part of season were witnessed. There was large variability in the dates of advance of monsoon over different regions and also the quantum and distribution of rainfall which had a direct bearing on the sowing operations of kharif and subsequent crops over many regions of the country. Late arrival, slow progress towards northwest and frequent breaks, persistent deficiency till end of July, 3 weeks late withdrawal, cloud bursts and floods in late September were the main characteristics of the 2012 south-west monsoon. Although rainfall deficiency was somewhat made up during second week of July, normal rainfall area had progressed from 10 to 36 % till end July.

Eight meteorological sub-divisions namely west UP; Saurashtra and Kutch; Punjab; Gujarat region; Haryana, Chandigarh and Delhi; Himachal Pradesh; north interior Karnataka and south interior Karnataka were most affected and did not witness more than 5 weeks normal to excess rainfall during this period. However two sub-divisions namely Punjab and Haryana, Chandigarh & Delhi were most seriously affected throughout the entire season with only 2 and 3 weeks witnessing normal to excess rainfall, respectively.

Most of the sowing operations for the kharif crops start in mid June and continue up to mid of August, thus making this period most important for crop production activities. Twelve meteorological sub-divisions namely Saurashtra & Kutch; Haryana, Chandigarh & Delhi; Punjab; west Rajasthan; east Rajasthan; east UP; west UP; Gujarat region; Himachal Pradesh; Madhya Maharashtra; south interior Karnataka and Kerala were most seriously affected.

Out of these 12 sub-divisions (except west Rajasthan which received normal rainfall in the first week i.e. week ending 13th June), in first six meteorological sub-divisions the rainfall was deficient to scanty up to week ending 15th of August when most of the crops are sown. Similarly other sub-divisions also suffered from deficient rains. In some of the sub-divisions where normal rains occurred during second week of August, it again followed a long dry spell thereby affecting the crop sowing.

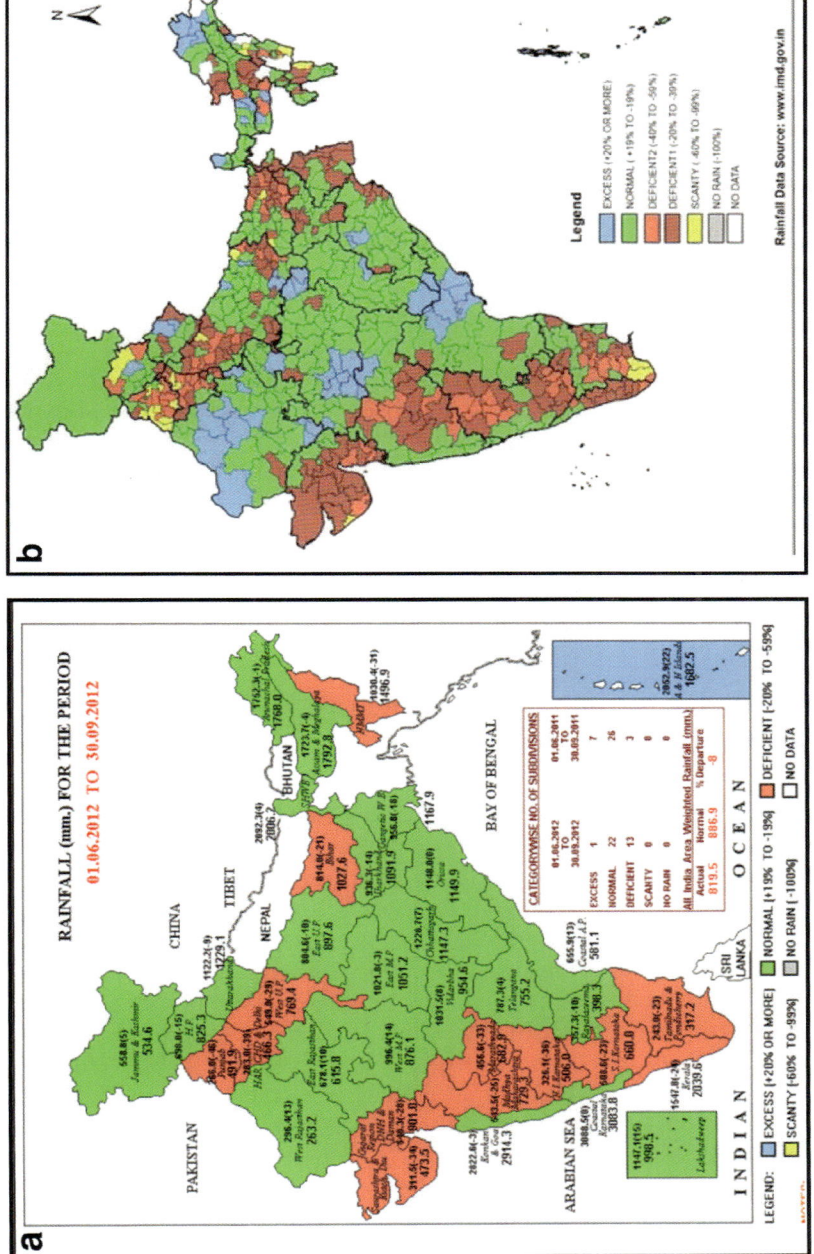

Fig. 3 (**a**) Sub-division-wise rainfall (mm) and deviations (*Source*: IMD) (**b**) District-wise rainfall deviations during June–September (*Source*: MNCFC; *Data Source*: IMD)

3.2 Aridity Anomaly

The bi-weekly Aridity Anomaly Map for 2012 developed by India Meteorological Department is presented in Fig. 4. It showed that in the month of June a major part of the country was under severe and moderate arid condition, because of the delay in monsoon. The condition improved in July and August. However parts of Gujarat, Maharashtra and Rajasthan remained under moderate/severe arid condition.

3.3 Remote Sensing Based Assessments

Based on various remote sensing derived products and meteorological data the monthly agricultural drought assessment was carried out for 2012. The October drought assessment map is presented in Fig. 5 and the number of districts under each state is given in Table 2. As can be seen from the map, the moderate drought condition was prevalent in Rajasthan, Gujarat, Maharashtra, Andhra Pradesh and Karnataka. The total number of districts under moderate condition was 51, while that under mild drought conditions were 43. Mandal level drought assessment had also been carried for Andhra Pradesh state (Fig. 6), which showed 164 mandals in moderate drought condition and 203 mandals in mild drought condition, while the worst hit districts were Anantpur and Kadapa.

4 Operational Agromet Advisory Service for Management of Drought During Monsoon

Integrated Agromet Advisory Service scheme under the aegis of India Meteorological Department (IMD), Ministry of Earth Sciences (MoES) is operating in the country with an objective to serve the farming community at different parts of the country. Agro-meteorological service is a step to contribute to weather information-based crop/livestock management strategies and operations dedicated to enhancing crop production and food security even in extreme weather conditions like drought, flood, cyclone etc. Under this project IMD has started issuing quantitative district level (612 districts) weather forecast upto 5 days from 1st June, 2008. These forecasts are communicated to 130 AgroMet Field Units (AMFUs) located at State Agriculture Universities (SAUs), institutes of Indian Council of Agriculture Research (ICAR) etc. for preparation of district level agromet advisories twice a week.

Advisories based on medium range weather forecast are valid for 5 days. The forewarning of weather helps the farmers to decide the timing of farm operations, e.g., land preparation, tillage, planting, transplanting, thinning, weeding, irrigation,

Agricultural Drought Assessment: Operational Approaches in India

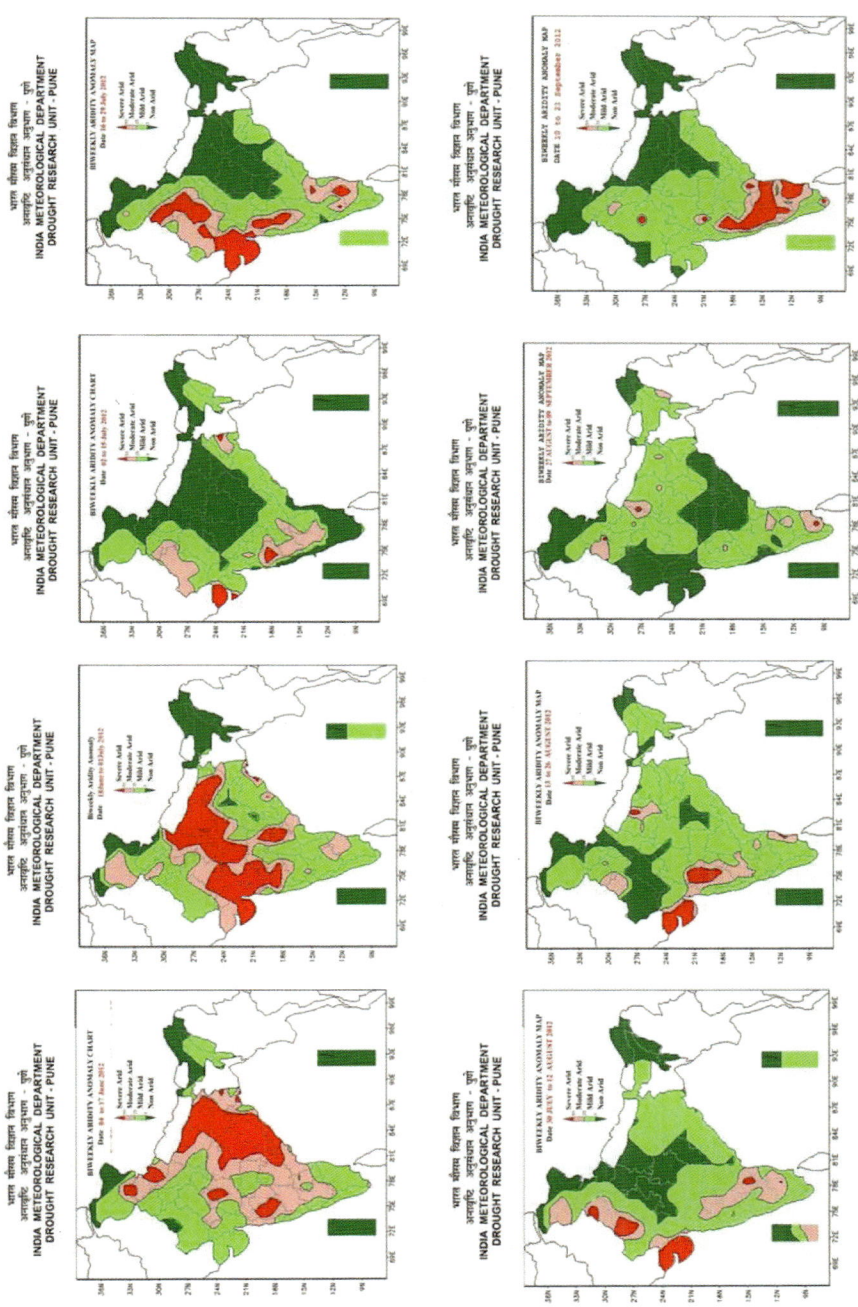

Fig. 4 Bi-weekly aridity anomaly map of 2012 developed by India Meteorological Department

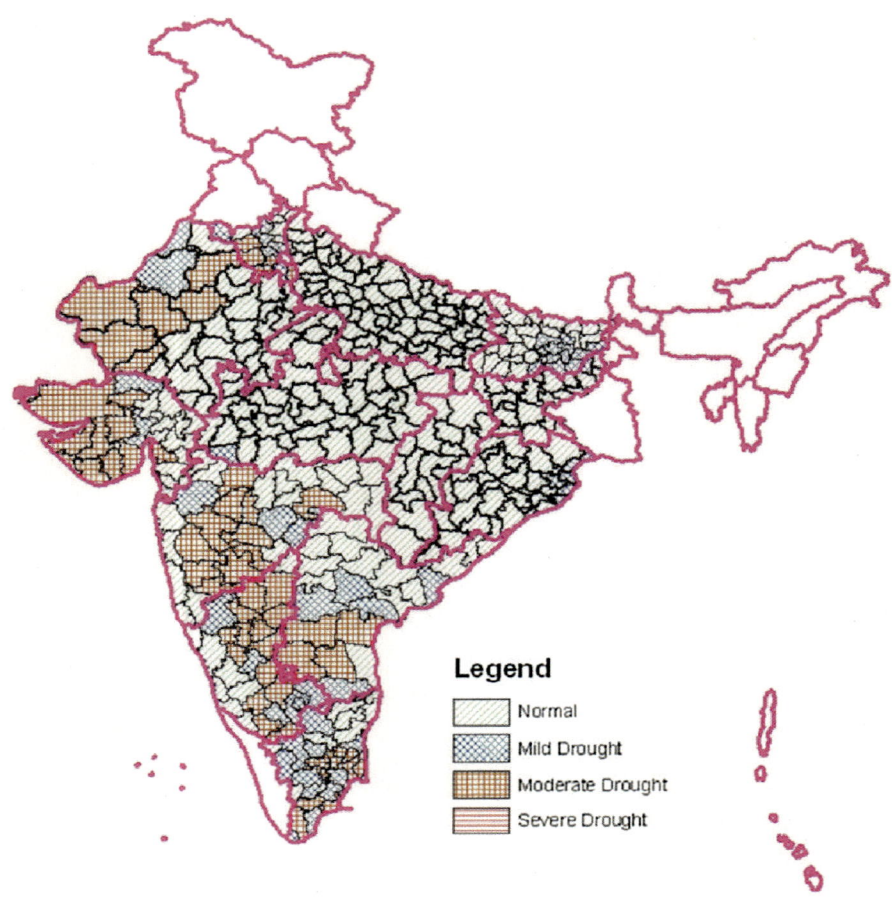

Fig. 5 Agricultural drought assessment for kharif, 2012 carried out under NADAMS programme

Table 2 Number of districts under various agricultural drought conditions during kharif, 2012 as assessed under NADAMS programme

State	Normal	Mild drought	Moderate drought
Andhra Pradesh	13	5	4
Bihar	31	6	0
Chhattisgarh	15	0	0
Gujarat	13	3	9
Haryana	8	5	7
Jharkhand	22	0	0
Karnataka	8	7	15
Maharashtra	18	5	10
Madhya Pradesh	45	0	0
Odisha	30	0	0
Rajasthan	24	2	6
Tamil Nadu	20	10	0
Uttar Pradesh	69	0	0
Total	**316**	**43**	**51**

Agricultural Drought Assessment: Operational Approaches in India 361

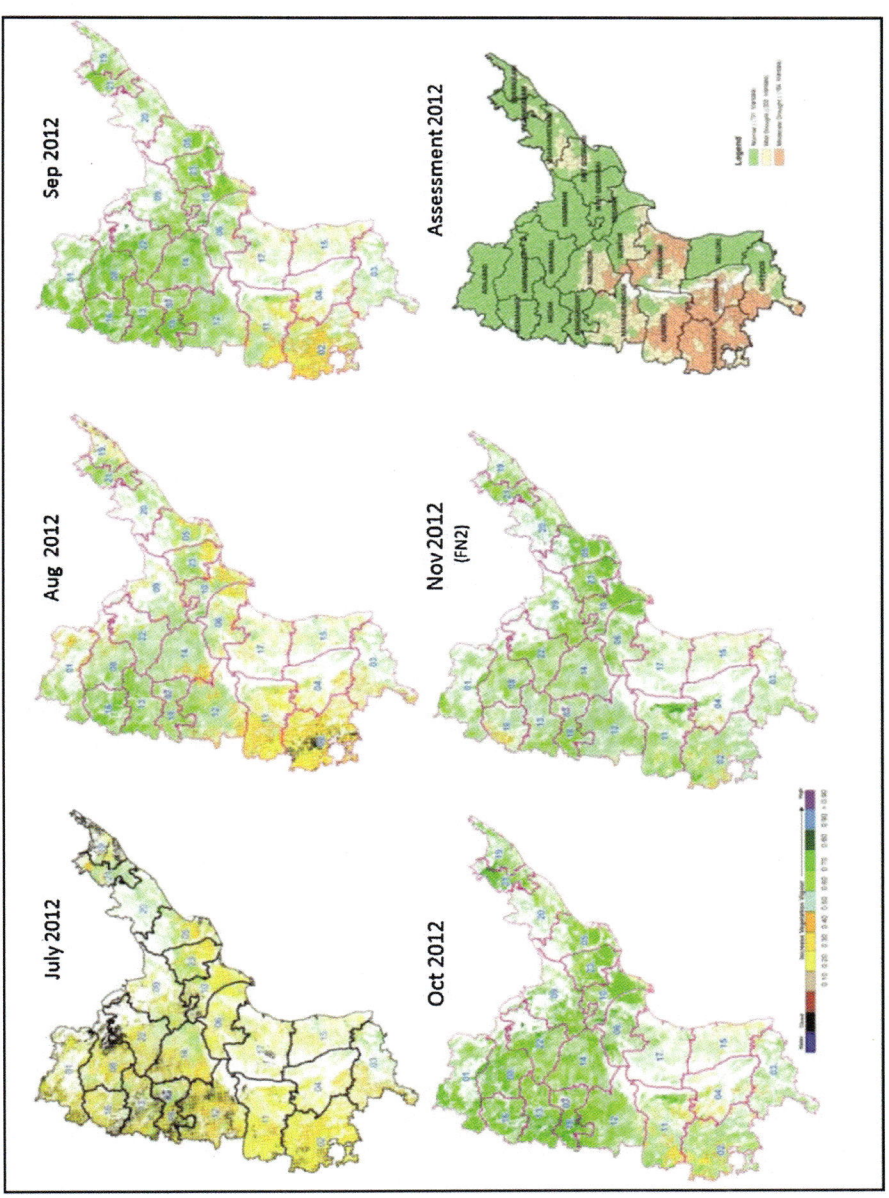

Fig. 6 Mandal-level agricultural drought assessment for Andhra Pradesh state during kharif, 2012 carried out under NADAMS programme

harvesting, application of insecticides, herbicides, fungicides, and fertilizer; type of tillage, depth of planting, density of planting; and choice of crops and crop varieties. Special advisories are being generated under drought condition based on the realised weather, weather forecast and contingent planning specific to the crops at a particular location. Some of the examples of advisories issued under drought conditions in 2002 are given below.

In north interior Karnataka the normal date of onset of monsoon was delayed by 2–6 days and upto 18th July the monsoon was deficient by −37 %. In central dry zone and north transition zone, early sown crops were under moisture stress due to insufficient rainfall. Under such condition, sowing of short duration varieties of tobacco, groundnut, maize, cowpea and sunflower was suggested instead of normal crops. In north dry zone, sowing of contingent crops viz. onion, cucumber, ridge gourd, sesamum, maize, sunflower, niger, castor, bajra, groundnut (spreading), pigeon pea, horse gram or cowpea were recommended. In south interior Karnataka, rainfall deficiency of −50 % was noticed. Crops which were sown earlier (20 % area), during first fortnight of May, were withering and in remaining 80 % of the normal sown area, advisories were communicated for postponing of sowing. Sowing of medium duration varieties like red gram, finger millet, maize and sunflower was suggested instead of groundnut after revival of monsoon. In Maharashtra and Marathawada, in view of deficient rain during earlier part of the season farmers were advised to undertake sowing of *kharif* bajra, tur, sunflower, castor, soybean, sesamum instead of regular kharif crops, with the help of available moisture in the field. In Bihar, farmers were advised to avoid transplanting in upland areas and to undertake sowing of til (var. Krishna), intercropping of maize (Suwan) + urd (T-9, Pant U-31, Pant U-19) instead of that. Also transplanting of short duration varieties in medium land and medium duration varieties in low land situation were suggested. In Gujarat, due to delay in onset by 7–26 days, deficient rainfall was received in Gujarat region (−49 %) and Saurashtra and Kutch (−72 %). In north Gujarat zone, contingency measures like adoption of short duration varieties or early maturing varieties of guar, cluster bean, cow pea, mung and green gram were suggested to farmers. In Saurashtra and Kutch, farmers were suggested to undertake rice cultivation with SRI techniques with irrigation facilities and select short duration varieties of castor and sesame and fodder crops like forage maize. In semiarid eastern plain zone of Rajasthan, sowing of short duration varieties of oilseed and pulses was suggested in place of cereals. In Haryana, as a contingency measure, sowing of short duration varieties of bajra, sorghum, guar, green gram, urd, arhar and moth was suggested. In west Rajasthan farmers were advised to undertake sowing of fodder bajra, cluster beans and pulses in place of cereals with receipt of rainfall.

Conclusion

Drought is a major disaster for India, considering majority of Indian agriculture is rainfed. Because of the complexities in its initiation and impact, it is difficult to characterize drought with a single indicator/index. It has been tried to study drought using large number of agro-meteorological and remote sensing based indices. 2012 was a rainfall deficient year. The assessment carried out under NADAMS programme (using an integrated approach) and the India Meteorological Department (using Aridity Anomaly Index) showed a good agreement in the overall agricultural drought situation and had highlighted the areas under various agricultural drought conditions. India being a vast country with diverse climate, in general, the country experiences drought every season in some part of the country. The other two recent years that India faced the agricultural drought were 2002 and 2009, which experienced moisture stress during the sowing period. Different variants of drought have different impacts on crop growth, ultimately impacting the food security of the nation. Hence, there is a need to study the drought in an integrated manner and use a composite index for drought assessment.

Acknowledgements Authors are grateful to Sh. Sanjeev Gupta, JS(IT), DAC, Dr. V.K. Dadhwal, Director, NRSC and Dr. L.S. Rathore, Director General, IMD for their kind support. The authors wish to thank Smt. S. Mamatha and Shri Karan Choudhary of MNCFC for providing inputs for this article.

References

DAC (2009) Manual for drought management. Department of Agriculture & Cooperation, New Delhi

Gao BC (1996) NDWI—a normalized difference water index for remote sensing of vegetation liquid water from space. Remote Sens Environ 58:257–266

Hardisky MA, Klemas V, Smart RM (1983) The influences of soil salinity, growth form, and leaf moisture on the spectral reflectance of Spartina alterniflora canopies. Photogramm Eng Remote Sens 49:77–83

Jackson RD, Reginato RJ, Idso SB (1977) Wheat canopy temperature: a practical tool for evaluating water requirements. Water Resour Res 13:651–656

Khanna S, Palacios-Orueta A, Whiting ML, Ustin SL, Riaño D, Litago J (2007) Development of angle indexes for soil moisture estimation, dry matter detection and land-cover discrimination. Remote Sens Environ 109:154–165

Kogan FN (1995) Drought of late 1980s in the USA as derived from NOAA polar orbiting satellite data. Bull Am Meteorol Soc 76:655–668

Kogan FN (2001) Operational space technology for global vegetation assessment. Bull Am Meteorol Soc 82(9):1949–1964

Minamiguchi N. (2005) The application of Geospatial and Disaster Information for food insecurity and agricultural drought monitoring and assessment by the FAQ GIEWS and Asia FIVIMS. Workshop on reducing food insecurity associated with natural disaster in Asia and Pacific, Bangkok, Thailand

Moran MS, Clarke TR, Inoue Y, Vidal A (1994) Estimating crop water deficit using the relation between surface-air temperature and spectral vegetation index. Remote Sens Environ 49:246–263

Murthy CS, Sesha Sai MVR (2011) Agricultural drought monitoring and assessment. In: Roy PS, Dwivedi RS, Vijayan D (eds) Remote sensing applications. NRSC, Hyderabad

Murthy CS, Sesha Sai MVR, Bhanuja Kumari V, Prakash VS, Roy PS (2007) Agricultural drought assessment at disaggregated level using AWiFS/WiFS data of Indian Remote sensing satellites. Geocarto Int 22:127–140

Pai DS, Bhan SC (2012) Monsoon 2012: a report. India Meteorological Department, New Delhi

Palacios-Orueta A, Khanna S, Litago J, Whiting ML, Ustin SL (2006) Assessment of NDVI and NDWI spectral indices using MODIS time series analysis and development of a new spectral index based on MODIS shortwave infrared bands. In: Proceedings of the 1st international conference on Remote Sensing and Geoinformation Processing. Trier, Germany, http://ubt.opus.hbz-nrw.de/volltexte/2006/362/pdf/03-rgldd-session2.pdf

Palmer WC (1965) Meteorological drought. Research paper no. 45. U.S. Weather Bureau, Washington, DC

Thenkabail PS, Gamage MSDN, Smakhtin VU (2004) The use of remote sensing data for drought assessment and monitoring in Southwest Asia. Research report 85. International Water Management Institute, Colombo

Thornthwaite CW, Mather JR (1955) The water balance. Clim Lab Climatol 8(1):1–104

Trends in Extreme Temperature Events over India During 1969–2012

A.K. Jaswal, Ajit Tyagi, and S.C. Bhan

1 Introduction

Intergovernmental Panel on Climate Change (IPCC) in their recent report has stated that effects of climate change will lead to more extreme temperatures and more hot days than cold days (IPCC 2013). It has now become increasingly evident that one of the major impacts of global warming on the earth's surface will be in the form of increasing frequency and intensity of extreme weather events. According to IPCC (2007), the global mean surface temperatures have increased by 0.76 ± 0.18 °C over the last 100 years (1906–2005). A great number of research papers focusing on extreme temperature on global, regional and national scales have been written by many researchers worldwide (e.g. Zhai and Pan 2003; Griffiths et al. 2005; Klein Tank et al. 2006). The percentile-based temperature indices have been used to analyze changes in temperature extremes for various parts of the world (e.g. Peterson et al. 2002; Vincent et al. 2005; Zhang et al. 2005; Klein Tank et al. 2006; Alexander et al. 2006). Tebaldi et al. (2006) have shown that the twenty-first century would bring global changes in temperature extremes consistent with a warming climate.

Regional studies of extreme temperature variations for India are very few. Kothawale (2005) studied the temperature extremes in India for the period 1970–2002 and noted that the number of hot days is maximum over central part of India and minimum along the west coast of India during the summer season. Rao et al. (2005) have reported that 80 % stations in peninsular India and 40 % stations

---

A.K. Jaswal (✉)
India Meteorological Department, Shivajinagar, Pune 411005, India
e-mail: jaswal4@gmail.com

A. Tyagi
Ministry of Earth Sciences, Lodhi Road, New Delhi 110003, India

S.C. Bhan
India Meteorological Department, Lodhi Road, New Delhi 110003, India

© Capital Publishing Company 2015
K. Ray et al. (eds.), *High-Impact Weather Events over the SAARC Region*,
DOI 10.1007/978-3-319-10217-7_25

in northern India showed increasing trend in the days with critical extreme maximum temperature while about 80 % of the stations in northern India showed increasing trend in the extremes in night temperatures during 1971–2000. Kothawale et al. (2010) have found widespread increasing trend in the frequency of occurrence of hot days and hot nights and widespread decreasing trend in those of cold days and cold nights in pre-monsoon season. Revadekar et al. (2012) have found widespread warming with increase in intensity and frequency of hot events and decrease in frequency of cold events in India. The main purpose of this study is to analyze monthly temperature extremes over India during summer and winter seasons using daily temperature data for 1969–2012.

2 Data and Methodology

2.1 Methodology

Available literature shows that there are usually two criteria to define high temperature days (DeGaetano and Allen 2002; Zhai and Pan 2003). One is relative, using temperature percentiles and the other is absolute temperature values. However, temperature percentile indices have now been used widely in many regional studies world over. Considering the serious effects on human health caused by high temperatures, we have used both criteria in defining extreme temperature events (number of days) in this study. In relative sense, we have worked out extreme temperature events on the basis of percentile threshold of long-period data (1971–2000) of each station by comparing daily maximum temperature (D_{max}) and daily minimum temperature (D_{min}) of the station. On the basis of percentiles and absolute values, the three relative/absolute hot events are hot days (HD) ($D_{max}>$90th percentile; $D_{max}>$35 °C, respectively), very hot days (VHD) ($D_{max}>$95th percentile; $D_{max}>$40 °C, respectively) and extremely hot days (EHD) ($D_{max}>$98th percentile; $D_{max}>$43 °C, respectively), while the three relative/absolute cold events are cold nights (CN) ($D_{min}>$10th percentile; $D_{min}<$10 °C, respectively), very cold nights (VCN) ($D_{min}<$5th percentile; $D_{min}<$5 °C, respectively) and extremely cold nights (ECN) ($D_{min}<$2nd percentile; $D_{min}<$2 °C, respectively). The percentile and absolute value based data series of HD, VHD, EHD and CN, VCN, ECN are prepared for all 227 stations for summer and winter months respectively for the period 1969–2012 which are further analyzed.

2.2 Data Used

The data used in this study are the daily surface maximum temperatures and minimum temperature obtained from the database of India Meteorological Department (IMD), National Data Centre (NDC) located at Pune, where all climatological data of India are processed, quality checked and archived. Since daily values of maximum temperature are available from 1969 only, the period of study is restricted to 1969–2012. Initially we have selected more than 300 surface meteorological stations from IMD's network, which are having long data series of daily maximum and minimum temperatures. First we checked the availability of daily maximum temperature values for each year and stations with too many continuous missing values were dropped from the selected data. Finally, a dataset belonging to 227 meteorological stations covering most of the climate types found in the country was selected for monthly extreme temperature analysis in India. The geographical coverage of stations utilized in this study is shown in Fig. 1.

As the high and low temperature events that can cause large-scale discomfort to human being occur mainly in summer and winter seasons, we have conducted our study for hot days in March, April, May and June months and cold nights in December, January and February months only. Therefore for each station, data series of relative and absolute HD, VHD and EHD for March, April, May and June

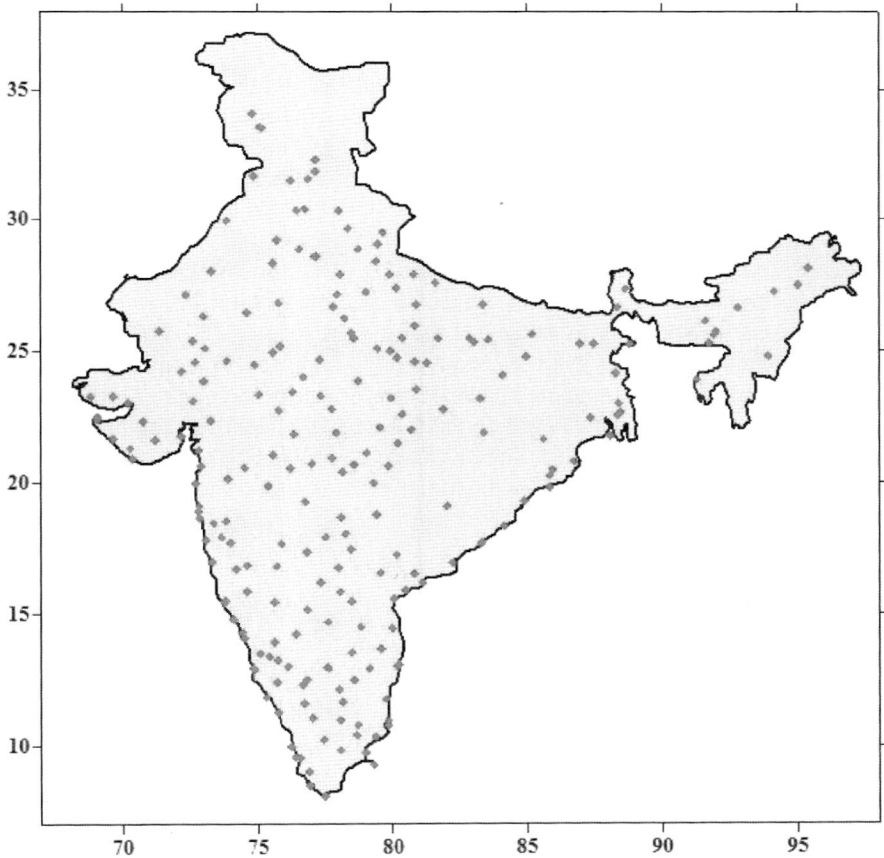

Fig. 1 Geographical locations of 227 surface meteorological stations used in the study

months and CN, VCN and ECN for December, January and February months were prepared for the period of 1969–2012. In order to investigate trends in monthly number of relative/absolute HD, VHD, EHD, CN, VCN and ECN, a linear regression was fitted to each time series by using a least square regression for each station. Two-sided Student's *t*-tests were used to check the statistical significance levels at 95 % level of significance. Statistics of numbers of stations showing positive or negative trends in relative/absolute HD, VHD, EHD, CN, VCN and ECN is given in Tables 1 and 2. The spatial patterns of linear trends are shown in Figs. 2, 3, 4, 5, 6, 7, and 8.

3 Results and Discussion

3.1 Number of Hot Days Trends in Summer Months

3.1.1 March

The trend results for the extreme temperature indices of 227 stations in this study indicate overall increase in HD, VHD and EHD over India during 1969–2012. The number of stations showing positive trends in HD, VHD and EHD are 149, 159 and 166 respectively out of which almost 40 % are statistically significant at 95 % level (Table 1a). The percentile based HD, VHD and EHD trends are coherently significantly positive over extreme north, west, south and northeast India as shown in Fig. 2a, c, and e. The trends in percentile based HD, VHD and EHD are significantly negative over Indo-Gangetic plains and east India. Out of total 227 stations under study, the number of stations having absolute value based HD, VHD and EHD trends in March are 209, 136 and 29, respectively (Table 1b). Stations having absolute value based positive trends in HD, VHD and EHD for the month of March are 136, 96 and 20, respectively. Absolute value based HD and VHD are also having similar trends as percentile based. Stations having absolute EHD are very few as not many stations report $D_{max} > 43$ °C in the month of March as shown in Fig. 2f.

3.1.2 April

The trend analysis of both percentile as well as absolute value based data series of April month indicates general increase in HD, VHD and EHD over India during 1969–2012. Out of 227 stations analyzed in this study, number of stations showing positive trends in HD, VHD and EHD are 155, 146 and 145 respectively out of which almost 35 % are statistically significant at 95 % level (Table 1a). The percentile based HD, VHD and EHD trends are coherently significantly positive over almost entire country except east where many stations are having negative trends as shown in Fig. 3a, c, and e. The trends in percentile based HD, VHD and EHD are significantly positive over coastal stations particularly on the west coast of India

Table 1 Number of stations having positive or negative trends in monthly hot days (HD), very hot days (VHD) and extremely hot days (EHD) based on percentiles and absolute methods for March to June over India during 1969–2012

	March			April			May			June		
	HD	VHD	EHD	HD	VHD	EHD	HD	VHD	EHD	HD	VHD	EHD
(a) Number of stations based on percentiles												
Positive	149	159	166	155	146	145	150	149	149	171	175	181
Positive significantly	67	62	58	61	55	42	62	58	47	45	47	34
Negative	78	68	61	72	81	82	77	78	78	56	52	46
Negative significantly	19	17	12	12	13	13	14	13	11	7	10	4
(b) Number of stations based on absolute values criterion												
Positive	136	96	20	118	102	82	133	95	91	123	103	95
Positive significantly	43	24	4	32	29	23	30	13	17	21	12	15
Negative	73	40	9	97	64	45	82	76	54	92	58	35
Negative significantly	14	5	1	18	10	7	12	11	11	8	3	2

Table 2 Number of stations having positive or negative trends in monthly cold nights (CN), very cold nights (VCN) and extremely cold nights (ECN) based on percentiles and absolute methods for December, January and February over India during 1969–2012

	December			January			February		
	CN	VCN	ECN	CN	VCN	ECN	CN	VCN	ECN
(a) Number of stations based on percentiles									
Positive	73	76	72	80	95	103	64	68	79
Positive significantly	14	20	26	27	26	28	14	14	11
Negative	154	151	155	147	132	124	163	159	148
Negative significantly	69	68	60	63	53	40	78	66	51
(b) Number of stations based on absolute values criterion									
Positive	47	38	22	60	43	40	47	19	16
Positive significantly	12	10	6	12	10	7	4	5	2
Negative	116	71	34	106	73	40	110	82	45
Negative significantly	52	22	7	31	25	15	58	36	13

where trend values are higher at many stations. Out of total 227 stations under study, the number of stations having absolute value based HD, VHD and EHD trends in April are 215, 166 and 127, respectively (Table 1b). Stations having absolute value based positive trends in HD, VHD and EHD for the month of April are 118, 102 and 82, respectively. Patterns of absolute value based HD and VHD trends for April are similar to percentile based trends except over south India where many stations are having opposite trends (negative) as compared to percentile based trends (Figs 3b, d).

3.1.3 May

In the month of May, trend analysis of both percentile as well as absolute value based data series indicates overall increase in HD, VHD and EHD over India during 1969–2012 except over Indo-Gangetic plains. The number of stations showing positive trends in HD, VHD and EHD are 150, 149 and 149 respectively out of which almost 35 % are statistically significant at 95 % level (Table 1a). The percentile based HD, VHD and EHD trends are coherently significantly positive over almost entire country except Indo-Gangetic plains where many stations are having significant negative trends as shown in Fig. 4a, c, and e. The trends in percentile based HD, VHD and EHD are significantly positive over coastal stations particularly on the west coast of India where trend values are higher at many stations. Out of total 227 stations under study, the numbers of stations having absolute value based HD, VHD and EHD trends in May are 215, 171 and 145, respectively (Table 1b). Stations having absolute value based positive trends in HD, VHD and EHD for the month of May are 133, 95 and 91, respectively. The patterns of absolute value based HD, VHD and EHD trends are similar to the percentile based trends except over south

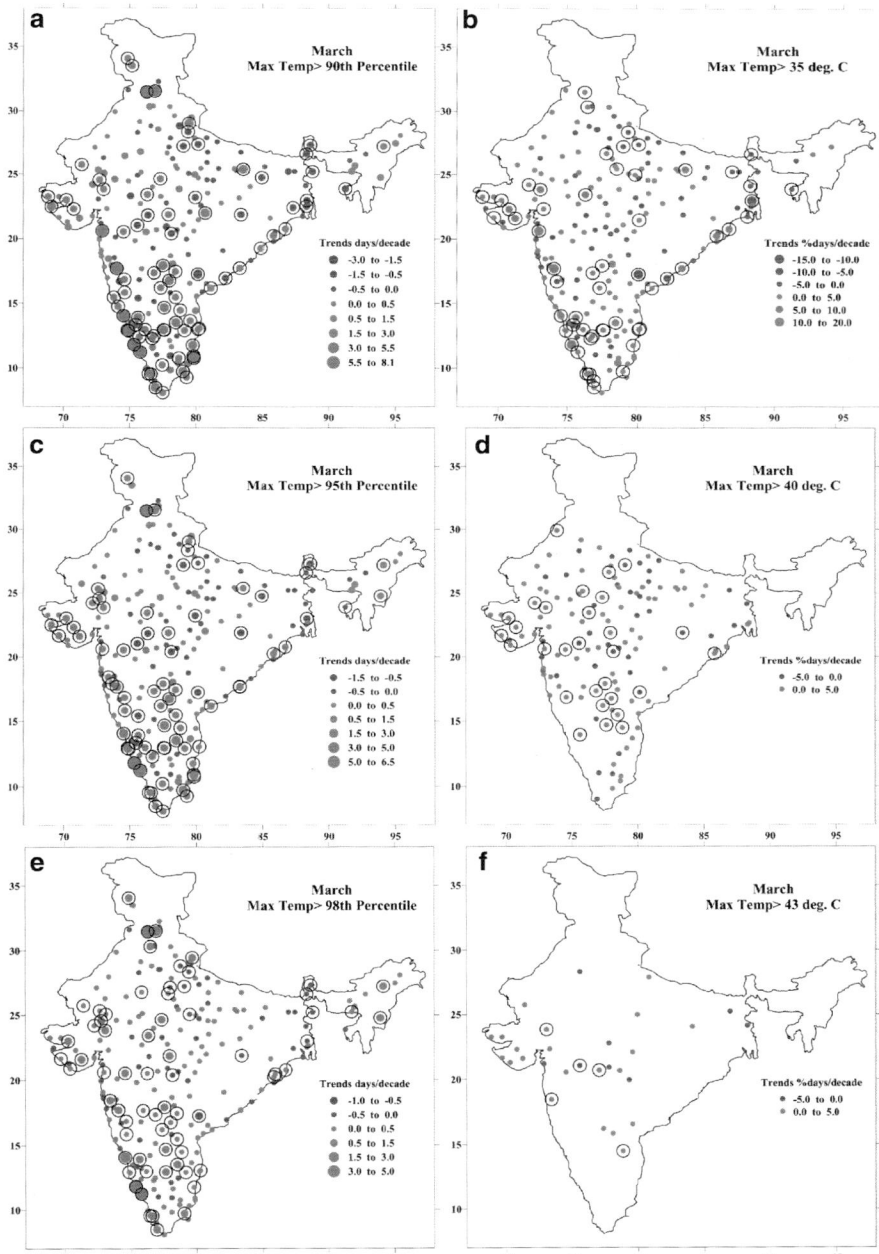

Fig. 2 Spatial distribution of trends in frequencies of daily maximum temperature above 90th, 95th and 98th percentile and daily maximum temperature above 35, 40 and 43 °C in the month of March during the period 1969–2012

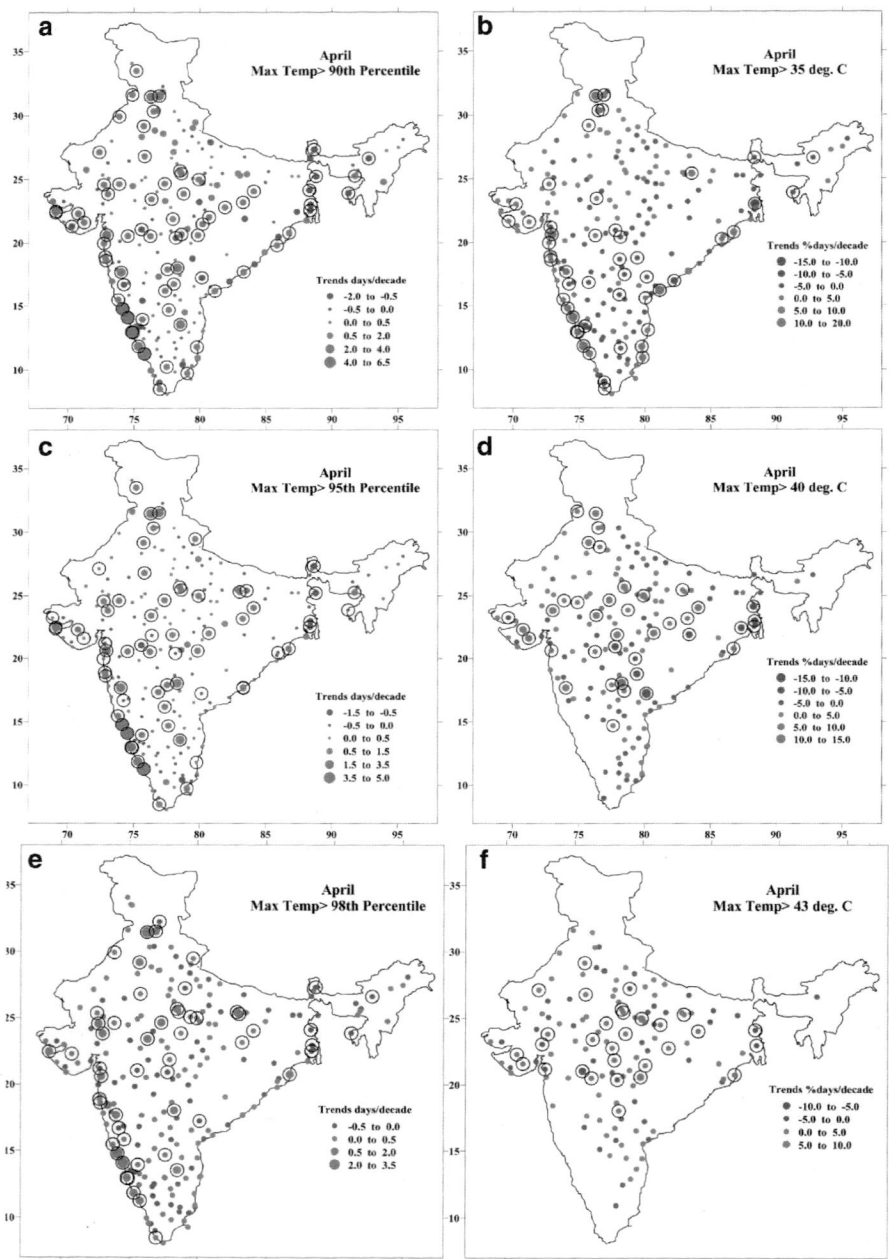

Fig. 3 Spatial distribution of trends in frequencies of daily maximum temperature above 90th, 95th and 98th percentile and daily maximum temperature above 35, 40 and 43 °C in the month of April during the period 1969–2012

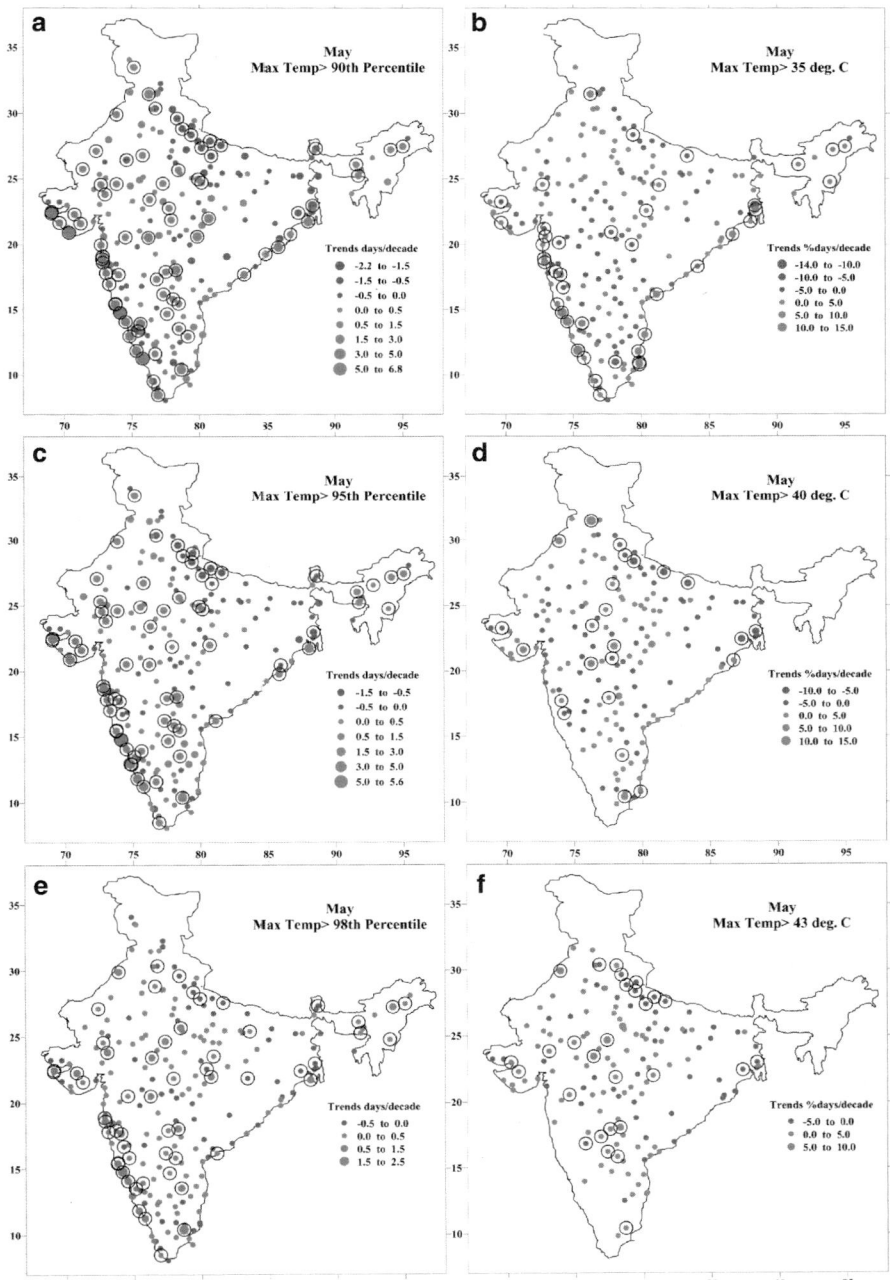

Fig. 4 Spatial distribution of trends in frequencies of daily maximum temperature above 90th, 95th and 98th percentile and daily maximum temperature above 35, 40 and 43 °C in the month of May during the period 1969–2012

India where many stations are having opposite trends (negative) as compared to the percentile based trends as shown in Fig. 4b, d, and f.

3.1.4 June

The trend analysis of both percentile as well as absolute value based data series of HD, VHD and EHD in the month of June indicates overall increase over many regions during 1969–2012 except over Indo Gangetic plains and a few pockets along east coast of India. The number of stations showing positive trends in percentile based HD, VHD and EHD are 171, 175 and 181 respectively out of which almost 30 % are statistically significant at 95 % level (Table 1a). The percentile based HD, VHD and EHD trends are coherently significantly positive over almost entire country except over some pockets along east coast and over Indo Gangetic plains where many stations are having significant negative trends as shown in Fig. 5a, c, and e. The trends in percentile based HD, VHD and EHD are significantly positive over west coast of India where trend values are also higher at many stations. Out of total 227 stations under study, the number of stations having absolute value based HD, VHD and EHD trends in June are 215, 161 and 130, respectively (Table 1b). Stations having positive trends in absolute value based HD, VHD and EHD for the month of June are 123, 103 and 95, respectively. The patterns of absolute value based HD, VHD and EHD trends are similar to the percentile based trends except over central India where many stations are having opposite trends (negative) in absolute value based HD as compared to the percentile based as shown in Fig. 5b.

3.2 Number of Cold Nights Trends in Winter Months

3.2.1 December

The trend results for the extreme temperature indices of 227 stations in this study indicate overall decrease in CN, VCN and ECN over India during 1969–2012 suggesting warming in night temperature. The number of stations showing negative trends in percentile based CN, VCN and ECN are 154, 151 and 155 respectively out of which almost 40 % are statistically significant at 95 % level (Table 2a). The percentile based CN, VCN and ECN trends are coherently significantly negative over the country as shown in Fig. 6a, c, and e. The magnitude of negative trends in percentile based CN, VCN and ECN are higher over the coastal stations suggesting large scale warming of night temperatures in December. Out of total 227 stations under study, the number of stations having absolute value based CN, VCN and ECN trends in December are 163, 109 and 56, respectively (Table 2b). Stations having absolute value based negative trends in CN, VCN and ECN for the month of December are 116, 71 and 34, respectively. Spatial patterns of absolute value based CN and VCN trends (Fig. 6b, d) are similar to percentile based trends (Fig. 6a, c).

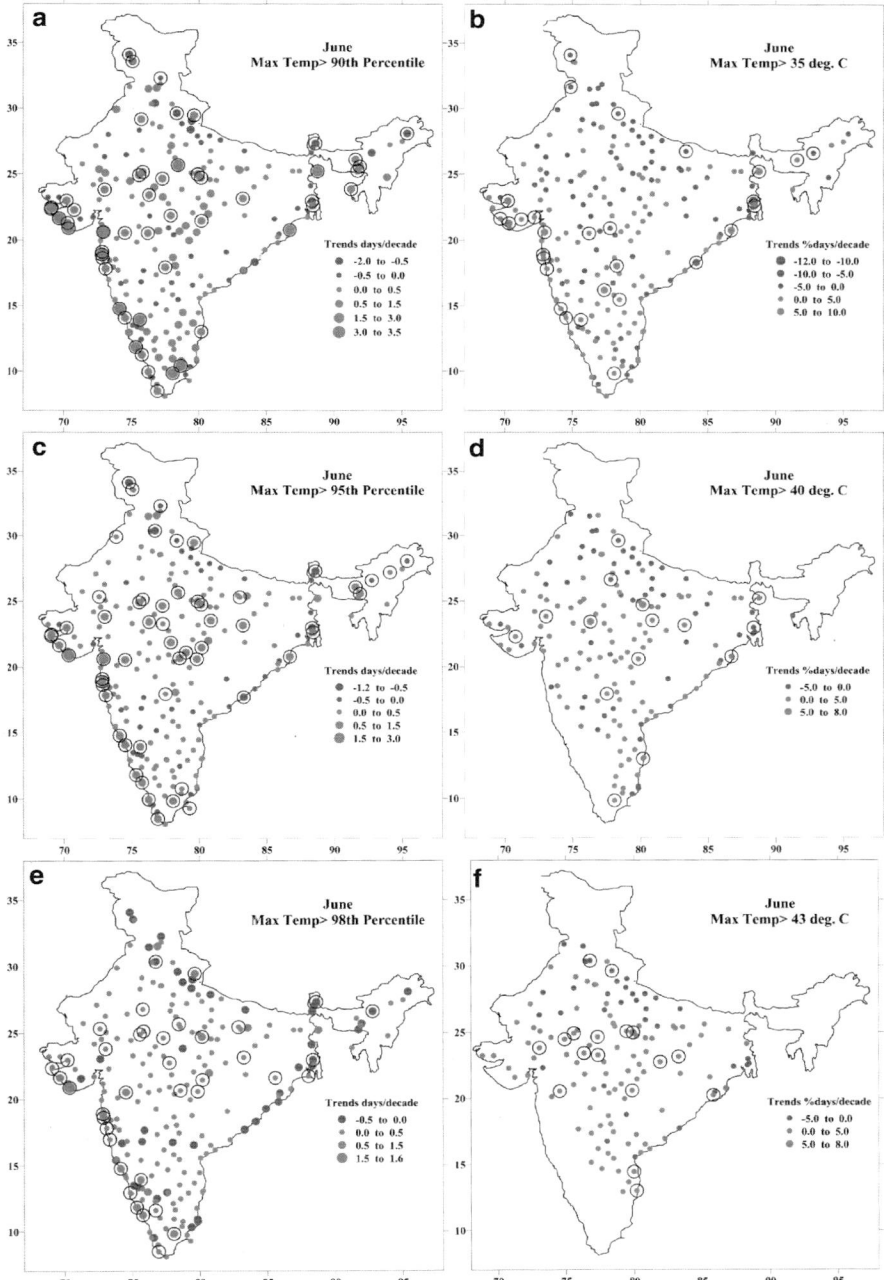

Fig. 5 Spatial distribution of trends in frequencies of daily maximum temperature above 90th, 95th and 98th percentile and daily maximum temperature above 35, 40 and 43 °C in the month of June during the period 1969–2012

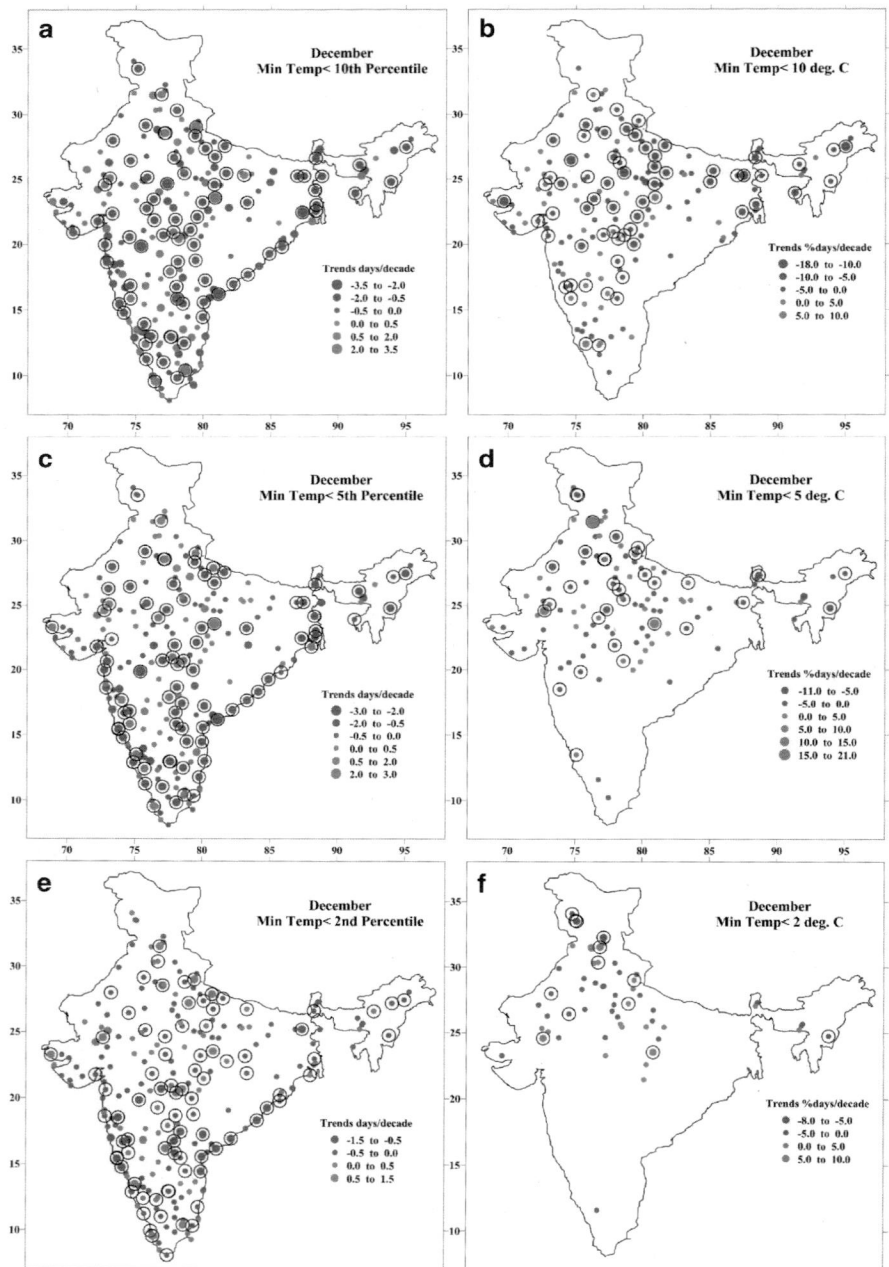

Fig. 6 Spatial distribution of trends in frequencies of daily minimum temperature below 10th, 5th and 2nd percentile and daily minimum temperature below 10, 5 and 2 °C in the month of December during the period 1969–2012

Though stations having absolute value based ECN trends are very few in December (many stations do not report $D_{min} < 2$ °C), the spatial patterns indicate warming in night temperature over north and northwest India as shown in Fig. 6f.

3.2.2 January

The results of trend analysis of CN, VCN and ECN data series of January are mixed. There are pockets of warming and cooling in night temperature over India during 1969–2012. Out of 227 stations, numbers of stations showing negative trends in percentile based CN, VCN and ECN are 147, 132 and 124 respectively out of which almost 40 % are statistically significant at 95 % level (Table 2a). The spatial patterns of percentile based CN, VCN and ECN trends show pockets of significantly negative as well as pockets of significantly positive trends over the country as shown in Fig. 7a, c, and e. The magnitude of negative trends in percentile based CN, VCN and ECN are higher over the coastal stations as well as in south India suggesting strong warming of night temperatures in January. Out of 227 stations under study, the number of stations having absolute value based CN, VCN and ECN trends in January are 166, 116 and 80, respectively (Table 2b). Stations having absolute value based negative trends in CN, VCN and ECN for the month of January are 106, 73 and 40, respectively. The geographical patterns of absolute value based CN, VCN and ECN trends are similar to percentile based trends.

3.2.3 February

The trend analysis of extreme temperature indices of 227 stations for the month of February indicates overall decrease in CN, VCN and ECN over north India and increase/decrease in south India during the study period 1969–2012. The number of stations showing negative trends in percentile based CN, VCN and ECN are 163, 159 and 148, respectively out of which almost 45 % are statistically significant at 95 % level (Table 2a). The spatial patterns of percentile based CN, VCN and ECN trends are coherently significantly negative over north India and positive over south India as shown in Fig. 8a, c, and e. The magnitude of negative trends in percentile based CN, VCN and ECN are higher over north India suggesting large scale warming of night temperatures in February which agrees with the earlier reported study by Jaswal (2010). Out of total 227 stations under study, the number of stations having absolute value based CN, VCN and ECN trends in February are 157, 101 and 61, respectively (Table 2b). Stations having absolute value based negative trends in CN, VCN and ECN for February are 110, 82 and 45, respectively. Spatial patterns of absolute value based CN and VCN trends are similar to percentile based trends.

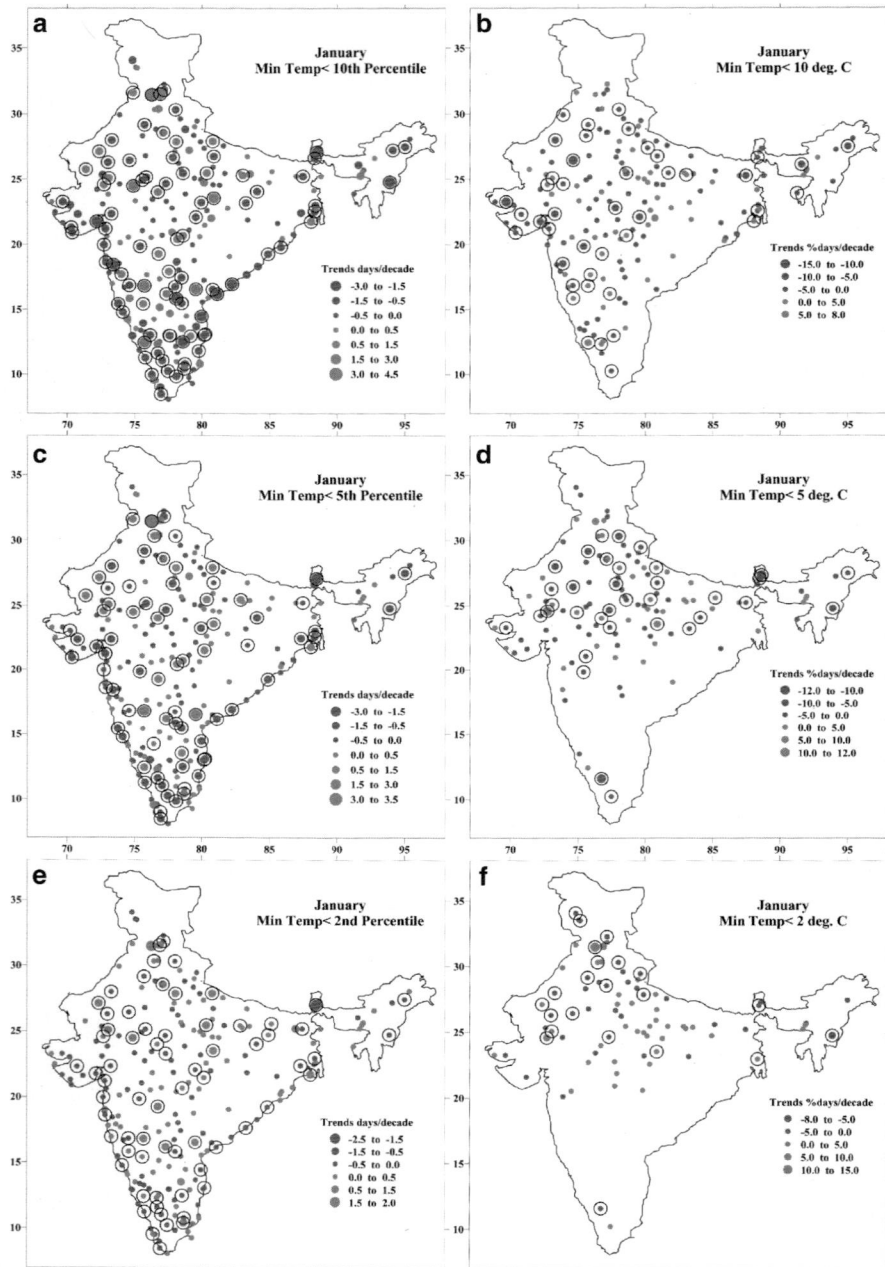

Fig. 7 Spatial distribution of trends in frequencies of daily minimum temperature below 10th, 5th and 2nd percentile and daily minimum temperature below 10, 5 and 2 °C in the month of January during the period 1969–2012

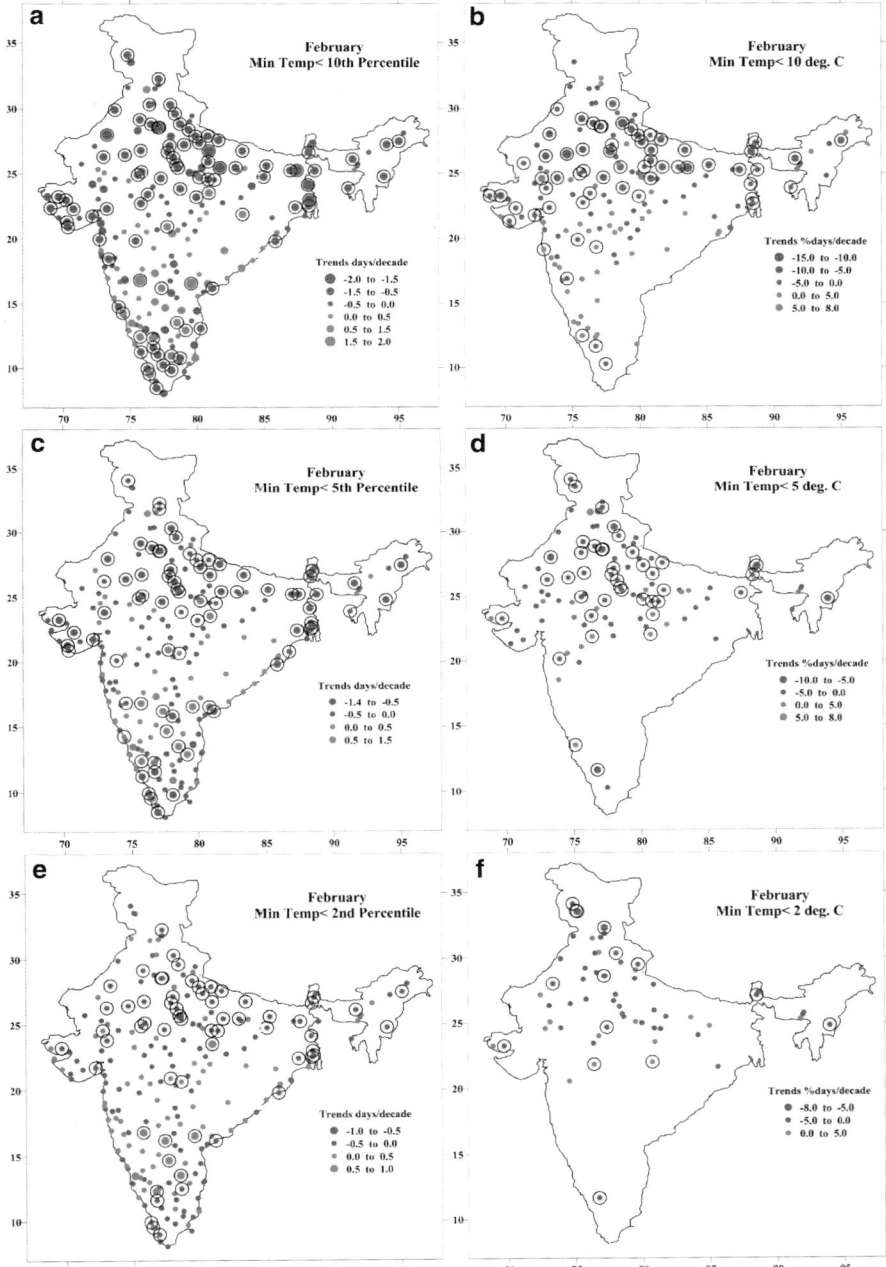

Fig. 8 Spatial distribution of trends in frequencies of daily minimum temperature below 10th, 5th and 2nd percentile and daily minimum temperature below 10, 5 and 2 °C in the month of February during the period 1969–2012

4 Discussion

This study is undertaken in the backdrop of recent IPCC report stating effects of climate change that lead to more extreme temperatures and more hot days than cold days (IPCC 2013). Unusually hot or cold temperatures can result in prolonged extreme weather events like summer heat waves or winter cold spells. Heat waves can lead to illness and death, particularly among older adults, the very young and other vulnerable groups. The impacts of most extremes are typically felt at a local or regional scale and so regional studies of climate extremes are important for assessing potential climate impacts. The great economic and social impacts of extreme events support increasing attention to studies on such issues. Therefore, in this study we have examined the statistics and behaviour of monthly extreme climate events based on percentile as well as absolute value methods over India for summer and winter months for the period of 1969–2012. The data analysis shows overall widespread increase in HD, VHD and EHD for March over India during 1969–2012 but more coherently over west and south India. In the month of April, positive trends in HD, VHD and EHD are well spread over the country but more over north, west, central India and also along the west coast. Our results are in line with earlier studies by Kothawale (2005), Jaswal (2010), Kothawale et al. (2010) and Revadekar et al. (2012). The geographical patterns of HD, VHD and EHD trends are positive almost all over the country in May and June except over the Indo-Gangetic plains where the trends are negative. The data analysis of winter months CN, VCN and ECN for December and January shows general negative trends over the country during 1969–2012 suggesting increase in night temperature.

The geographical distribution of CN, VCN and ECN trends for February indicates more coherent significantly negative trends in northern half of the country while the trends in southern half are mixed. Possible reasons for having such patterns might be due to global warming and/or urbanization which need further investigation. Number of stations showing positive trends in HD, VHD and EHD are above 65 % in most of the summer months both on relative and absolute basis. The highest number of stations with positive trends is for EHD in June with 80 % stations on relative basis and 73 % stations on absolute basis suggesting strong daytime warming over the country which may be due to late onset of monsoon over the country. Similarly, number of stations showing negative trends in CN, VCN and ECN are above 65 % in most of winter months both on relative and absolute basis. February month is having highest number of stations with negative trends for CN on relative basis (72 %) and for VCN on absolute basis (81 %) suggesting strong warming in night temperatures which is consistent with the study by Jaswal (2010). The positive trends in summer months hot days and negative trends in winter months cold nights over India during the period of study corroborates well with the projections given by IPCC (2007) and potential future assessment in changes in climate extremes given by Tebaldi et al. (2006).

Conclusions

An attempt has been made in the present study to analyse observed summer and winter months' temperature extremes over India during 1969–2012. Following are the conclusions of this study:

(a) On all-India scale, hot, very hot and extremely hot days are increasing in all summer months suggesting hot days have become more common now.
(b) Geographical distribution of trends suggests significant increase in hot days in north India in April and south India in March. Also hot, very hot and extremely hot days are significantly increasing over west coast of India in all summer months.
(c) There is significant decrease in hot, very hot and extremely hot days over Indo-Gangetic plains in the month of May.
(d) Stations having extremely hot days are the highest in June when 80 % of the total number of stations are having positive trends which may be due to erratic behaviour of onset of monsoon over India.
(e) All-India trend analysis of cold, very cold and extremely cold nights suggests significant decreasing trend in December, mixed trend in January while in February, there is significant decreasing trend in north India and mixed trend in south India.

Acknowledgements Authors are thankful to the reviewers for suggesting constructive suggestions that helped to improve the manuscript.

References

Alexander L, Zhang X, Peterson TC et al (2006) Global observed changes in daily climate extremes of temperature and precipitation. J Geophys Res 111:D05109. doi:10.1029/2005JD006290

DeGaetano AT, Allen RJ (2002) Trends in twentieth-century temperature extremes across the United States. J Clim 15(22):3188–3205

Griffiths GM, Chambers LE, Haylock MR et al (2005) Change in mean temperature as a predictor of extreme temperature change in the Asia-Pacific region. Int J Climatol 25:1301–1330. doi:10.1002/joc.1194

IPCC (2007) Climate change 2007—the physical science basis. Contribution of Working Group I to the fourth assessment report of the IPCC. Cambridge University Press, Cambridge

IPCC (2013) Climate change 2013—the physical science basis. Intergovernmental Panel on Climate Change. https://www.ipcc.ch/report/ar5/wg1/

Jaswal AK (2010) Recent winter warming over India—spatial and temporal characteristics of monthly maximum and minimum temperature trends for January to March. Mausam 61(2):163–174

Klein Tank AMG, Peterson TC, Quadir DA et al (2006) Changes in daily temperature and precipitation extremes in Central and South Asia. J Geophys Res 111, D16105. doi:10.1029/2005JD006316

Kothawale DR (2005) Surface and upper air temperature variability over India and its influence on Indian monsoon rainfall. Ph.D. thesis, University of Pune

Kothawale DR, Revadekar JV, Rupa Kumar K (2010) Recent trends in pre-monsoon daily temperature extremes over India. J Earth Syst Sci 119(1):51–65

Peterson TC, Taylor MA, Demeritte R et al (2002) Recent changes in climate extremes in the Caribbean region. J Geophys Res 107(D21):4601. doi:10.1029/2002JD002251

Rao GSP, Murthy MK, Joshi UR (2005) Climate change over India as revealed by critical extreme temperature analysis. Mausam 56(3):601–608

Revadekar JV, Kothawale DR, Patwardhan SK, Pant GB, Rupakumar K (2012) About the observed and future changes in temperature extremes over India. Nat Hazards 60(3):1133–1155

Tebaldi C, Hayhoe K, Arblater JM, Meehl GA (2006) Going to the extremes: an intercomparison of model-simulated historical and future changes in extreme events. Clim Change 79:185–211. doi:10.1007/s10584-006-9051-4

Vincent LA, Peterson TC, Barros VR et al (2005) Observed trends in indices of daily temperature extremes in South America 1960–2000. J Clim 18:5011–5023

Zhai P, Pan X (2003) Trends in temperature extremes during 1951–1999 in China. Geophys Res Lett 30:1913. doi:10.1029/2003GL018004

Zhang X, Aguilar E, Sensoy S et al (2005) Trends in middle east climate extremes indices during 1930–2003. J Geophys Res 110(D22):104. doi:10.1029/2005JD.006181

Analysis of Extreme High Temperature Conditions over Uttar Pradesh, India

Ramesh Chand and Kamaljit Ray

1 Introduction

The state of Uttar Pradesh (Fig. 1) is in the heart of Indo-Gangetic plain with river Ganges flowing right through it, Himalayas to the north and the Chota Nagpur plateau and the Vindhya hills to the south. The climate of Uttar Pradesh varies from temperate in eastern Uttar Pradesh, extreme in western UP to semi arid in Bundelkhand and Agra zone. Therefore it is extremely difficult to categorize it in a particular climatic frame. Nevertheless, summers are very hot and winters are chilly. Generally heat waves develop in the north west parts of India or over the northern parts of Pakistan and expand into the neighbouring subdivisions of the country including Uttar Pradesh. Heat wave may also develop in situ over the region in case of development of an upper-air anticyclone circulation. The temperatures start to increase over Uttar Pradesh in March and by April the land area becomes hot with day time maximum temperature reaching about 40 °C at many locations. During this season the range of maximum and minimum temperatures is found to be more than 15 °C at many stations. Maximum temperatures rise sharply exceeding 45 °C by the end of May and early June, resulting in harsh summers particularly over SW-U.P. (Bundelkhand). Heat waves form when high pressure aloft (3,000–7,600 m) strengthens and remains over a region for several days to several weeks. Under high pressure, the air subsides (sinks) toward the surface. This sinking air acts as a dome capping the atmosphere. This cap helps to trap heat instead of allowing it to lift. This results in a continuous build-up of heat at the surface that is experienced as a heat wave.

Extreme positive departures from the normal maximum temperature result in heat waves during the summer season (De et al. 2005). These are called heat waves,

R. Chand • K. Ray (✉)
India Meteorological Department, Lodhi Road, New Delhi 110003, India
e-mail: kamaljit_ray@rediffmail.com

Fig. 1 Map of Uttar Pradesh

because the spells of hot weather are often seen to move from one region to another (Chaudhury et al. 2000). Over India, heat waves are prominent extreme temperature events occurring during the pre-monsoon season (April–June). During the period 1901–2010, around 23 severe heat wave conditions have occurred in India, resulting in death of 8,869 persons and a damage of 144 million US dollar (EMDAT 2011). One of the predicted outcomes of the global warming is the increase in frequency of extreme events, including heat waves, etc. across different parts of the world on varying scale/intensity (Srivastava et al. 2001; Ray et al. 2009; Manton 2010). Hence, understanding the pattern and frequency of heat waves is essential for better forecasting and management of the extreme temperature conditions.

Attri and Tyagi (2010) in climate profile of India analysed data for the period 1901–2009 and suggested that annual mean temperature for the country as a whole has risen by 0.56 °C (Fig. 2) over the period. It may be mentioned that annual mean temperature has been generally above normal (normal based on period, 1961–1990) since 1990. This warming is primarily due to rise in maximum

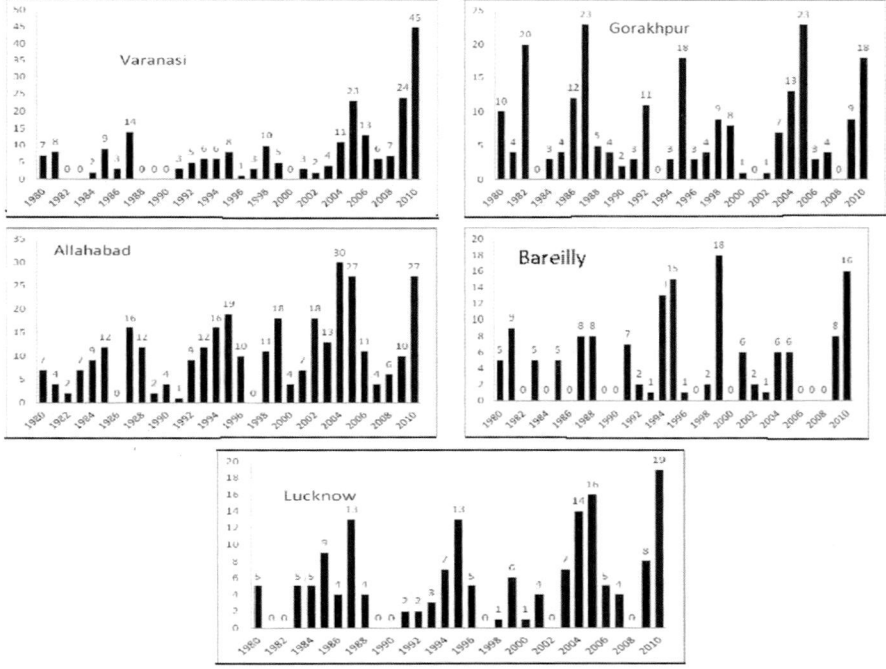

Fig. 2 Year-wise number of heat waves in various stations of Uttar Pradesh

temperature across the country, over larger parts of the data set. However, since 1990, minimum temperature is steadily rising and rate of its rise is slightly more than that of maximum temperature (IMD Annual Climate Summary 2009). Warming trend over globe of the order of 0.74 °C has been reported by IPCC (2007). Climate variability and occurrence of extreme weather events are the major concerns linked to global warming. In a study, Hingane et al. (1985) had prepared an all-India mean series of seasonal and annual surface air temperatures for a long-term trend studies, using data for 1901–1982. The analysis indicated a significant warming of 0.4 °C per 100 years in mean annual temperature of the country as a whole. The gridded 30 years moving averages of mean maximum temperatures over Gujarat state (Ray et al. 2009) also indicated an increase of 0.11 °C during the period 1969–2005. Ray et al. (2013) studied heat waves over Gujarat state and observed that the number of moderate heat wave days and severe heat wave days were highest in the last decade (2001–2010) as compared to earlier three decades. In this paper an attempt has been made to study heat wave during March, April, May and June for five stations of Uttar Pradesh. This study was undertaken to analyze the decadal climatology of extreme high temperature conditions over Uttar Pradesh. The stations chosen were Allahabad, Gorakhpur, Varanasi, Lucknow and Bareilly.

2 Data and Methodology

Maximum temperature data for summer months (March, April, May and June) for the period 1980–2010 was obtained from National Data Centre of ADGM(R) Pune office for five stations in Uttar Pradesh. For synoptic analysis, synoptic charts were downloaded from NCEP website (www.esrl.noaa.gov/psd/data/reanalysis/reanalysis.shtml).

Heat wave is considered only when the maximum temperature of a station reaches at least 40 °C for plains and at least 30 °C for hilly regions. It has been defined in following three categories:

1. *First category:* When normal maximum temperature of a station is less than or equal to 40 °C, heat wave (HW) may be considered, if the maximum temperature departure from normal is 5–6 °C and severe heat wave (SHW) may be considered if the departure from normal is 7 °C or more.
2. *Second category:* When normal maximum temperature of a station is more than 40 °C than heat wave (HW) may be considered if maximum temperature departure from normal is 4–5 °C and severe heat wave (SHW) if the departure from normal is 6 °C or more.
3. *Third category:* When actual maximum temperature remains 45 °C or more irrespective of normal maximum temperature, heat wave should be declared.

As per the above criterion heat waves days and severe heat wave days were calculated for five stations in Uttar Pradesh for 30 years period (1980–2010).

3 Results and Discussion

3.1 Study of Decadal Pattern

The total number of heat waves and severe heat wave days in the three decades for five IMD stations are shown in Table 1. The decadal analysis indicates that the number of moderate heat wave days are highest in the last decade as compared to earlier

Table 1 Decadal frequency of MHW and SHW for five stations of Uttar Pradesh

Stations/decades	Moderate heat wave (MHW)			Severe heat wave (SHW)		
	1981–1990	1991–2000	2001–2010	1981–1990	1991–2000	2001–2010
Allahabad	58	92	145	12	12	7
Varanasi	27	46	121	6	2	17
Gorakhpur	59	56	78	19	4	3
Lucknow	39	40	72	6	0	5
Bareilly	28	58	42	8	2	3

two decades in most of the stations. The districts like Allahabad, Gorakhpur, Varanasi and Lucknow show increase in moderate heat waves in the last decade and Varanasi indicates appreciable increase in severe heat wave days in the last decade (2000–2010). Bareilly does not indicate an increase in the heat waves in the last decade. If we see the total heat wave days (MHW + SHW), then except for Bareilly it is highest in the last decade for all stations. The frequency of yearly heat wave days was calculated from total heat wave days in the 30-year period. It was seen that out of the five stations considered Allahabad has the highest frequency of heat waves (11 days per year) followed by Gorakhpur and Varanasi (7 days per year) and then Lucknow and Bareilly (5 days each per year). This shows plains and eastern parts of Uttar Pradesh are more prone to heat waves.

Table 2 gives the year-wise frequency of MHW and SHW days in the last decade (2001–2010). In the year 2010, heat wave prevailed on 45 days in Varanasi, which was the highest number of heat wave days recorded in the past 30 years (1980–2010). Moderate heat wave prevailed on 36 days and severe heat wave on 9 days. Similarly in Allahabad heat wave prevailed on 29 days in 2004 followed by 27 days in 2005 and 2010. Lucknow recorded MHW on 19 days in the year 2010; Bareilly recorded MHW on 16 days and Gorakhpur on 18 days. In the last decade maximum heat wave days were recorded in 2010 followed by 2005 and 2004 for Uttar Pradesh as a whole. Varanasi recorded highest number of severe heat wave days (nine) in 2010, followed by 5 days in 2005. Figure 2 shows the number of heat wave days in last 30 years for all five stations. Highest numbers of heat wave days were realized in 2010 for Varanasi and Lucknow in the past 30 years. Allahabad recorded highest heat wave days in 2004 followed by 2005 and 2010. Except for Bareilly rest all stations show an increase in heat wave days in the last decade.

The damage potential of the heat waves depend upon the duration of the event; therefore cases of continuous days of MHW were also calculated. Table 3 gives the instances of continuous heat wave days in the past 30 years. Allahabad recorded continuous spell of heat wave for 17 days from 1 to 16 June, 1995, Gorakhpur recorded heat wave for continuous 11 days from 9 to 19 June, 1995 and during the same year, Bareilly recorded heat wave spell of 8 days (14–21 June). Heat wave prevailed over the state of Uttar Pradesh during the entire month in various regions travelling from east to west. Table 4 shows the severe heat wave for continuous 4 days or more. Most of the cases of continuous SHW were found to be in the first decade, with 90 % of the cases in 1987. Maximum frequency of heat waves over most parts of the state was found to be in the month of June. The continuous SHW prevailed in Varanasi from 22 to 27 March 2010, when maximum temperatures were around 42–43 °C as against normal of 34 °C and in the year 2005, it prevailed from 15 to 20 June when maximum temperatures were around 45–47 °C as against normal of 39 °C.

In the past 30 years the longest continuous severe heat wave (7 days) realized in any station in Uttar Pradesh was in Allahabad from 24 to 30 June in 1987. Next higher number of SHW of 6 days was recorded in Gorakhpur and Varanasi, during the same period when temperatures crossed 45 °C (Table 4). Bareilly and Lucknow also had SHW conditions on 4–5 days in 1987. Table 5 shows the highest

Table 2 Year-wise frequency of MHW and SHW days in the last decade (2001–2010)

Year	2001		2002		2003		2004		2005		2006		2007		2008		2009		2010	
Station	MHW	SHW	MHW	SHW	MHW	SHW	MHW	SHW	MHW	SHW	MHW	SHW	MHW	SHW	MHW	SHW	MHW	SHW	MHW	SHW
Gorakhpur	0	0	1	0	7	0	10	3	23	0	3	0	4	0	0	0	9	0	18	0
Allahabad	7	0	18	0	13	0	26	3	27	0	11	0	4	0	6	0	8	2	25	2
Varanasi	3	0	2	0	4	0	11	0	18	5	13	0	6	0	7	0	21	3	36	9
Lucknow	4	0	0	0	7	0	14	0	12	4	5	0	4	0	0	0	7	1	19	0
Bareilly	5	1	2	0	1	0	6	0	4	2	0	0	0	0	0	0	8	0	16	0

Table 3 Days of continuous heat waves in the past 30 years

Station	Duration	Period
Allahabad	17 days	1 June–16 June 1995
Gorakhpur	11 days	9 June–19 June 1995
Varanasi	13 days	21 March–2 April 2010
Lucknow	11 days	10 June–20 June 2005
Bareilly	8 days	1 May–8 May 1995 & 14 June–21 June 1995

Table 4 Severe heat wave days for continuous 4 days and more

Station	Duration	Period	Temperatures[a]
Allahabad	7	24th–30th June 1987	45–47 (37 °C)
Gorakhpur	6	25th–30th June 1987	44–46 (36 °C)
Varanasi	6	25th–30th June 1987,	45–46 (36 °C)
		15th–20th June 2005,	45–47 (39 °C)
		22th–27th March 2010	42 (34 °C)
Lucknow	5	26th–30th June 1987	44–46 (36 °C)
Bareilly	4	27th–30th June 1987	44–45 (36 °C)

[a]Figures within brackets are normals

Table 5 Highest temperatures recorded in various observatories in the past 30 years

1970–2010		
Stations	Highest temp.	Date
Allahabad	48.4	30 May, 1994
Varanasi	46.8	1 June, 1998
		23 May, 1998
Gorakhpur	45.5	25 May, 1995
Bareilly	46.6	2 June, 1995
Lucknow	46.5	31 May, 1995

temperatures recorded in the last 30 years. All stations have recorded highest temperatures in the 1991–2000 decade. The highest temperature of 48.4 was recorded in Allahabad on 30 May in 1994 followed by 46.8 recorded in Varanasi on 1 June and 23 May in 1998. In order to analyze the increase in heat waves in the last decade the heat wave spells of 2, 4, 6 and more than 8 days for all stations are shown in Table 6. It was found that the 2-day spell of heat waves have increased significantly in March and April in the last decade as compared to earlier decades for all stations except Bareilly. The 4-day spells have also increased in the last decade as compared to earlier ones and the rise has been in the months of March and April particularly.

Some of the earlier studies have also indicated a relation between El-Nino and heat waves (De and Mukhopadhyay 1998; Chaudhury et al. 2000; Ray et al. 2013). During the period 1980–2010, it was seen that the maximum number of heat wave days and human lives lost were comparatively large during the years preceded by warm ENSO years. In this study it was seen that in the past 30 years, the years with

Table 6 Spells of heat wave for continuous 2, 4, 6 and 8 days or more for different stations

Station	Continuous spell of heat wave	Decade	March	April	May	June	Total (March–June)
Allahabad	2 days or more	Decade 1 (1981–1990)	1	4	4	9	18
		Decade 2 (1991–2000)	4	3	6	9	22
		Decade 3 (2001–2010)	10	11	3	9	33
	4 days or more	Decade 1 (1981–1990)	0	0	1	4	5
		Decade 2 (1991–2000)	0	0	0	4	4
		Decade 3 (2001–2010)	3	4	1	4	12
	6 days or more	Decade 1 (1981–1990)	0	0	0	2	2
		Decade 2 (1991–2000)	0	0	0	3	3
		Decade 3 (2001–2010)	3	2	0	2	7
	8 days or more	Decade 1 (1981–1990)	0	0	0	0	0
		Decade 2 (1991–2000)	0	0	0	2	2
		Decade 3 (2001–2010)	0	1	0	1	2
Gorakhpur	2 days or more	Decade 1 (1981–1990)	2	1	1	7	11
		Decade 2 (1991–2000)	1	0	2	6	9
		Decade 3 (2001–2010)	8	4	0	4	16
	4 days or more	Decade 1 (1981–1990)	0	0	0	4	3
		Decade 2 (1991–2000)	0	0	1	2	2
		Decade 3 (2001–2010)	2	1	0	2	5
	6 days or more	Decade 1 (1981–1990)	0	0	0	4	4
		Decade 2 (1991–2000)	0	0	1	2	3
		Decade 3 (2001–2010)	1	0	0	2	2
	8 days or more	Decade 1 (1981–1990)	0	0	0	1	1
		Decade 2 (1991–2000)	0	0	1	0	1
		Decade 3 (2001–2010)	0	0	0	2	2

Analysis of Extreme High Temperature Conditions

Varanasi	2 days or more	Decade 1 (1981–1990)	0	1	0	6	7
		Decade 2 (1991–2000)	0	2	4	7	13
		Decade 3 (2001–2010)	9	11	2	7	29
	4 days or more	Decade 1 (1981–1990)	0	0	0	3	3
		Decade 2 (1991–2000)	0	0	1	2	3
		Decade 3 (2001–2010)	3	3	0	3	9
	6 days or more	Decade 1 (1981–1990)	0	0	0	1	1
		Decade 2 (1991–2000)	0	0	0	0	0
		Decade 3 (2001–2010)	2	2	0	2	6
	8 days or more	Decade 1 (1981–1990)	0	0	0	0	0
		Decade 2 (1991–2000)	0	0	0	0	0
		Decade 3 (2001–2010)	1	1	0	2	4
Lucknow	2 days or more	Decade 1 (1981–1990)	2	1	1	7	11
		Decade 2 (1991–2000)	1	0	2	6	9
		Decade 3 (2001–2010)	8	4	0	4	16
	4 days or more	Decade 1 (1981–1990)	0	0	0	3	3
		Decade 2 (1991–2000)	0	0	0	2	2
		Decade 3 (2001–2010)	2	1	0	2	5
	6 days or more	Decade 1 (1981–1990)	0	0	0	1	1
		Decade 2 (1991–2000)	0	0	0	1	1
		Decade 3 (2001–2010)	2	0	0	2	2
	8 days or more	Decade 1 (1981–1990)	0	0	0	0	0
		Decade 2 (1991–2000)	0	0	0	0	0
		Decade 3 (2001–2010)	0	0	0	1	1

(continued)

Table 6 (continued)

Station	Continuous spell of heat wave	Decade	March	April	May	June	Total (March–June)
Bareilly	2 days or more	Decade 1 (1981–1990)	0	2	2	3	7
		Decade 2 (1991–2000)	2	3	3	4	12
		Decade 3 (2001–2010)	2	5	0	2	9
	4 days or more	Decade 1 (1981–1990)	0	0	0	3	3
		Decade 2 (1991–2000)	0	0	3	2	5
		Decade 3 (2001–2010)	0	1	0	2	3
	6 days or more	Decade 1 (1981–1990)	0	0	0	2	2
		Decade 2 (1991–2000)	0	0	1	1	2
		Decade 3 (2001–2010)	0	0	0	2	2
	8 days or more	Decade 1 (1981–1990)	0	0	0	1	1
		Decade 2 (1991–2000)	0	0	1	0	1
		Decade 3 (2001–2010)	0	0	0	0	0

high number of heat wave days were either El-Nino years or years preceded by an El-Nino year. The year 1983, 1987, 2004, 2005 and 2010 were years when heat wave prevailed for a longer period in Uttar Pradesh. The years 1982, 1986, 1987, 1991, 1994, 1997, 2002, 2004, 2006 and 2009 were El-Nino years during last 30 years. Thus three El-Nino years and 2 years succeeding El-Nino years were with maximum heat wave days in the last three decades.

The study indicates an appreciable rise in extreme high temperature episodes over Uttar Pradesh in the last decade, as compared to earlier decades. More detailed study needs to be initiated to obtain a relation between El-Nino and temperature extremes during summer.

3.2 Case Study of Severe Heat Wave Condition (24–30 June 1987)

During the period from March to May, normal temperature was quite high over the state. Moderate heat waves to severe heat waves prevailed over the region. Figure 3 indicates the mean temperature distribution during the SHW period in

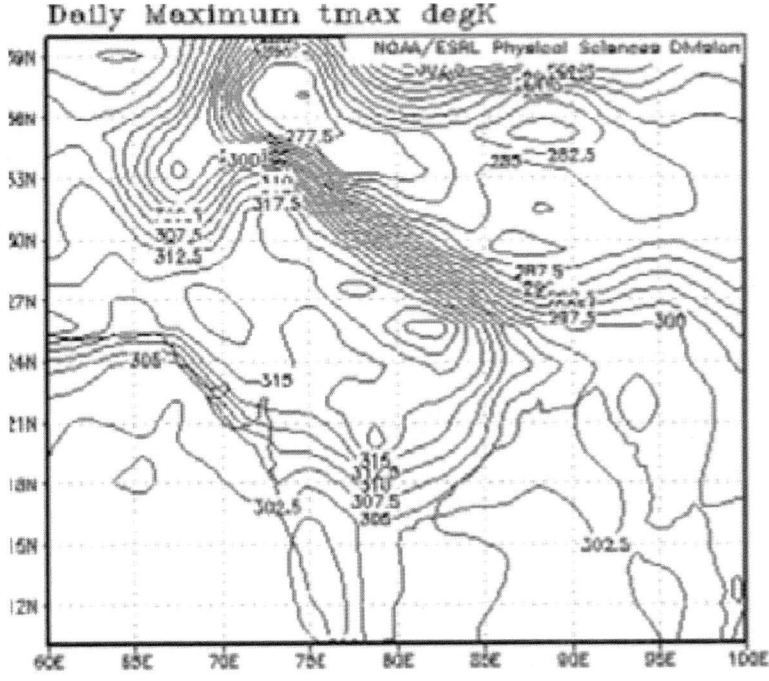

Fig. 3 Daily maximum temperature (°K) averaged during 26 June 1987 to 30 June 1987

June 26–30, 1987. A major part of Uttar Pradesh state was severely affected during the period. Allahabad recorded 45 °C and more from 25 to 30 June, with the highest recorded being 46.7 °C on 29 June. During the period (26th June to 30th June 1987) of severe heat wave, low level continental dry southerly to south-south easterly winds were blowing over the region as monsoon trough was likely to establish during this period. The extreme high temperature were also due to subsidence caused by an anticyclonic circulation over the region at 500 hPa level. Figure 4 shows mean air temperature during severe heat wave episode over Uttar Pradesh during 27 June 1987 to 30 June 1987. Pockets of temperature of more than 47 °C in plains of UP with steep decrease towards north in foothills of Himalaya were seen. Vast area were under the grip of severe heat wave. Maximum temperature averaged for the period 24th June 1987 to 30th June 1987, when severe heat wave prevailing over the region was more than 46 °C. The extreme high temperature was also due to subsidence caused by an anticyclonic circulation over the region at 500 hPa level which continued till 30 June, 1987 (Figs. 5 and 6).

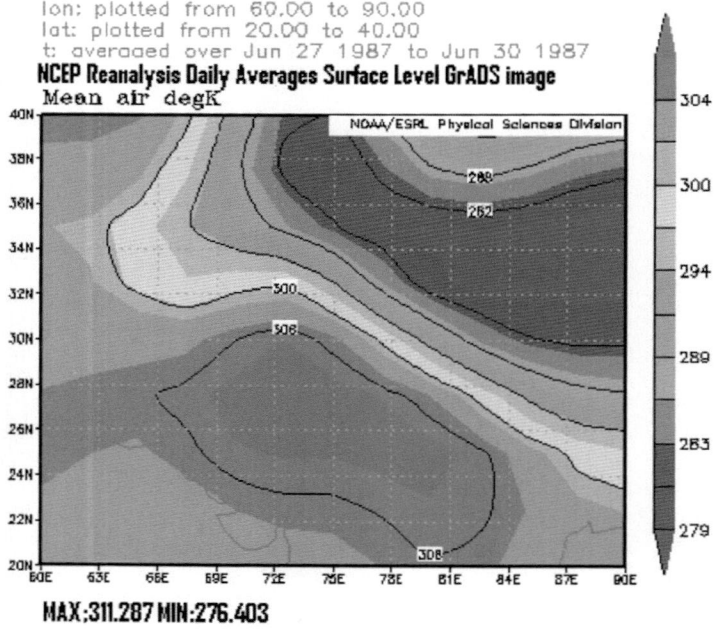

Fig. 4 Mean air temperature 27 June 1987 to 30 June 1987

Analysis of Extreme High Temperature Conditions

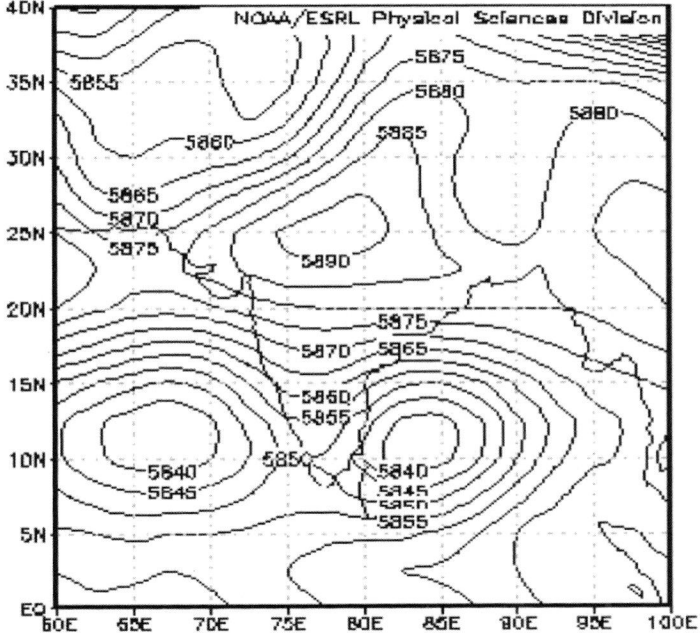

Fig. 5 NCEP reanalysis daily averages pressure level image at 500 hPa of 27 June 1987

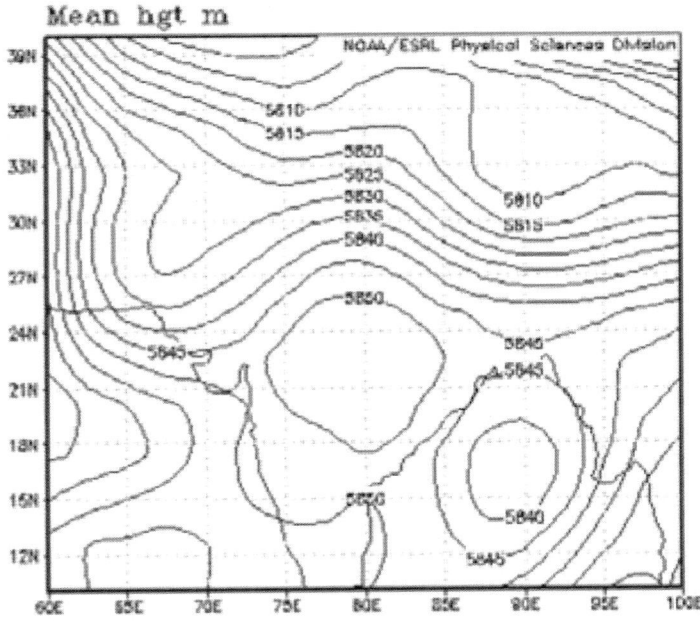

Fig. 6 NCEP reanalysis daily averages pressure level image at 500 hPa of 30 June 1987

Conclusions

This study was carried out to analyse the decadal pattern of extreme temperature conditions over different weather stations of Uttar Pradesh. The study indicates an appreciable rise in heat wave episode over U.P. in the last decade, as compared to earlier two decades. More detailed study is needed to obtain a relation between occurrence of heat wave and El-Nino and delayed onset of monsoon. Detailed study also needs to be initiated to correlate heat wave with loss of life (livestock) and loss of agriculture yields. The decadal analysis indicated that:

(a) The number of moderate heat wave and severe heat wave were highest in the month of June for all the stations considered.
(b) The total number of heat wave (moderate) is highest in the last decade (2001–2010) as compared to earlier two decades. In general there is a continuous increase in moderate heat wave from first decade to third decade whereas severe heat waves were highest in first decade.
(c) In last decade, analysis shows 2010 was warmest year followed by 2005 and 2004.
(d) Plains of east UP is more prone to heat wave (both moderate and severe) as compared to west UP.
(e) June is the month of longest duration of SHW in UP. No station reported severe heat wave for more than 7 days (Allahabad – 24th to 30th June 1987) in UP during last 30 years whereas heat wave for 17 days was recorded during 1 June to 16 June 1995.

References

Attri SD, Tyagi A (2010) Climate profile of India, Met monograph no. Environment Meteorology-01/2010
Chaudhury SK, Gore JM, Sinha Ray KC (2000) Impact of heat waves over India. Curr Sci 79:153–155
De US, Mukhopadhyay RK (1998) Severe heat wave over Indian subcontinent in 1998 in a perspective of global climate. Curr Sci 75:1308–1311
De US, Dube RK, Prakasa Rao GS (2005) Extreme Weather Events over India in the last 100 years. J Ind Geophys Union 9(3):173–187
EM-DAT (2011) The OFDA/CRED International Disaster Database. www.em-dat.net
Hingane CS, Rupa Kumar K, Ramana Murthy BV (1985) Long-term trends of surface air temperature in India. Int J Climatol 5:521–528
IMD Annual Climate Summary (2009) National Climate Centre, Pune
IPCC (2007) Climate change 2007, the physical science basis. Cambridge University Press, Cambridge

Manton MJ (2010) Trends in climate extremes affecting human settlements. Curr Opin Environ Sustain 2:151–155

Ray K, Mohanty M, Chincholikar JR (2009) Climate variability over Gujarat, India. In: ISPRS archives XXXVIII-8/W3 workshop proceedings: impact of climate change on agriculture

Ray K, Chincholikar JR, Mohanty M (2013) Analysis of extreme high temperature conditions over Gujarat. Mausam 64(3):467–474

Srivastava AK, Sinha Ray KC, Yadav RV (2001) Is summer becoming more uncomfortable over major cities of India? Curr Sci 81:342–344

Use of Remote Sensing Data for Drought Assessment: A Case Study for Bihar State of India During Kharif, 2013

K. Choudhary, Inka Goel, P.K. Bisen, S. Mamatha, S.S. Ray, K. Chandrasekar, C.S. Murthy, and M.V.R. Sesha Sai

1 Introduction

Agricultural drought can be defined as the deficiency of soil moisture leading to crop stress at any stage of the crop. Impact of moisture deficiency is more felt during period just after germination, flowering and milking/fruit formation stages of the crop. Moisture stress during these phenological stages of the crop will result in yield reduction, both quantitatively and qualitatively.

Observation of vegetation from satellite platform provides a unique vantage view and an insight into the dynamics of the vegetation, both temporally and spatially (Kogan 1997). There are two main optical domains characterizing the optical properties of plant, viz, the visible region and Near-Infrared Region (NIR). In the visible bands (0.4–0.7 μm), light absorption by leaf pigments dominate the reflectance spectrum of the leaf. The near-infrared spectral region between 0.7 and 3.0 μm has strong reflectance because of the spongy mesophyll cells, which mainly reflect light at cell/air space interface (Tucker and Sellers 1986).

Taking the advantage of this differential reflectance nature from the vegetation, the Normalized Difference Vegetation Index (NDVI) has been derived as

$$NDVI = (NIR - Red) / (NIR + Red) \qquad (1)$$

where NIR is reflectance in the near-infrared region and Red that of red region. NDVI has become the primary tool for description of vegetation changes and

---

K. Choudhary • I. Goel • P.K. Bisen • S. Mamatha • S.S. Ray (✉)
Department of Agriculture and Cooperatives, Mahalanobis National Crop Forecast Centre, New Delhi, India
e-mail: shibendu.ncfc@nic.in

K. Chandrasekar • C.S. Murthy • M.V.R. Sesha Sai
National Remote Sensing Centre, ISRO, Hyderabad, India

interpretation of the impact of environmental phenomena (Kogan 1990). NDVI is also effectively used for monitoring drought, estimating net primary production (NPP) of vegetation and crop yields, detecting weather impacts and other events important for agriculture, ecology and economics (Tucker et al. 1985; Hielkema et al. 1986; Kogan 1987, 1990; Rasmussen 1997).

However, NDVI as a vegetation index has limited success in estimating vegetation water content (VWC) (Chen et al. 2005). A potentially better way of estimating VWC is to use indices based on the longer wavelength reflective infrared range (1,240–3,000 nm) and in particular the shortwave infrared (SWIR) reflectance (1,300–2,500 nm) (Chen et al. 2005). The shortwave infrared is sensitive to vegetation cover, leaf moisture and soil moisture. The combination of NIR and SWIR bands has the potential of retrieving canopy water content (Ceccato et al. 2002). The combination of NIR and SWIR has different nomenclatures with different authors. Gao (1996) and Chen et al. (2005) defined NDWI (Normalized Difference Water Index) as:

$$\mathrm{NDWI} = (\mathrm{NIR} - \mathrm{SWIR}) / (\mathrm{NIR} + \mathrm{SWIR}) \tag{2}$$

where the NIR and SWIR are the reflectance values in near-infrared and shortwave infrared region.

The water indices using the 2,130 nm appeared more useful in extracting the vegetation water status and in drought detection and water sustainability studies (Hojin et al. 2004). Hence, in this study, the NDWI using the 2,130 nm is being used along with NDVI.

Vegetation Condition Index (VCI) provides information of how the current status of the vegetation compared with the historic maximum and minimum (Kogan 1997). Under ideal conditions of good rainfall, adequate nutrients and management inputs, the crop in a region could grow to its maximum, producing maximum NDVI/NDWI for that year. On the contrary in a drought year with less rainfall and inadequate inputs results in very low NDVI. The maximum and minimum NDVI are the conceivable limits of the vegetation ecology over the several years considered. When the current year NDVI is related to the maximum and minimum values, it helps in getting a fair idea of the present status of vegetation compared to the historic maximum and minimum. In this study, 12 years historic database (2001–2012) of NDVI and NDWI was used to derive the VCI of 2013. The VCI of NDVI and NDWI are defined as

$$\mathrm{VCI}(\mathrm{NDVI}) = (\mathrm{NDVI}_i - \mathrm{NDVI}_{min}) / (\mathrm{NDVI}_{max} - \mathrm{NDVI}_{min}) \times 100$$
$$\mathrm{VCI}(\mathrm{NDWI}) = (\mathrm{NDWI}_i - \mathrm{NDWI}_{min}) / (\mathrm{NDWI}_{max} - \mathrm{NDWI}_{min}) \times 100$$

where NDVI_i is the NDVI at time i in current year, NDVI_{min} is the historic minimum and NDVI_{max} is the historic maximum NDVI during same period.

Mahalanobis National Crop Forecast Centre (MNCFC), in collaboration with National Remote Sensing Centre (NRSC), carries out district level drought assessment based on remote sensing, meteorological and agricultural data under NADAMS (National Agricultural Drought Assessment and Monitoring System) programme. The year 2013 had a normal monsoon over the whole country, except for states like Bihar, Jharkhand and North Eastern States. Till end of September 2013, Bihar had a deficit rainfall of 30 %, with some districts getting more than 60 % deficit rainfall. The meteorological drought resulted in agricultural drought in many parts of the state causing reduction in cropped area and delayed sowing time.

In this context, this study attempts to analyse the agricultural situation during Kharif (rainy season) 2013, using remote sensing derived indices.

2 Materials and Method

2.1 Study Area

The area chosen for the study is the state of Bihar in India. The landform of Bihar is a vast stretch of fertile alluvial plain occupying the Gangetic Valley. Bihar is endowed richly with water resources, both the groundwater resource and the surfacewater resource. Not only by rainfall but it has considerable water supply from the rivers which flow within the territory of the state. Ganga is the main river which is joined by tributaries with their sources in the Himalayas. Around 61 % of crop area is irrigated. Agriculture is the biggest industry in the state. The major crops grown in the state are paddy, wheat, lentils, sugarcane and jute.

2.2 Satellite Data

NOAA AVHRR satellite data starting from 1 June 2013 to 30 October was used for creating NDVI images. Satellite image was georeferenced and NDVI was created for every single day during the above mentioned period. Fortnightly and monthly NDVI composites were created for every month. The MODIS Terra 16-day 1 km Vegetation Index Product (http://LPDAAC.usgs.gov) starting from first fortnight of June to second fortnight of October for the year 2013 was used for generating NDWI. These products were computed from the atmospherically corrected bi-directional surface reflectance that have been masked for water and heavy aerosols (http://tbrs.arizona.edu/project/MODIS/userguide-doc.php). Using the Eq. (2) the NDWI was derived from near-infrared and shortwave infrared reflectance. Forest and other non-agricultural areas of the state were masked using Land Use Land Cover Layer generated by NRSC. Figure 1 shows the NDVI and NDWI images of the state of Bihar during Kharif 2013. Further VCI of NDVI and NDWI at district level was derived using the historic NDVI and NDWI data, respectively.

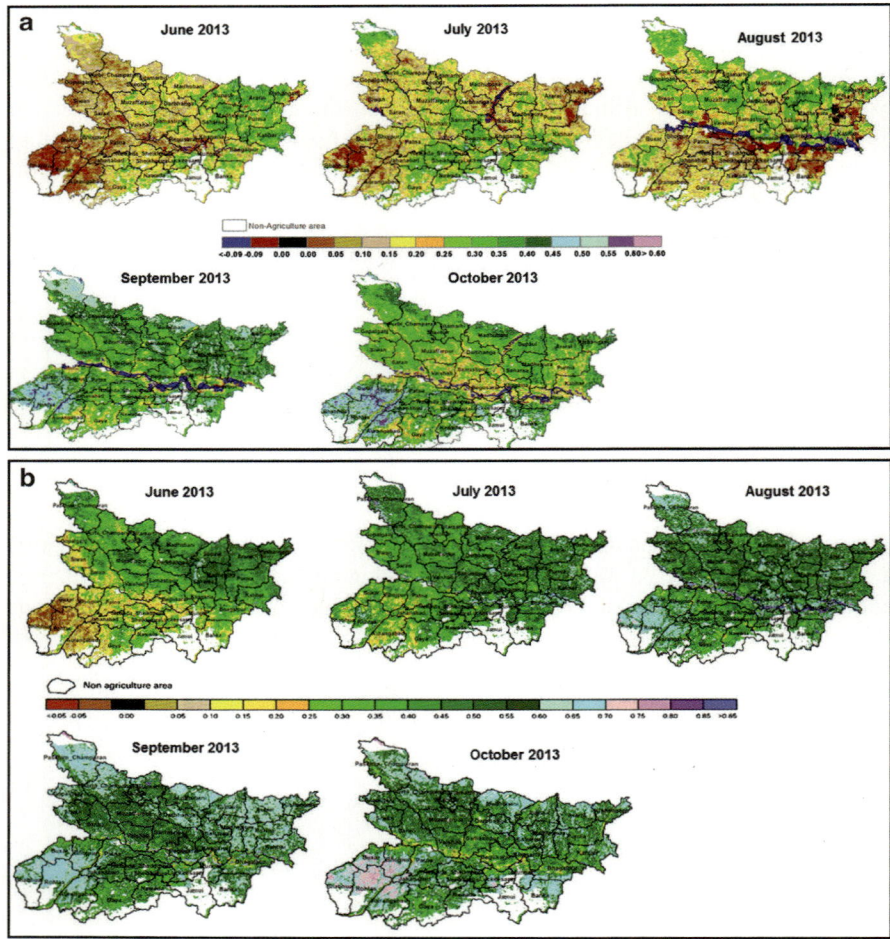

Fig. 1 (**a**) NDVI (**b**) NDWI images of Bihar state during kharif 2013

2.3 Rainfall Data

Rainfall is the most important factor for drought related studies. This study used the weekly rainfall deviation data at district level. This data is available at IMD website. The data was downloaded in excel format and further weekly and monthly district rainfall deviation maps were prepared in GIS environment (Fig. 2).

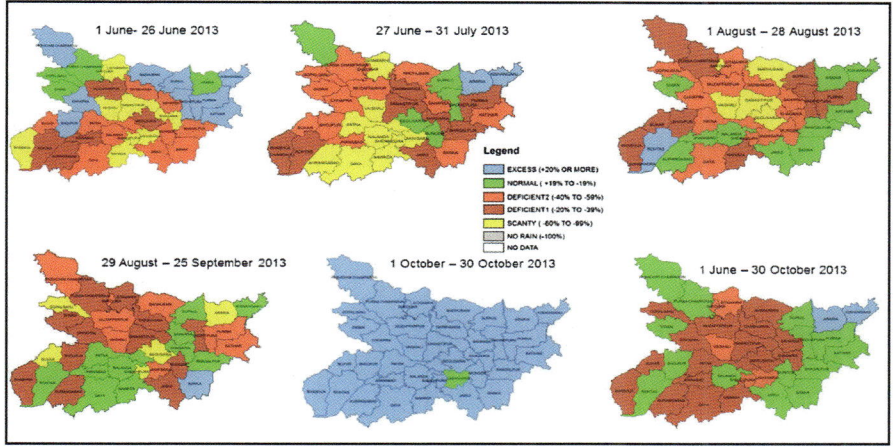

Fig. 2 Rainfall deviation maps of Bihar state during kharif 2013

2.4 Area Favourable for Crop Sowing (AFCS)

The Area Favourable for Crop Sowing (AFCS) was derived from (1) Shortwave Angle Slope Index (SASI) (Khanna et al. 2007) data integrated with ground data on cropping pattern, soil and irrigation availability and (2) Soil Moisture Index from spatial soil water balance model. AFCS reflects the agricultural area with significant surface wetness and hence favourable for crop sowing activity. The AFCS at district level has been derived and used with other indicators for drought assessment (Fig. 3).

3 Results and Discussion

During June 2013, with early advance, the monsoon rainfall pattern covered the whole country by 16 June, compared to the normal date of 15 July. However, for Bihar state there was deficient rainfall during the kharif season of 2013. In the year 2013, till end of September, Bihar had a deficit rainfall of 30 %, with some districts getting more than 60 % deficit rainfall. Nine out of 37 districts had more than 50 % deficit rainfall, while seven districts had normal or excess rainfall.

In kharif season (June to October) paddy was the major crop in Bihar. It can be observed in Fig. 1 that the peak NDVI was achieved during the month of September.

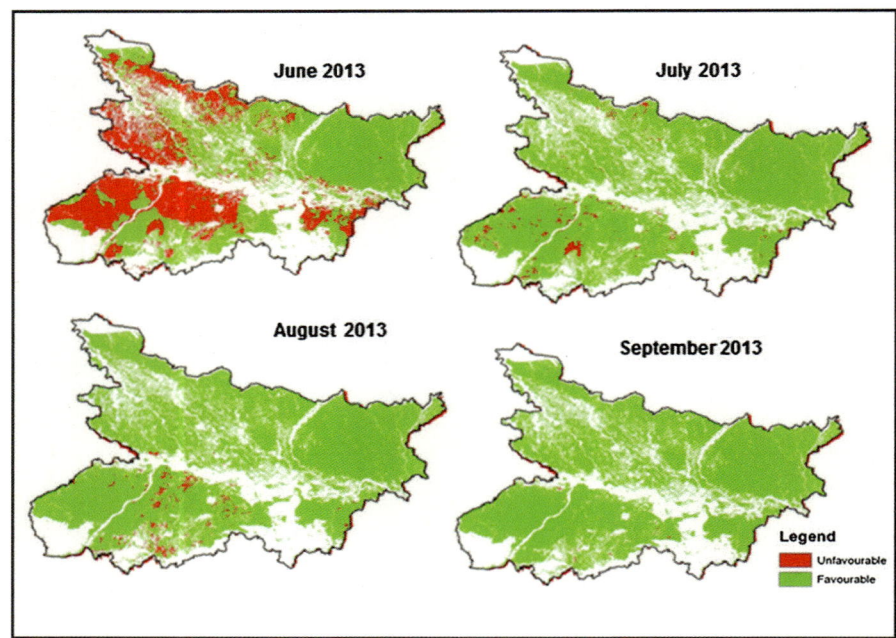

Fig. 3 Area Favourable for Crop Sowing for Bihar state of India during kharif 2013

In kharif 2012 agricultural condition was more or less normal in Bihar state. Comparison of NDVI for year 2012 and 2013 over few districts of Bihar are depicted in Fig. 4. For the districts Kishanganj, Araria, Supaul and Paschim Champaran 2013 NDVI values closely followed the 2012 NDVI values showing the close-to-normal agricultural condition. But, in case of Vaishali, Nawada, Gaya and Muzzafarpur, 2013 NDVI values showed lower values in the months of August, September and October highlighting the poor crop condition in these districts. Similar patterns were also found for NDWI values.

Less than normal NDVI and NDWI, signifying delayed crop sowing/reduced crop area/poor crop growth, are observed in central and southern districts of Bihar. More than 20 districts have received deficient rainfall up to 30 October 2013. Also, field visit has shown large number of fallow lands in Gaya and Nawada districts.

VCI of NDVI and NDWI at district level was derived using historic NDVI and NDWI values respectively. Other indicators such as rainfall deviation, area favourable for crop sowing and per cent irrigated area were integrated with VCI through a logical modelling approach to carry out the drought assessment for the kharif season 2013.

Under NADAMS programme, the districts are categorised as Normal, Watch and Alert for the months of June, July and August and for the months of September

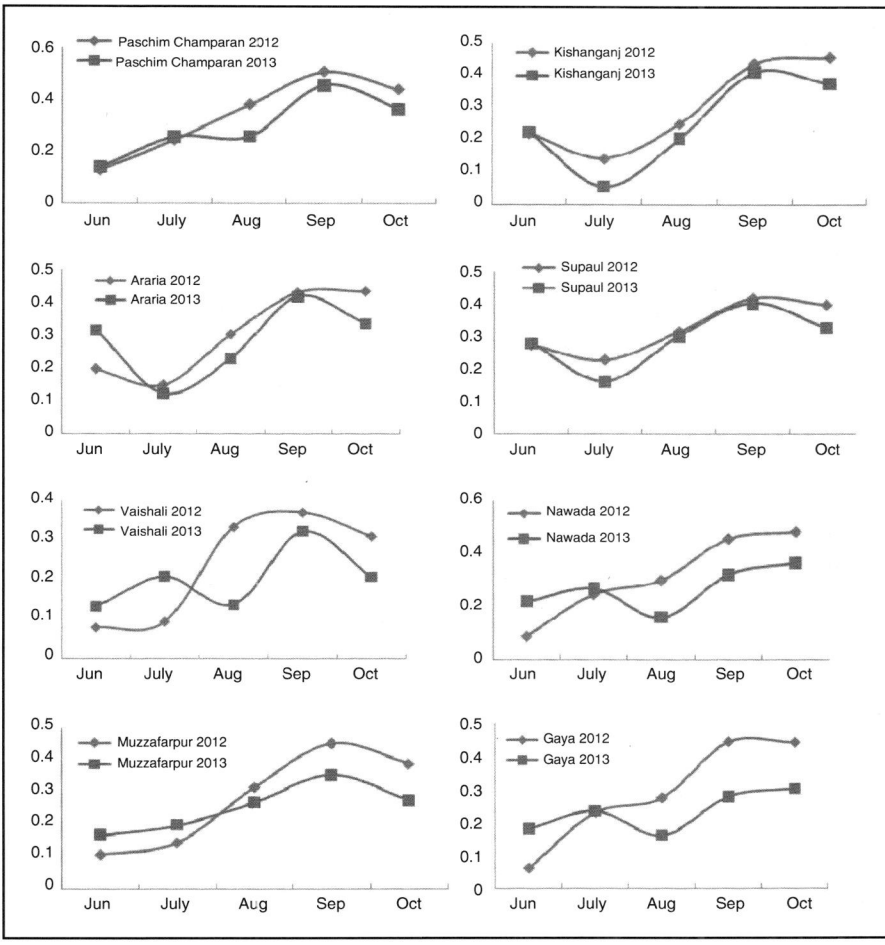

Fig. 4 NDVI profile of various districts showing comparison of agricultural condition in 2012 and 2013 kharif season

and October the districts are classified in Normal, Mild drought and Moderate drought category (Fig. 5). For the kharif 2013, considering all of the above mentioned factors five districts of Bihar (Gaya, Luckeesarai, Nawada, Sheikhpura, Vaishali) have been assessed under Moderate drought category, 12 districts (Begusarai, Darbhanga, Jahanabad, Katihar, Madhubani, Muzaffarpur, Patna, Saharsa, Samastipur, Saran, Sheohar, Sitamarhi) under Mild drought category and remaining 20 districts have been assessed under Normal category. This information has been regularly circulated to concerned national and state departments for necessary remedial action.

Fig. 5 Agricultural condition during various months of Kharif 2013

Conclusion

Among the different types of the droughts, agricultural drought poses challenges in monitoring and assessment. This study shows that satellite based indices can be effectively used for monitoring and assessing the agricultural drought. The analysis of NDVI and NDWI and ground observations showed that in the 2013 kharif cropping season, many districts in Bihar were having comparatively poor agricultural condition, though no district had severe drought condition.

Acknowledgement The authors wish to thank Dr. V.K. Dadhwal, Director, NRSC and Sh. Sanjeev Gupta, Joint Secretary, Government of India for their strong support for this work.

References

Ceccato P, Gobron N, Lassee S, Pinty B, Tarantola S (2002) Designing a spectral index to estimate vegetation water content from remote sensing data: Part 1, Theoretical approach. Remote Sens Environ 82:188–197

Chen D, Huang J, Jackson TT (2005) Vegetation water content estimation for corn and soybean; using spectral indices derived from MODIS near and short-wave infrared bands. Remote Sens Environ 98:225–236

Devenport ML, Nicholson SE (1993) On the relation between rainfall and Normalized Difference Vegetation Index for diverse vegetation types of East Africa. Int J Remote Sens 12:2369–2389

Gao BC (1996) NDWI—a Normalized Difference Water Index for remote sensing of vegetation liquid water from space. Remote Sens Environ 58:257–266

Hielkema JU, Prince SD, Astle WL (1986) Rainfall and vegetation monitoring in the Savannah zone of the Democratic Republic of Sudan using the NOAA-AVHRR. Int J Remote Sens 7:1499–1514

Hojin K, Huete AR, Pamela N, Ed Glenn, Emmerich W, Scott RL (2004) Monitoring riparian and semiarid upland vegetation using vegetation and water indices from the MODIS satellite sensor. In: Research insights in semiarid ecosystems (RISE) Symposium, 13th November, 2004, University of Arizona, Tucson, Marley Building

Khanna S, Orueta AP, Whiting ML, Ustin SL, Riaño D, Litago X (2007) Development of angle indexes for soil moisture estimation, dry matter detection and land-cover discrimination. Remote Sens Environ 109:154–165

Kogan FN (1987) Vegetation index for areal analysis of crop conditions. In: Proceedings of the 18th conference on agricultural and forest meteorology, AMS held in W. Lafayette, Indiana

Kogan FN (1990) Remote Sensing of Weather impacts on vegetation on non-homogeneous area. Int J Remote Sens 11:1405–1419

Kogan FN (1997) Global drought watch from space. Bull Am Meteorol Soc 78:621–636

Kogan FN (2000) Global drought detection and impact assessment from space. In: Wilhite DA (ed) Drought: a global assessment, vol 1. Routledge, London

Rasmussen MS (1997) Operational yield forecast using AVHRR NDVI data: reduction of environmental and inter-annual variability. Int J Remote Sens 18:1059–1077

Tucker CJ, Sellers PJ (1986) Satellite remote sensing of primary production. Int J Remote Sens 7:1395–1416

Tucker CJ, Vanpract C, Sharman MJ, Van Ittersum G (1985) Satellite remote sensing of total herbaceous biomass production in the Senegalese Sahel: 1980–1984. Remote Sens Environ 17:233–249

Xiao X, Zhang Q, Braswel B, Urbanski S, Boles S, Woty SC et al (2004) Modelling gross primary production of a deciduous broadleaf forest using satellite image and climate data. Remote Sens Environ 91:256–270

Index

Notes: Figures are denoted by an italic *f* and tables by an italic *t*

A
ADAS, cloud analysis component of, 25
Advanced research WRF (ARW-WRF), 24, 180
    dynamic solver, 51
    solver, 65
Adverse weather forecasting, 98–99
    gale wind, 98–99
    heavy rainfall, 98
    storm surge, 99
Adverse weather warning, 93
Agricultural drought, 350
    assessment, 351–355
Arakawa C-grid staggering, 148
Area average meridional winds, 12, 14, 15
Area average vorticity, 15
Area average zonal winds, 10–12
Area Cyclone Warning Centres (ACWCs), 104–107
Aridity anomaly index, 358
ARPS model, cloud analysis package, 24
ARPS3DVAR, 24, 33
    data assimilation technique, 25
    WRF, coupled, 26
Automatic Weather Stations (AWS), 91
Average divergence, 12–13
Azimuthally averaged radial velocity, 168, 169, 172

B
Bangladesh
    frequently occurring thunderstorms/nor'westers, 35
    Meteorological Department (BMD), 4
    STS in pre-monsoon season (March-May) of 2013, 404
    STS, 3–4
    thunderstorm climatology, 37–38
Baroclinic zones, 313
Bay of Bengal cyclones, initial and boundary conditions on mesoscale simulation, 179–188
    double nested two-way interactive ARW-WRF model, 180
Bay of Bengal TC disaster, 191
Betts-Miller-Janjic (BMJ), 148
Bihar, India Kharif season, 2013
    remote sensing data for drought assessment
    materials and method, 401–403
    results and discussion, 403–406
BMJ group, simulated rainfalls, 17
Brier score (BS), 123, 127

C
Central sea level pressure (CSLP), 148
Cherrapunji, 1871-2010
    rainfall pattern and heavy rainfall events, 335–344
    annual one-day, two-day and three-day maximum rainfall series, 343
    daily rainfall of different intensities, variation of, 338, 340–342

Cherrapunji (cont.)
    data used and methodology, 336–337
    monthly and seasonal rainfall trends, 337–21
Coastal hourly observations, 93
Convective available potential energy (CAPE), 3, 4, 9, 55, 68, 70f, 73, 76, 81f
    distribution, 7f
    field, 5
    mean and median values, 16
    simulated CAPE field, 242
    simulated maximum CAPE, 249–251,
    simulated mean CAPE, 45
    strong CAPE with temporal variation, 249
Convective parameters, 15–17
Conventional radar, 91
CPS, forecast performance of, 116–126
Crop moisture index (CMI), 352
Crop water stress index (CWSI), 353
Cyclone alert, 106
Cyclone disasters, reduction of, 87
Cyclone forecasting system, 95–103
    ensemble prediction system (EPS), 97
    Hurricane WRF model (HWRF), 96
    multi-model ensemble (MME), 97
    non-hydrostatic mesoscale modelling system WRFDA-WRF-ARW, 96
    NWP Models, 95–98
    Quasi-Lagrangian model (QLM), 96
Cyclone intensity, 89
Cyclone monitoring
    forecasting, standard operation procedure (SOP), 89
    observational network, 89–92
Cyclone warning, 106
Cyclone warning centres (CWCs), 105–107
Cyclone warning dissemination system (CWDS) stations, 107–108
Cyclone warning organisation, 103–107
Cyclonic storm Phailin (8–14)
    October 2013, 115
    genesis potential parameter (GPP), grid point analysis of, 116

D

Data assimilation system, 25
Data buoys, 91
Disaster management, 287
Doppler Weather Radar (DWR), 91
    data, 108
    data quality control, 27
    derived products, 72
    image, 23
    induction of, 95
    observation, 24, 67
    observational networks, 67
    products, 73
    reflectivity
        images, 54, 317, 320f
        of storm, 51–52
Double nested two-way interactive ARW-WRF model, 180
Drop size distribution (DSD), 25
Drought during monsoon, operational agromet advisory service, 23–362
DWR Agartala, 24, 26, 30

E

Elevated mixed layer (EML), 63
El Nino-Southern Oscillation (ENSO) phenomenon, 129
Error of representativeness, 65–67
ESSO-IMD, 108
Extreme rainfall (ER) events, 229–230

F

Final Reanalysis (FNL) data, 148–149
Flash floods, 207
FNL IBCs, 182–183, 186
Forecast and warning products, 100–101
Four-stage warning bulletin, 106

G

Genesis potential parameter (GPP), 111–113, 127
    cyclogenesis prediction, forecast skill, 116
    cyclone Phailin, analysis and fore-casts of area average GPP, 118f
    cyclone Phailin, grid point analysis of, 117f
    forecasts, deterministic verification of, 116f
    for the NIO, 98
GFS IBCs, 182–183, 187
Global forecast system (GFS), 26, 96, 97, 192, 197, 199
    high resolution GFS datasets, 180
    IMD-GFS, 111–113
    of NCEP, 74, 290–291
    NCEP-based GFS system, 202
    simulation, 181–187
GFS T574L50, 192, 200, 279
GFS T574, 200, 202, 279, 281, 282, 284

Index 411

Global mean surface temperatures, 365
Goddard cumulus ensemble (GCE) microphysics, 51–52
Grill schemes, 148
Gujarat, Western India
   heavy rainfall during September 2013, simulation of, 277–286
      22-29th September 2013, 278
      data and methodology, 279
      IMDGFS T574 model, 278
      Mesoscale Convective Systems (MCSs), 277–278
      results and discussion, 281–285
      synoptic features, 279–280
   heavy rainfall trend in last decade, 259–275
      3–7 September, 2011, 271–274
      15–20 September, 2008, 274–275
      24 September–28 September, 2013, 271
      data and methodology, 260
      results and discussion, 260–275

H
Hail formation, 49
Hail warning product (HWP), 77, 80, 79–80*f*
Hailstones, 49
   28 Mar 2013 hailstorm in Delhi, India, 49–60
      equivalent potential temperature (EPT), 58, 58*f*
      geopotential height, 58–59, 58*f*
      model simulated synoptic situation, 52–56
      numerical weather prediction model, 50–52
      TRMM observational dataset, 52–56
Hailstorms with squall, 73
Heat waves, 383–385
Hurricane track forecasts, 147
Hydrological drought, 350
Hydrometeors, 147

I
Ice mixing ratios, 56
Ice-phase microphysical scheme, 148
Ice precipitation, 56
IMDGFS T574, 278
India
   extreme rainfall events
      data used and methodology, 230–231
      long term trends, 229–230
      global ensemble forecasting system (GEFS), 97

India during 1871–2011
   rainfall concentration, changes in, 325–333
   data and methodology, 326–327
   results and discussion, 327–333
India, 1969–2012
   extreme temperature events, 365–381
   data and methodology, 366–368
   results and discussion, 368–380
India, 2012, agricultural drought assessment, 349–23
India Meteorological Department (IMD), 111
   cyclone warning organization, 103
Indian DWR data, 24
Indian voluntary observing fleet (IVOF), 91
INSAT satellite, 107
INSAT-3A, 90
Intensity forecasting accuracy, 102
Intensity prediction after landfall, 125
Intergovernmental Panel on Climate Change (IPCC), 385

K
Kain-Fritsch (KF) scheme, 148
"Kal Boishakhi" or nor'westers/thunderstorms of pre-monsoon season (March-May), 63
Kalpana-I, 90
Kessler-KF combination, 154
KI index, mean and median values, 16

L
Landfall forecast accuracy, 101
Land falling tropical cyclones (TCs), 179
Landfall point error (Phailin), 119
Landfall point forecast error, 101
Lin-KF combination, 153
Local analysis and prediction system (LAPS), 25
Local severe storm (LSS), 35

M
Mann Kendal non-parametric trend test, 208
Marshall-Palmer DSD, 25
Maximum (minimum) vorticity, 161
Maximum sustained surface wind (MSW), 87
Maximum temperature departure, 157–159
Maximum winds, 18–21
Mesoscale convective system (MCS), 277, 283
   stratiform, 242

Mesoscale convective systems (MCS), simulation of
   Bulk Richardson Number Shear (BRNSHR), 69
   CAPE, 68
   850 hPa wind vector and 10 m wind speed analysis, 71
   precipitation, 71
   SREH, 68–69
   3DVAR data assimilation, 63–72
   3DVAR system, 65–67
   assimilation thunderstorm forecasting, 69
   WRF ARW mode configuration, 66t
Mean forecast errors, 120, 125, 197
Mean intensity forecast errors, 127
Mean track forecast error, 127
Meteorological drought, 350
Microwave imageries, 90
MME track forecast error for Phailin, 116–119
Moisture Adequacy Index (MAI), 352
Moisture deficiency, 399
Moisture stress, 399
Monitoring of cyclone
   standard operation procedure, 92–95
      centre and intensity, determination of, 92–93
      genesis parameters, 92
      landfall point estimation error, 95
      location estimation error, 94
      radar techniques, 94–95
      satellite technique, 93–94
      synoptic technique, 93
Monthly thunderstorm records-data sources, 36

N

NADAMS programme, 404
National data buoy programme (NDBP), 91
NCEP-based GFS system, 202
NCEP FNL data, 148
NCEP reanalysis, 5
NCMRWF, 97, 202
Near IR (NIR) channels, 90
Negative vorticity, 15
NIO. See North Indian Ocean (NIO)
NOAA-AVHRR data, 353, 354
Non-monotonic or abrupt change point (climate jump), 208–209
NORIC experiment, 26, 29–33
Nor'westers, climatology of, 3–4

Normalized Difference Vegetation Index (NDVI), 353, 399–400
Northern Part of Bangladesh
   heavy rainfall events in monsoon season, NWP technique, 241–257
   19 September 2012, synoptic conditions, 246
   heavy rainfall prediction, skill scores of, 251–257
   methodology, 243–244
   results and discussion, 244–246
North Indian Ocean (NIO), 87–88, 93, 95, 176
   cyclones over
      climatological characteristics of, 89
      observational systems, 89–92
      satellite based observations, 90
      genesis potential parameter (GPP), 98
      Hurricane-WRF model for cyclone track and intensity forecast, 96
      real-time forecast, 97
      TC forecasting skill accuracy, 101–103
Numerical weather prediction (NWP)
   models, 36
   system, 65
NWP-based objective cyclone prediction system (CPS), 111–114
   dynamical-statistical model for cyclone intensity prediction (SCIP), 113
   genesis potential parameter (GPP), 112–113
   rapid intensification (RI) index, 113–114
   track prediction, multi-model ensemble (MME) technique, 113

O

Observed rainfall
   09 May 2013, Squall line event, 28
Ocean–atmosphere interaction, "memory" of, 136
Orography, 242
OSCAT-based surface winds, 90

P

Palmer Drought Index (PDI), 352
Palmer Drought Severity Index (PDSI), 352
Percentile-based temperature indices, 365
Polarization corrected temperature (PCT), 304–305, 307, 308f
Polar orbiting satellite, 91

Positive vorticity, 15
Precipitation Concentration Index (PCI), 325–333
 for monsoon rainfall, 330
 for post-monsoon rainfall, 331–332
Prolonged dry conditions, 350

Q

Quadrant wind radii forecasting, 101

R

Radar data assimilation (DA), 23
Radar observation
 09 May 2013, squall line event, 26–29
Rapid intensification prediction, forecast skill of RI-Index, 125
Rapid intermittent assimilation cycle, 29
Reflectivity forecast, 29–32
 hook shaped echo, 30
 observed reflectivity plot, 30
RIC experiment, 30
 East-West oriented reflectivity field, 31

S

SAC-ISRO, 108
Satellite based observations, 90
S-band radar stations, 91
S-band weather radar, 64
Scatterometer-based satellites, 90
SCIP intensity forecast, 120
 errors, 120
Sea-surface temperatures (SSTs), 129
Severe thunderstorms (STS), 3, 4, 8, 10, 15–17, 19
 observed weather, 404
 surface pressure, wind and relative humidity, 10
Ship observations, 91
Short-Wave IR (SWIR) channels, 90
Short wave perturbations, 313
Simulated instability indices, 18t
Simulated maximum wind, 19
 spatial distribution, 19
Simulated rainfall distribution, 17–18
Simulated vertical structure, 157
Southern India, severe convective weather events, 73
 doppler weather radar (DWR), 74
 hailstorm events, 75–83
 WRF-ARW model, 74
Southwest Monsoon over India

 heavy rainfall, observational analysis, 207–226
 annual one-day maximum rainfall, 211–215
 for Bangaldesh and Nepal, 222–224
 monotonic trends, 208
Squall line, 23, 31, 32
 in Doppler Weather Radar (DWR), 23
 model domain, 26
 over Indian subcontinent, 23
 precipitation, impact on, 32–33
 reflectivity forecast, impact on, 29–32
Squall Line event, 09 May 2013, 26–29, 33
Standardized precipitation index (SPI), 352
Standard Operation Procedure (SOP), 89, 92
 forecasting and decision support system (DSS), 99–100
Storm Relative Environmental Helicity (SREH), 8–10, 15, 68, 69, 70f
Synoptic conditions, 28
 09 May 2013, squall line event, 29

T

TC monitoring, high wind speed recorders (HWSRs), 92
TC/Typhoon, 87
Third-order Runge–Kutta time integration, 24
26 April 2010, thunderstorm event, 38–47
 data assimilation model, 45
 observed features, 39–42
 STORM field observations, assimilation of, 42–47
3DVAR systems, 23
TIGGE Cyclone XML (CXML) data, 111
Tornadic thunderstorms forecasting, 69
Tornadoes, 303
Total precipitable water (TPW), 321
Track forecast
 accuracy, 101–102
 cone of uncertainty (COU), 100–101
 products, 100
Track prediction, performance of MME, 116–120
TRMM observational dataset, 52
Tropical cyclone (TC), 87
 adverse weather forecasting, 98–99
 early warning system, 87–88
 formation, 191
 monitoring and forecasting process of, 88f

Tropical cyclone (TC) activity
  out-of-phase interannual oscillation, 130
  Western North Pacific (WNP), 129
Tropical cyclone (TC) activity over the
      Arabian Sea (AS), 129–144
  data and methodology, 131–133
  interdecadal variations, 136–137
  rainfall anomaly in Southern Pakistan,
      142–143
  seasonal variation, 137–138
  spatial distribution, 138–141
  tropospheric biennial oscillation (TBO), 136
Tropical cyclone (TC) intensity
  eddy angular momentum fluxes
    cyclone season of October-December
      viz., Jal (04–08 November 2010)
      and Thane(25–31 Dec 2011),
      165–176
    data and methodology, 166
    SCS Jal (04–08 Nov 2010), 167, 172–176
    VSCS Thane (25–31 Dec 2011),
      167–172
Tropical Cyclone Sidr
  intensity of, 150
  Mean Sea Level Pressure (MSLP), 154–155
  simulated maximum updrafts, 160, 160f
  simulated vertical temperature departure
    (°C) profile, 157–158
  synoptic situation, 149
  track error, 153–154
  track of, 150–153
  vertically integrated space averaged
    vorticity, time variation of, 161
  vertical velocity, 159–160, 160f
  vorticity, 161–162
  wind, 155–157
Tropical Pacific SST anomalies, 130
Tropical Rainfall Measuring Mission
    (TRMM), 21, 52, 304, 308f, 309f
TRRM images, 315
TRRM satellite, 29, 32–33
2D axi-symmetric nonhydrostatic model, 148

U

Uttar Pradesh, India
  exceptionally heavy rainfall, convergence
    of synoptic and dynamical
    conditions, 313–324
    data source and methodology, 315
    dynamical and thermodynamical
      conditions, 319–323
    precipitation during
      15–17 June 2013, 315

    pre-monsoon trough phase, 313
    synoptic conditions, 317–319
  extreme high temperature conditions,
    383–396
    24–30 June 1987, 393–395
    data and methodology, 386
    results and discussion, 386–395
  rainfall simulation using WRF-ARW
      Model AIRS profiles, 287–300
    data and methodology, 288–289
    results and discussion, 292–299

V

Vegetation Condition Index (VCI), 400
Vertically integrated moisture flux
    convergence (VIMFC), 304
Vertical profile divergence, 16f
Vertical profile vorticity, 17f
Vertical wind shear (VWS), 3–4, 8, 49, 58,
    130, 165, 167, 172–176
Very severe cyclonic storm (VSCS), 87
VSCS 'Phailin' over North Indian Ocean
  global forecast system, 191–202
    central MSLP with maximum
      sustainable wind, 199
    cyclone intensity forecast, 197–199
    observed heavy rainfall, 199–201
VSCS Sidr, 149

W

Warning Dissemination Mechanism, 107–108
Western disturbances (WD), 54, 304, 314,
    318, 323
Wind gusts, 303
WMO/ESCAP panel countries, 108
WRF model, and ARPS3DVAR assimilation
    system, 24
WRF-ARW model, 7
  convective available potential energy
    (CAPE), 9
  Convective Inhibition (CIN) 9
  convective parameters, 8
  Energy Helicity Index (EHI), 9–10
  equivalent potential temperature (θE), 10
  K Index, 8
  Lifted Index (LI), 8
  precipitable Water (PW), 9
  sea level pressure (SLP), 8
  Storm Relative Environmental Helicity
    (SREH) index, 9
  Total Total Index (TTI), 8
WSM6 microphysical processes, 147

Printed by Printforce, the Netherlands